Optimal Experimental Design with R

Optimal Experimental Design with R

Dieter Rasch
Jürgen Pilz
Rob Verdooren
Albrecht Gebhardt

CRC Press
Taylor & Francis Group
Boca Raton London New York

CRC Press is an imprint of the
Taylor & Francis Group, an **informa** business
A CHAPMAN & HALL BOOK

CRC Press
Taylor & Francis Group
6000 Broken Sound Parkway NW, Suite 300
Boca Raton, FL 33487-2742

First issued in paperback 2019

ISBN-13: 978-1-4398-1697-4 (hbk)
ISBN-13: 978-0-367-38276-6 (pbk)

Visit the Taylor & Francis Web site at
http://www.taylorandfrancis.com

and the CRC Press Web site at
http://www.crcpress.com

Contents

List of Figures

List of Tables

Preface

Experimental design is the stepchild of applied and mathematical statistics. In hundreds of text books and monographs about basic and advanced statistics, nothing is said about planning a survey or a design—statistics is understood there as a collection of methods for analysing data only. As a consequence of this situation, experimenters seldom think about an optimal design and the necessary sample size needed for a precise answer for an experimental question. This situation consequently is reflected in statistical program packages—they mainly are packages for data analysis. This is also the case for the S- or R-packages and for more than ninety books describing statistics by R (see http://www.r-project.org).
By reading the title of most of these books, we can already see that they deal with data analysis only.
The present book has the following goals:

1. to introduce experimenters into the philosophy of experimentation and the need for designing experiments and data collection – these readers can skip the more theoretical parts of the text and Appendix A,

2. to give experimenters and consulting statisticians an easy process for constructing optimum experimental designs and calculating the size needed in experimentation by using R-programs,

3. to show by examples how the R-programs should be used,

4. to give mathematicians interested in the theoretical background of experimental designs the theoretical background of the programs throughout the text and in the short theoretical Appendix A.

Together with this book the R-program package OPDOE has been developed. It follows the structure of the book, and it will be available as soon as the book is published. When the book is printed not all the programs are mentioned in our book, but the program package OPDOE with missing programs will be available online under `www.crcpress.com/product/isbn/9781439816974`.
This book is not a text book which introduces the reader systematically into the theory of experimental design. It should instead be understood as a manual for the R-package OPDOE. For instance Chapter three, "Size of Experiments in Analysis of Variance Models", is more a collection of recipes showing how to determine the sample size in different situations than a chapter which could be read through with pleasure from the beginning to the end.

Vienna, Klagenfurt and Wageningen, Spring 2011

Acknowledgments

The authors wish to express their thanks to:

Dr. Minghui Wang, ORTEC Finance BV, Rotterdam, The Netherlands, who wrote most of the programs in Chapter 3;

Dr. Bernhard Spangl, University of Natural Resources and Applied Life Sciences, Vienna, for Table 3.9 in Chapter 3;

Dr. Petr Šimeček, Institute of Animal Science, Prague, who was in charge of the initial steps of the writing of this book, and

Dr. Ben Torsney, University of Glasgow, for many helpful comments and proposals.

Because the authors are not native English speakers, we are happy to have found help from:

Sandra Almgren, Kremmling, Colorado, USA, who we have to thank for many helpful suggestions and corrections.

We give further thanks to:

Beate Simma, Alpen Adria Universität Klagenfurt, for carefully retyping the whole manuscript in LaTeX.

1

Introduction

In this chapter we give an introduction both to basic ideas of the philosophy of experimentation, including basic definitions of designing experiments, and into the R-programs. Here especially we give a description of the R-package OPDOE, which was developed during the process of writing this book.

1.1 Experimentation and empirical research

Empirical research: we understand this term to mean the acquisition of knowledge either through passive observations (as in surveys) or from experiments, where we actively influence or manipulate the research material and then make observations. Given the steadily increasing financial problems facing research institutions, how can empirical research be conducted as efficiently (cost-effectively) as possible? This book seeks to make a contribution to answering this question. The methods are in existence, but they are not widely used. The reasons for this stem on the one hand from the traditions of experimental research, and also from the behavioural patterns of the researchers themselves. Optimal research design implies that the objective of the investigation is determined in detail before the experiment or survey is carried out and that the precision requirements for the type of analysis planned for the data using the chosen statistical model are formulated, and that all possible things which could have a negative influence on the research work or could bias or disturb the results are considered. This all makes work, takes time, and is much more difficult than simply starting with the practical investigation and formulating the objective of the study after the data have been collected and then seeing what reliability the results have. Another reason that design methods are not used as much as they could or should be is that the designs are spread throughout a huge number of articles and books and they are sometimes not easy for the experimenter to understand.

Even if many things are equal for surveys and experiments, the two fields of empirical work differ in many aspects or in terminology. What is called "blocking" by experimenters is called "stratification" in the survey area.
To avoid confusion, this book uses the language of experimentation only. But

many sample size problems discussed here can be used in surveys as well. By offering R-programs for the construction of experimental designs and the determination of the minimum sample size needed for some precision, we hope that even experimenters with limited knowledge about the theory of experimental designs are able to improve the efficiency of their experimentation by using these programs.
On the other hand, statisticians and experimenters who are interested in the background needed for constructing designs will find this in Appendix A and in the theoretical parts of some chapters—they can be skipped by those who wish to run the programs only.
The necessity of designing experiments is also not reflected in the statistical R-literature. All statistical R-books edited by Chapman and Hall and by Springer up to 2008 do not deal with experimental design. Some of them understand "statistical computing" as analysing data. In *A First Course in Statistical Computing with R*, by W.J. Brown and D.J. Murdoch (2008) we find: "Statistical computing involves doing computation to aid in statistical analysis. For example, data must be summarized and displayed." There is no word in the entire book about planning the data collection.

We assume here that the reader is familiar with the statistical methods for analysis, and for these methods we provide procedures to determine the size of data to be sampled to do the analysis with a predetermined precision. We further provide methods for constructing optimal designs for regression and ANOVA models. How to analyse the data will not be demonstrated here, as there are already enough books handling just this aspect.

In many areas experiments take place, different feeds and fertilising agents are used, different therapies are compared. In industrial research one must frequently investigate the influence of various factors on the quality or quantity of an end product or adjust production conditions to achieve an optimum. This can require the systematic variation of relevant factors. In all these areas there are basic principles for selecting efficient (optimal) designs, which we will discuss in this book. The most important thing here is that the problem of the empirical investigation should be properly formulated. It is certainly true, and this will be clarified in examples, that this is often not a simple task and can require days or months of work. Since an experiment often lasts far longer than its analysis, one can quickly alter a false analysis of correctly obtained data, but often the most sophisticated analysis cannot save an insufficiently thought-through experiment, which must therefore be repeated at consequently greater cost. Concerning the design of experiments besides books cited elsewhere in the text we refer to Atkinson, Donev and Randall (2007); Bailey (2008); Beth, Jungnickel and Lenz (1999); Bose (1941); Cox (1958, 2009). This book will only discuss experiments, but part of the results can be used in surveys too. The process of gaining knowledge in the empirical sciences can be considered as follows:

1. formulation of the problem,

2. fixing the precision requirements,

3. selecting the statistical model for planning and analysis,

4. construction the (optimal) design of the experiment,

5. performing the experiment,

6. statistical analysis of the observed results, and

7. interpreting results.

(1–4—pre-experimental statistics, 6–7—post-experimental statistics)

This book deals mainly with the pre-experimental statistics. We hope that scientists after reading the book will no longer collect data first and then afterwards put the question "What can I do with my data and how should I analyse them?" Too often we are confronted in our consultation with just such a strange situation. Clients come to us with a set of collected data and ask what to do with these data.

Each statistical analysis has a design counterpart. The design tells the researcher how data should be collected and how many observations are needed. After a rough problem formulation, we must think about the statistical analysis we plan to do after data collection is completed. Only within the statistical model underlying the future analysis can precision requirements be formulated. If we plan to test a hypothesis about an expectation, the precision is given by the risk of the first and the second kind of the corresponding statistical test and by what is often called the effect size. It is the standardised minimum difference from the null hypothesis value of the expecation which is of practical interest—for just this value the risk of the second kind should have a given value. But if we plan to construct a confidence interval, the precision is defined by the confidence coefficient and the expected width of the interval. Therefore the kind of the future analysis is part of the exact formulation of the research question.

But to be able to find an optimum design, the model for the future observations and the analysis needed to answer the experimental question must be fixed already in the planning phase. If possible, it should not be changed after experimentation.

A different situation occurs if the aim of the experimentation is the model choice. So the method of analysing future data may be "regression analysis," for instance, "growth curve analysis." When we do not know the function in growth research between the growing character and the age of individuals, model selection must be done first. In such cases, we can try to find optimal designs for discriminating between the elements of a set of given growth functions and later design experiments for estimating the parameters of the selected function. We find some hints to that in the section on non-linear regression.

Sometimes, however, it is not research problems which need optimal experimental designs but production processes. Examples are the mixture designs (discussed in Chapter ten) where an optimum mixture of ingredients for a production process or the optimum combination of fertilisers in agricultural production is needed. Thus the design of experiments (here it is often abbreviated by DOE) plays an important role in industry and in agriculture, horticulture, fisheries (optimum fishing strategies), forestry, ecology and other areas.

1.2 Designing experiments

The *statistical planning* of an experiment includes the construction of an optimum statistical experimental design and the determination of the minimum sample size which is necessary to achieve predetermined precision requirements in the context of a chosen statistical model for the analysis of the results. How this can be done is shown in Chapter two.

The optimum design determines the *structure* of the experiment as well as the size of the experiment and often depends upon the statistical model for the analysis. Thus the model and the method of analysis planned for the data to be collected must be fixed before the experiment starts and should not be selected when the data are already collected.

A *statistical model* $\boldsymbol{y}$ for the result y of a statistical experiment is defined by a model equation (perhaps with side conditions) and by the distributional assumptions made about the random components of $\boldsymbol{y}$. (Random variables are in bold print.)

We will explain this and the following terms and definitions by an example which we took from Rasch et al. (2007b). This was a real life problem which was put to the first author during the socialist era in East Germany. At this time, everything had to be planned. Cooperatives had to know, for instance, their gasoline consumption for the next year and received only as much gasoline as they planned (and sometimes less). To help the managers of cooperatives, an agricultural university decided to determine by experimentation how high the gasoline consumption of a special type of lorry (at this time there was only one) is when used for harvest transport. The petrol consumption data given by the producer of the cars were useless; they were valid for paved streets but not for farm land where the soil was often muddy. The lady responsible for this experimentation asked us how many lorries we had to test and how often we would travel with each of them. This seems to be an easy question, but as we will soon see, it is not very easy to answer. Of course the same question arises to the producers of any kind of car if they wish to determine the fuel consumption of their cars on normal streets.

Another problem leads to the same type of model: how can we estimate the average yearly milk production in a country (the Czech Republic, for example)? We know that the sires as well as the cows have an influence (by their genotype) on the yearly milk production of a heifer or daughters in the following lactations. Then the analogue question is: how many sires should be taken into the analysis and from how many of their daughters are yearly milk records to be collected? There are many fields of applications with an analogue problem. But we will explain the theory behind it with the lorry example and show by this example how the 7 steps in Section 1.1 must be handled.

Example 1.1 For a certain type of lorry the average fuel consumption (in litres per 100 kilometres) used by the lorries for harvest has to be determined. Is the research question adequately formulated? Clearly not (although at least the units of measurement of fuel consumption are given). It seems to be quite clear what we wish to know, but really we are far from a precise formulation of the problem (stage 1 in Section 1.1) to be solved. But before we can make the formulation more precise we first have to choose a statistical model (stage 3 in Section 1.1). First we need to define the population from which the sample will be drawn. This is the total number of lorries of the type in question which have been or will be produced under the same conditions. This number is unknown, but is very large so that it can be modelled by an infinite set (models are only approximate representations of reality), and we model the fuel consumption y_i of the i-th lorry, as a random variable $\boldsymbol{y}_i$. What do we want to determine (or estimate, in statistical terminology)?

Certainly not $\boldsymbol{\mu}_i = E(\boldsymbol{y}_i)$. This expectation—the average fuel consumption of the i-th lorry only interests the owner of this particular lorry (whose journeys are indeed his universe or population.).

Instead, we require a general statement applicable to all farmers who use this type of lorry to bring in the harvest. So we continue modelling the situation described above. We assume that the average fuel consumption $\boldsymbol{\mu}_i$ of the i-th car deviates by the random variable $\boldsymbol{a}_i$ from the overall mean fuel consumption μ. This deviation arises mainly from random variations of technical parameters of components of the car during production. Individual records $\boldsymbol{y}_{ij}$ of the i-th car in the j-th journey again vary randomly around $\boldsymbol{\mu}_i$ by $\boldsymbol{e}_{ij}$ (a random error variable which depends, for instance, on the speed, temperature, wetness of the ground in the field, the experience of the driver etc.).

This leads us to the following model for the individual observations (measurements) $\boldsymbol{y}_{ij}$:

$$\boldsymbol{y}_{ij} = \mu + \boldsymbol{a}_i + \boldsymbol{e}_{ij} \quad (i = 1, \ldots, a;\ j = 1, \ldots, n_i) \tag{1.1}$$

In (1.1) it is assumed that a number a of the cars selected from the production will be tested in such a way that from the i-th car n_i measurements of the petrol consumption are available. Further the conditions

$$E(\boldsymbol{a}_i) = E(\boldsymbol{e}_{ij}) = 0, \quad var(\boldsymbol{a}_i) = \sigma_a^2, var(\boldsymbol{e}_{ij}) = \sigma^2$$

have to be fulfilled, and the different random variables on the right hand side of (1.1) are assumed to be stochastically independent. The model (1.1) together with the side conditions are a model II (with a random treatment factor) of simple analysis of variance.

After our first stage of a model choice, we can give a more precise description of the aim of our experiment.

Under special further assumptions like those about the distribution of the $\boldsymbol{a}_i$ and the $\boldsymbol{e}_{ij}$, a confidence interval for μ in model (1.1) will be constructed. Also the required precision of such an estimate has to be defined. This can be done in several ways. In our example we require that, given a confidence coefficient of $1 - \alpha = 0.95$, the half expected length of the confidence interval is not larger than a value δ in litres/hundred kilometres given in advance. The aim of an optimal statistical design may now be the determination of positive integers $a; n_1, \ldots, n_a$ in such a way that the total costs of the experimentation are minimised, but the precision required is guaranteed. For this we denote by c_1 the costs of a lorry to be used in the experiment and by c_2 the costs of a measurement (y_{ij}).

In this way stages 1 to 3 of Section 1.1 have been largely dealt with. Next we need to determine the most cost-effective experimental design, but for this we need techniques which are not dealt with until later chapters, and we will defer the determination of the optimal values of $(a; n_1, n_1, \ldots, n_a)$ until Chapter two, where we will continue with this example.

1.3 Some basic definitions

We call an experimental design optimal if it is constructed to minimise the value of a given optimality criterion while keeping some requirements of precision. For brevity we call such designs *optimal* designs. See, for instance, Shah and Sinha (1989).

In an experimental design N experimental units have to be assigned to the $a = v$ treatments either completely randomly or in a restricted random manner.

An experimental design is p-factorial or factorial if each treatment is a factor level combination of $p > 1$ factors. An experimental design is called a block design if a noise factor is eliminated by blocking. The levels of the noise factor are then called blocks. The number k_i of experimental units in the i-th block is called the block size. Block designs are discussed in Section 1.4 and Chapter six.

If an experiment involves only one treatment factor, its levels are called treatments. If there are several treatment factors, the combination of the levels of the treatment factors which are applied to a particular sampling unit is called a treatment combination and the experiment is called a factorial experiment. Factorial designs are discussed in Chapter seven.

In cases where the experimental units are randomly allocated to the treatments (or treatment combinations in factorial designs) without any restriction, we have a so-called completely randomised design. The sample size problems for these designs are discussed in Chapter two.

Experimental results are dependent on the influence of the experimental conditions defined by the *factors*. In planning an experiment, the non-random factors whose values can be determined by the research workers before or during the experiment are of special interest, because often the optimal choice of these values is a part of planning the experiment. The possible values of a factor are called *factor levels* or more succinctly, *levels*. If a factor is fixed during the experiment, or if it has only one level, we call it a *constant factor*. It is then a part of the background or the basic conditions of the experiment. If it is the objective of an experiment to investigate the influence of a factor or factors, and if the research worker varies the factor(s) systematically in the experimental design, then we call the factor(s) *design factor(s)*. If the evaluation of possible effects of some design factors is a part of the problem formulation of the experiment, then these are called *treatment factors*. The remaining design factors are called *block factors*. All the other factors that cannot systematically be varied but might have some influence on the experimental results are subsumed by the term *residual factors*. The residual and the block factors are the noise factors. Qualitative and quantitative factors are defined analogously to qualitative and quantitative characters. However, if we have a large number of factors, not all the combinations of the factor levels can be applied in a factorial experiment. Then so-called *fractional factorial* designs are in use. Their construction is shown in Chapter seven.

A *statistical experimental design* is a specification for the layout of a statistical experiment. It must ensure that no aspects which are important for the analysis or for the model choice and the attainable precision are neglected. To explain the last statement in the definition, let us consider a block design. As we will see below, a block design defines which treatment occurs in which block and the distribution of the experimental units over the blocks and the random allocation of treatments to these units. By treatments we mean the levels of a treatment factor, but in the block designs in Chapter six, treatments can also be understood as the relevant combinations of the levels of several treatment factors as they occur in Chapter seven.

Other important questions like who will do the experimental work, on which day, when and where the experiment is performed and other problems do not belong to statistical experimental design and will not be considered in this book.

Having said this, we will nevertheless use the term *experimental design* in place of *statistical experimental design.* Furthermore we would emphasize that this text deals exclusively with such designs having a non-negligible stochastic component in the observed random variables.

In this book we will consider only the design problems in the pre-experimental statistics. The analysis of the designs considered here can be found in a huge number of excellent books, some of them discussing/featuring R-programs and therefore the R-programs for the analysis do not need to be repeated here.

1.4 Block designs

If it turns out during the planning phase of an investigation that a factor (not used as a treatment factor) could influence the outcome of an experiment or survey, then we should find ways of eliminating its effect as far as is possible.

In mathematical statistics there are standard procedures for at least partly eliminating the effect of a nuisance factor.

Firstly it can be done, provided that the nuisance factor is quantitative and the dependency of the measurement of interest on it is at least approximately known by using the function describing this relation, and doing an analysis of covariance.

Even more generally applicable is randomisation. It is less effective, but does not require knowledge of what the nuisance factors might be.

Blocking (in experiments) and the analogous stratification in surveys on the one hand are similar to randomisation and on the other are further principles of experimental design and will be treated in greater detail below.

In experiments, the grouping together of experimental units which have the same level of a possible nuisance factor is called blocking; in sample surveys it is called stratification. The levels are called blocks or strata. If the factor has in reality little effect on the results, the loss of degrees of freedom in the analysis of variance of such designs may not be compensated by a decrease in the residual variance.

Blocking is especially useful when the nuisance factor is qualitative, since analysis of covariance is not then available, but blocking is also possible when the nuisance factor is quantitative. In this case we can group its values into classes and use these classes to define the levels of the nuisance factor.

Of course we need *a priori* information in order to choose the size and number of the blocks before the experiment begins (which should be the rule). In agricultural field experiments such information can be obtained from so-called uniformity trials. These are experiments which are run using only one level of the factor being investigated (for example, using only one variety of plant) in order to test whether the chosen levels of the noise factor lead to significant differences in the factor of interest. For details of field experiments, see Thomas (2006).

To construct a block design we proceed as follows. First we must specify the $b \geq 2$ levels of the nuisance factor which define the blocks. It must be possible to allocate each experimental unit unambiguously to a block. In the case of naturally occurring blocks, such as pairs of twins, the pairs of eyes of different patients, or litters of piglets, there is no problem, but when blocks (or strata in surveys) are formed artificially, care must be taken to ensure that they do not overlap and that there are no experimental units which belong to none of the blocks. The number b of the blocks is determined as part of the experimental design process. In this book we assume that all b blocks have the same block size k.

Example 1.2 In an ophthalmic hospital the effects of two kinds of eye drops A and B have to be tested. It is expected that the effects of the eye drops vary more between the individual patients than between the left and the right eye of each patient. Ten patients are available for the experiment. The research worker defines the factor "patient" as a noise factor. Each patient is considered as a block with $k = 2$ experimental units, namely the two eyes. Further we assume that the position of the eye (left or right) does not have any influence on the experimental result. The block size is two (assuming that each patient can see with both eyes). It is important to realise that the 10 patients are not the experimental units, rather, their $N = 20$ eyes. In the experimental design, one of the eyes of each of the patients has to be treated at random with one of the medicines, say by A, and the other eye with the medicine B. The reactions of the treatments are measured and in this way the difference in the effects of the eye drops can be estimated based on 10 individual comparisons.

In this example $\nu = 2$ treatments (A, B) are allocated in $b = 10$ blocks (patients) with the block size $k = 2$ (eyes). Because each treatment occurs in each block, the design is called a *complete* block design. Unless stated otherwise we assume that a treatment occurs once at most in the same block.

Besides complete block designs we will often use incomplete block designs, in which not all treatments can occur in each block. The need for such designs stems from the fact that the number of treatments v can be larger than the block size k. Suppose for example that we have to compare the effects of 5 different eye drops A, B, C, D, E. Since only two treatments can be applied to any one patient, this experiment has to be laid out as an incomplete block design (the block size is limited to $k = 2$). Such a design can be found in Table 1.1.

TABLE 1.1
Incomplete Block Design with $\nu = 5$ Treatments A, B, C, D, E in $b = 10$ Blocks with Size $k = 2$.

Patient No.	Treatments
1	A,B
2	A,C
3	A,D
4	A,E
5	B,C
6	B,D
7	B,E
8	C,D
9	C,E
10	D,E

An incomplete block design is called balanced, if each treatment is replicated the same number r of times in the experiment (here $r = 4$) and if furthermore the number λ of the occurrences of each of the possible treatment pairs in a block is constant (here 10 pairs; $\lambda = 1$). The fact that $\lambda = 1$ means that only one direct comparison between any two eye drops is possible. If we denote the results of the experiment, i.e., the effects of treatment T (T equals A, B, C, D, or E), for the i-th patient by $w_i(T)$, then the difference between, say, A and B can be estimated directly by $w_1(A)$—$w_1(B)$. We can also further compare A and B via blocks containing A and B together with the same treatment, for example,

$$\{w_2(A) - w_2(C)\} - \{w_5(B) - w_5(C)\}$$
$$\{w_3(A) - w_3(D)\} - \{w_6(B) - w_6(D)\}$$
$$\{w_4(A) - w_4(E)\} - \{w_7(B) - w_7(E)\}$$

The full analysis is described in several books like Hinkelmann and Kempthorne (1994, 2005) and many others. The construction of balanced incomplete block designs is described and demonstrated by R-programs in Chapter six.

We gather together the terms used in the above example in the following definition.

Definition 1.1 Block designs are special experimental designs for reducing the influence of nuisance factors by blocking. A relatively homogeneous group of experimental units is called a *block*. We say that a block is *complete* if each of the v treatments occurs at least once in this block; otherwise the block is called *incomplete*. A block design is called complete if all its blocks are complete, and is called *incomplete* if at least one of its blocks is incomplete.

The construction of complete block designs is no problem if they are balanced, in the sense that each treatment occurs equally often in any block and the blocks have—as already assumed—the same size k. Therefore, complete block designs will not be considered in this book (to repeat: the analysis of designs is not our task).
Thus only incomplete block designs are of interest concerning their construction. Especially of interest are designs with just one block factor and amongst them the so-called balanced designs. Designs with two and more block factors will not be considered in this book. Such designs like Latin Squares, Latin Rectangles, Greco-Latin Squares, or Change-Over Designs (Group–Period Designs) are not the scope of this book.
But regarding balanced incomplete design concerns, we will restrict ourselves to completely balanced incomplete block designs (BIBD) which are discussed in Chapter six. Another group—the partially balanced incomplete block designs—are also not considered here. A BIBD has blocks of equal size k and each treatment occurs just r times. Further every treatment pair occurs equally often (λ times) in the blocks of the design.
A partially balanced incomplete block design (PBIBD) is like a BIBD, but some treatment pairs occur λ_1 times, λ_2 times, ..., λ_k times. Then the PBIBD is called a design with k associate classes or a PBIBD(k). Thus a BIBD is a PBIBD(1). While the construction of PBIBD(k) for $k > 1$ and given association schemes is mathematically very interesting, eventually PBIBD(2) are sometimes used in practical experimentation (especially in agricultural field experiments) but relatively seldom. We therefore restrict ourselves in this book on the construction of BIBD.

1.5 About the R-programs

The statistical environment R is an implementation on the S programming language originally designed at Bell Laboratories between 1975 and 1976. See, for instance, Venables and Ripley (2003). The capabilities of R are further extended through R-packages that allow users to add specialised statistical methods as well as call procedures from C and Fortran. There exist several public repositories of R-packages; most popular is the CRAN repository hosted by R Development Team. The number of CRAN packages grows exponentially and reached the 1000 threshold in April 2007. At the time of finishing this manuscript there were more than 2200 packages.

There are many reasons for the success of R (see Fox, 2008), but this success is particularly related to its relative ease of use, and it offers great capability at the same time. The R-program and the vast majority of packages are freely available (under GNU licence, see www.gnu.org) for Windows, Linux and Apple platforms. If you have no experience with R, see Dalgaard (2008). It is hard to find something that is not yet implemented in any R-package. One of these omissions is the statistical design of experiments and sample size computation. There are a few packages for creating experimental designs; see Grömping (2009) for an overview. Furthermore in the package "agricolae" cited by Grömping (2009), we find an unsatisfactory solution like `design.bib`. We explain this statement further in Chapter six. The purpose of our R-package OPDOE is to fill existing gaps.

The organisation of the package follows the organisation of this book. Manual pages for each function refer to the page in the book, and the name of the parameters correspond to the formula in the book. We welcome suggestions for collaboration and further improvements.

The R-functions in this book and those appended in the OPDOE package follow the general naming convention: purpose.explanation.model, e.g., `power.prop_test.one_sample` determines the power (or optimal sample size, according to choice) for testing the proportion in a given sample.

Not all the programs are mentioned in our book. But the program package OPDOE with the missing programs will be available online under `www.crcpress.com/product/isbn/9781439816974`.

Part I

Determining the Minimal Size of an Experiment for Given Precision

Why is it so important that the size of a non-sequential experiment be determined before the experiment starts? The answer is that an experiment with a size chosen arbitrarily or by guessing what may be enough could lead to very high probabilities of incorrect decision. We demonstrate this by an example by which can be seen what may occur if no advance planning has been done.

Negative Example:

The following text in italics was taken from an information of the Ministry of Health in Austria entitled:

SCIENTIFIC ARGUMENTS FOR AN IMPORT BAN OF GENETICALLY MODIFIED MAIZE MON 863 (Zea mays L., line MON 863) OF MONSANTO (NOTIFICATION C/DE/02/9)

On 8th August 2005 the Decision (2005/608/EC) concerning the placing on the market, in accordance with Directive 2001/18/EC of genetically modified maize MON 863 was adopted by the Commission. The product may be placed on the market and put to the same uses as any other maize, with the exception of cultivation and uses as or in food. On 13th January 2006 the placing on the market of foods and food ingredients derived from genetically modified maize line MON 863 as novel foods or novel food ingredients under Regulation (EC) No 258/97 was authorised. With regard to the ***studies on nutritional equivalence assessment in farm animals****, which are quoted in HAMMOND et al. (2006) as scientific proof for the safety of maize MON 863, a lot of shortcomings have been detected: In this document, the scientific arguments, which are justifying the Austrian import ban of this GMO, are described. They focus particularly on the toxicological safety assessment and the antibiotic resistance marker (ARM) gene nptII, which is contained in maize MON 863, but also on the given risk management measures to prevent accidental spillage.* ***Summarizing*** *the evaluation of the* ***toxicological safety assessment*** *of the dossier, it can be stated that a lot of deficiencies are obvious: With regard to the* ***studies on nutritional equivalence assessment in farm animals****, which are quoted in HAMMOND et al. (2006) as scientific proof for the safety of maize MON 863, a lot of shortcomings have been detected: Concerning the experimental design it has to be criticised that reference groups are often contributing 60-80% of the sample size. Statistically significant differences between test and control groups are therefore often masked because group differences between iso- and transgenic diets fall into the broad range of reference groups. An important factor is also the sensitivity of the animal model: HAMMOND et al. (2006) described the use of an outbred rat model. The study compared a high number of different lines of maize, among them MON 863. The data vary considerably in and between the groups. That would allow the assumption that only effects with great deviations from the control would have been detectable with the chosen trial setup.* (Verbot des Inverkehrbringens von GV Mais Zea Mays L., MON863 in BGBL. II Nr. 257/2008)

Let us consider the design of Hammond et al. (2006) without the reference

groups. Then we have two control groups of rats fed with grain free of MON 863 and two groups fed with MON 863, one of them with 11% and one with 33% MON 863. The size of each group was $n = 20$.
Now let us assume that we wish to test the null hypothesis that there is no decrease in fecundity of rats in the treatment groups compared with the control groups. That means that there is no argument (with regard to fecundity) not to use MON 863. Let us further assume that we use a risk of the first kind of 0.1 or alternatively 0.05 (the probability that we reject the null hypothesis if it is true, the producer risk of Monsanto) and at first a risk of the second kind of 0.01 (the probability that there is a decrease of fecundity even it is present) if the decrease is 2% or larger. Then we will find out by the method described in Section 2.4.2.2., assuming that the usual fecundity is around 0.8, that in each of two groups to be compared, the group size should be 10894. An overview about the association between risk of the first and second kind, group size and minimum detectable decrease (from 0.8) as calculated by R is given below (one-sided alternative hypothesis):

Risk β	0.8 - δ	Size of each group; $\alpha = 0.1$	Size of each group; $\alpha = 0.05$
0.01	0.78	10894	13177
0.01	0.70	504	607
0.01	0.60	143	171
0.05	0.78	7201	9075
0.05	0.70	339	423
0.05	0.60	98	121
0.10	0.78	5548	7202
0.10	0.70	265	339
0.10	0.60	78	98
0.20	0.78	3838	5227
0.20	0.70	188	251
0.20	0.60	56	74

With 20 animals and with $\alpha = 0.1$ (one-sided) and $p_1 = 0.8; p_2 = 0.78$, the difference can be detected with probability 0.13 only. In the case $p_1 = 0.8; p_2 = 0.7$, the larger difference can be detected with probability 0.29 which is also not acceptable.

2

Sample Size Determination in Completely Randomised Designs

This chapter deals with the determination of the size of samples in completely randomised experiments when predetermined precision requirements should be fulfilled by just this size. That means we are looking for the smallest integer sample size by which this predetermined precision can be guaranteed. In completely randomised designs, we have further randomly assigned each experimental unit to one of the treatments when there are more than one and selected randomly the experimental units from the universe of elements.

2.1 Introduction

We start with a random variable $\boldsymbol{y}$ which follows a distribution with expectation (mean) μ and variance σ^2 (and thus standard deviation σ). If the n components $\boldsymbol{y}_i$ of the vector $(\boldsymbol{y}_1, \boldsymbol{y}_2, \ldots, \boldsymbol{y}_n)$ are mutually independent and all distributed exactly as $\boldsymbol{y}$, then the vector is called an *abstract random sample* of size n or an i.i.d. (identically and independently distributed) vector. To derive statistical procedures, distributional assumptions often have to be made. Most of the procedures for quantitative characters are derived under the assumption of normally distributed $\boldsymbol{y}$. But due to robustness results published in Rasch and Guiard (2004) and Rasch et al. (2007a), we do not need these assumptions for practical applications, especially not in sample size determination. An abstract random sample is a model for the random sampling process in selecting experimental units from the universe of units which will be defined later. If the components of an abstract random sample are distributed with mean μ and variance σ^2, then the arithmetic mean of those components is a random variable, whose mean is also μ and whose variance is σ^2/n.

2.1.1 Sample size to be determined

Our philosophy is that the size of an experiment (or of a survey) should be determined following precision requirements fixed in advance by the researcher

or a group of researchers. What can be done if the size calculated in this way is unrealistic will be discussed in Section 2.1.2.
To determine the sample size for the **point estimation** of the expectation μ of a random sample so that its variance is smaller than a given constant C means that

$$\frac{\sigma^2}{n} \leq C \text{ or } n = \left\lceil \frac{\sigma^2}{C} \right\rceil \tag{2.1}$$

where $\lceil x \rceil$ is the smallest integer larger or equal to x. Things in point estimation are very simple, but we have to know σ^2. But this is usually not the case, and therefore we introduce the following procedure.

Procedure I:

1. We look for some *a priori* information (an experiment in the past, results from the literature) $\tilde{\sigma}^2$ about σ^2 and use this as a rough estimate of σ^2. Of course such an approach gives only an estimate of the sample size n we are looking for.

2. If step 1 is not possible because we cannot get such information, then we can obtain an estimate of the unknown variance by guessing the largest possible and the smallest possible observations and calculating the difference between them. We divide this difference by 6 to estimate σ.

3. If even step 2 will not work or seems to be too heuristic, we can get a better solution for this problem by using a sequential procedure based on a two-stage method of Stein (1945)—see also Chapter five. In the first stage of an experiment with between 10 and 30 units, the variance σ^2 is estimated. The estimate is then used to calculate the total sample size. If the total sample size is smaller than or equal to the number used in the first step, then our chosen size for the first stage is sufficient; otherwise the rest of the experimental units must be used to generate a second sample in the procedure.

Procedure I will be used in other sections also. The abstract random sample defined above as an i.i.d. vector of random variables is a model for an exact random sample in applications. A random sample in reality is to be understood as a sample obtained by a random sampling procedure taking a set of n elements from a larger universe which often is assumed to be of very large size N (in theory to be infinite). The letter N in this book will be used twice. On the one hand it is the size of the universe. But on the other hand, if we have more than one sample in an experiment with sample sizes n_i; $i = 1, \ldots, a$, the total size of an experiment is also denoted by $N = \sum_{i=1}^{a} n_i$.

Definition 2.1 An exact unrestricted random sample of size n (without replacement) is defined as a sample obtained by a unrestricted random sampling procedure, which is defined as a procedure where each element of the N elements of the universe has the same probability to become an element of the sample with the additional property that each of the $\binom{N}{n}$ possible subsets has the same probability to become the sample. Sampling with replacement (where drawn elements are brought back to the population and thus can be drawn again) will not be considered in our book. If the population is infinite or N very large, both sampling procedures are identical. The base R-function `sample` allows us to draw a random sample. For this we assume that the elements of the universe are numbered from 1 to N and that we will randomly draw a sample of size n.

Example 2.1 Random sampling. The following example demonstrates how to use R to draw a random sample without replacement.

We will play Lotto. In Germany, twice every week 6 numbers are selected at random without replacement from 49; in Austria, from 45 numbers (the first nonzero integers). This is done by drawing balls from a sphere. Whether this is really done in a way that each ball has the same probability $\frac{1}{49}$ or $\frac{1}{45}$ to become an element of the subset of size 6 of winning numbers is not quite clear. Our R-program certainly fulfills the random selection.

We use the `sample` function to draw a random sample of size $n = 6$ from a universe of size $N = 49$.

```
> set.seed(123)
> sample(49, 6)

[1] 15 38 20 41 43  3
```

Remark: Using the same function, we can also draw a random sample with replacement. Default is drawing without replacement. If one wants to draw a sample of size 6 from 49 with replacement the command is: `sample (49, 6, replace = TRUE)`.

2.1.2 What to do when the size of an experiment is given in advance

There are cases where a limited number of experimental units is available or the resources for a research project are limited and allow for investigation of a given number of experimental units only. In such cases, our R-program must be able to calculate the precision achievable by samples of a given size. The researcher then has the choice of *performing an experiment with a low but perhaps still acceptable precision* or of saving money and manpower by not *investigating the problem under such circumstances.* The sample size commands in OPDOE can be used in any direction. Either we insert values for the precision and calculate the size or we insert the size and calculate the precision.

In the following sections, we will use examples of different situations to show how the R-programs can be used to obtain information about the precision achievable by limited sizes.

2.2 Confidence estimation

In confidence estimation, we define a random region in the parameter space of some distribution which covers the parameter with probability $1-\alpha$. We restrict ourselves in this chapter to cases where the parameter space is the real line or part of it, and therefore the regions are intervals.

In confidence estimation for real location parameters, we wish to obtain an interval as small as possible for a given confidence coefficient. But what does this mean if the interval is one-sided and one bound is infinite or if, as usual, the bounds are random? In case of two-sided intervals, we consider the half-length of the interval, and in case of one-sided intervals, the distance of the finite bound from the estimator for the location parameter. If length or distance is random, we consider the expected length (or distance), and our precision requirement is now defined by the confidence coefficient $1-\alpha$ (the probability that the random interval covers the corresponding parameter) and by an upper bound for the half expected length in the two-sided case or the expected distance of the random bound from the estimator of the parameter discussed in the one-sided case. In case of a scale parameter, the length or distance is replaced by a ratio, for instance, of one bound to the midpoint of the interval.

2.2.1 Confidence intervals for expectations

Expectations are location parameters; therefore, we work with distances and lengths.

2.2.1.1 One-sample case, σ^2 known

Suppose we have a random sample in form of the vector $(\boldsymbol{y}_1, \boldsymbol{y}_2, \ldots, \boldsymbol{y}_n)$ from a particular normal distribution $N(\mu, \sigma^2)$, and suppose that σ^2 is known. Let $\overline{\boldsymbol{y}}$ be the arithmetic mean of this sample. We need a confidence interval for μ. Writing u_P or sometimes $u(P)$ for the P-quantile of the standard normal distribution, a two-sided $(1-\alpha)$-confidence interval is given by

$$\left[\boldsymbol{\bar{y}} - u\left(1 - \frac{\alpha}{2}\right)\frac{\sigma}{\sqrt{n}}; \boldsymbol{\bar{y}} + u\left(1 - \frac{\alpha}{2}\right)\frac{\sigma}{\sqrt{n}}\right] \tag{2.2}$$

and the two one-sided $(1 - \alpha)$-confidence intervals for μ are

$$\left[-\infty; \boldsymbol{\bar{y}} + u(1 - \alpha)\frac{\sigma}{\sqrt{n}}\right] \tag{2.3}$$

and

$$\left[\boldsymbol{\bar{y}} - u(1 - \alpha)\frac{\sigma}{\sqrt{n}}; \infty\right] \tag{2.4}$$

respectively.

It can be shown that under the assumptions above, interval (2.2) has the smallest width of all $(1 - \alpha)$-confidence intervals. If we split α otherwise into two parts α_1, α_2 so that $\alpha_1 + \alpha_2 = \alpha$, we obtain intervals of larger length.

The following precision requirements can be used for all confidence intervals for location parameters.

Precision requirement Type A

Two-sided intervals: Given the confidence coefficient $(1-\alpha)$, the half expected width of a two-sided interval should be smaller than δ.

One-sided intervals: Given the confidence coefficient $(1 - \alpha)$, the difference between the random finite bound of a one-sided interval and the estimator of the parameter (expectation) should be smaller than δ.

The half expected width *E(**H**)* of (2.2) is

$$E(\boldsymbol{H}) = u\left(1 - \frac{\alpha}{2}\right)\frac{\sigma}{\sqrt{n}}$$

From

$$u\left(1 - \frac{\alpha}{2}\right)\frac{\sigma}{\sqrt{n}} \leq \delta,$$

it follows that for two-sided intervals,

$$n = \left\lceil u^2\left(1 - \frac{\alpha}{2}\right)\frac{\sigma^2}{\delta^2}\right\rceil \tag{2.5}$$

is the minimum sample size.

For one-sided intervals we obtain the minimum sample size as

$$n = \left\lceil u^2 (1-\alpha) \frac{\sigma^2}{\delta^2} \right\rceil \tag{2.6}$$

In the sequel we only deal with two-sided confidence intervals. When readers like to construct one-sided intervals they simply should replace $1-\frac{\alpha}{2}$ by $1-\alpha$.

Precision requirement Type B

If on the other hand the precision requirement states that both confidence limits must not differ by more than δ from the parameter μ with given probability $1-\beta$, then, following (Bock, 1998) we must choose

$$n = \left\lceil \frac{\sigma^2}{\delta^2} \left[u\left(1-\frac{\alpha}{2}\right) + u\left(1-\frac{\beta}{2}\right) \right]^2 \right\rceil \tag{2.7}$$

in the two-sided case. The R-program and an example of its use will be explained in the more realistic case of unknown variance but it is applicable for the case of known variances also.

2.2.1.2 One-sample case, σ^2 unknown

In the case of unknown variance, we have to use ***Procedure I*** above, and we have to replace the formulae (2.2)–(2.7). First we need the half expected width of the corresponding two-sided confidence interval. Writing $t(f;P)$ for the P-quantile of the central t–distribution with f degrees of freedom, the interval is given by:

$$\left[\overline{\boldsymbol{y}} - t\left(n-1; 1-\frac{\alpha}{2}\right)\frac{\boldsymbol{s}}{\sqrt{n}};\ \overline{\boldsymbol{y}} + t\left(n-1; 1-\frac{\alpha}{2}\right)\frac{\boldsymbol{s}}{\sqrt{n}} \right]$$

and its half expected length is:

$$E(\boldsymbol{H}) = t\left(n-1; 1-\frac{\alpha}{2}\right)\frac{E(\boldsymbol{s})}{\sqrt{n}} = \frac{t\left(n-1; 1-\frac{\alpha}{2}\right)\Gamma\left(\frac{n}{2}\right)\cdot\sqrt{2}}{\sqrt{n}\,\Gamma\left(\frac{n-1}{2}\right)\sqrt{n-1}}\sigma$$

$E(\boldsymbol{H}) \le \delta$ leads to the equation for n:

$$n = h(n) = \left\lceil t^2\left(n-1; 1-\frac{\alpha}{2}\right)\frac{2\cdot\Gamma^2\left(\frac{n}{2}\right)\cdot\sigma^2}{\Gamma^2\left(\frac{n-1}{2}\right)(n-1)\delta^2} \right\rceil \tag{2.8}$$

To solve this and other analogue equations with an implicit n we need the following procedure.

Procedure II

Choose a certain starting value n_0 and calculate iteratively the values

$n_1, n_2, n_3, \ldots$ by the recursive relation $n_{i+1} = h(n_i)$ until either $n_i = n_{i-1}$ or $|n_i - n_{i-1}| \geq |n_{i-1} - n_{i-2}|$. In the first case $n = n_i = n_{i-1}$ is the solution. In the second case we can find a solution by systematic search starting with $n = \min(n_{i-1}, n_{i-2})$. This algorithm often converges to the solution after one or two steps; we will use it in subsequent sections and chapters and of course in OPDOE.

Remember: Procedure I is already given in Section 2.1.1. When we start with $n = \infty$, but use approximately $\frac{2\cdot\Gamma^2\left(\frac{n}{2}\right)}{\Gamma^2\left(\frac{n-1}{2}\right)(n-1)} = 1$, we start with equation (2.5). In the one-sided case, the sample size formula is analogous to (2.8):

$$n = \left\lceil t^2(n-1;\ 1-\alpha)\frac{2\cdot\Gamma^2\left(\frac{n}{2}\right)\cdot\sigma^2}{\Gamma^2\left(\frac{n-1}{2}\right)(n-1)\delta^2}\right\rceil \tag{2.9}$$

The calculation can be done by the R built-in function `power.t.test`.

Example 2.2 We calculate the sample size needed to construct a two-sided confidence interval for $\alpha = 0.05$ and $\delta = 0.25\sigma$.
We find for $n_0 = \infty$; $t(\infty; 0.975) = 1.96$, and from (2.8) we obtain

$$n_1 = \left\lceil \frac{1.96^2}{0.25^2} \right\rceil = \lceil 61.47 \rceil = 62$$

Next we look up $t(61; 0.975) = 1.9996$, and this gives

$$n_2 = \left\lceil \frac{1.99996^2}{0.25^2} \right\rceil = \lceil 63.97 \rceil = 64$$

Finally with $t(63; 0.975) = 1.9983$, n_3 becomes

$$n_3 = \left\lceil \frac{1.9983^2}{0.25^2} \right\rceil = \lceil 63.89 \rceil = 64$$

and thus $n = 64$ is the solution. The operator $\lceil\ \rceil$ accelerates the convergence of the iteration algorithm.

With a slight modification we can use the R built-in function `power.t.test`, developed for testing problems and described in Section 2.4. The modification is to use power $= 1 - \beta = 0.5$ then formula (2.33) is for large n identical with formula (2.8) because then $t(n-1;\ 1-\beta) = 0$.

With the R built-in function `power.t.test` we calculate:

```
> power.t.test(type = "one.sample", power = 0.5, delta = 0.25,
+       sd = 1, sig.level = 0.05, alternative = "two.sided")
```

```
     One-sample t test power calculation

              n = 63.4024
          delta = 0.25
             sd = 1
      sig.level = 0.05
          power = 0.5
    alternative = two.sided
```

For convenience we provide a call wrapper `size.t.test` which returns the desired integer size parameter:

```
> size.t.test(type = "one.sample", power = 0.5, delta = 0.25,
+     sd = 1, sig.level = 0.05, alternative = "two.sided")

[1] 64
```

Hence we will choose $n = 64$.

If we took a random sample of size 20, for example, then by our call wrapper to the R-built-in-function `delta.t.test` the half expected width for a given α can be calculated in the following way:

```
> delta.t.test(type = "one.sample", power = 0.5, n = 20, sd = 1,
+     sig.level = 0.05, alternative = "two.sided")

[1] 0.4617063
```

If, on the other hand, the precision requirement states that both confidence limits must not differ by more than δ from the parameter μ with given probability $1-\beta$, then, in analogy to (2.7), we must choose in the two-sided case

$$n = \left\lceil \left[t\left(n-1;\, 1-\frac{\alpha}{2}\right) + t\left(n-1;\, 1-\frac{\beta}{2}\right)\right]^2 \frac{2\cdot\Gamma^2\left(\frac{n}{2}\right)\cdot}{\Gamma^2\left(\frac{n-1}{2}\right)(n-1)} \frac{\sigma^2}{\delta^2} \right\rceil \quad (2.10)$$

Example 2.2 – continued

We calculate the sample size needed to construct a two-sided confidence interval using the precision requirement type B:

```
> size.t.test(type = "one.sample", power = 0.95, delta = 0.25,
+     sig.level = 0.05, alternative = "two.sided")

[1] 210
```

2.2.1.3 Confidence intervals for the expectation of the normal distribution in the presence of a noise factor

The mean μ of a normally distributed variable may have to be estimated when the observations depend on a noise factor.

Example 1.1 – continued

Occasionally the average fuel consumption of a certain type of a lorry has to be estimated using data on the performance obtained from different vehicles which may have come from different production runs. The factor "lorry" is thus a random noise factor (or blocking factor) and should therefore be eliminated as far as possible. For the investigation, a random sample of vehicles is to be drawn from the population of all such lorries produced. An adequate model for this situation is the model equation of a Model II one-way analysis of variance (see end of Section 3.2).

Planning this experiment thus means determining the number a of cars as the levels of the noise factor (blocking factor) and the number n_i of journeys by vehicle $i(i = 1, ..., a)$. The required precision for the estimation of the mean μ is defined by the half expected width δ of a confidence interval for μ and the confidence coefficient $(1 - \alpha)$.

For an optimal design of experiment, we need estimates s_a^2 and s_R^2 of the variance components σ_a^2 (between the cars) and σ_R^2 (residual variance) or an estimate $\hat{\theta}$ of the intraclass-correlation coefficient $\theta = \sigma_a^2/(\sigma_a^2 + \sigma_R^2)$.

If no costs are taken into account, we define a design as optimal if it minimises $N = \sum n_i$. Without any restrictions on n_i and a, this leads to $n_i = n = 1$ and N defined by

$$N = a = \left\lceil \frac{t^2\left(N-1;\ 1-\frac{\alpha}{2}\right)}{c^2} \right\rceil \tag{2.11}$$

where

$$c = \frac{\delta}{\sqrt{s_a^2 + s_R^2}} \tag{2.12}$$

If, on the other hand a lower bound, n_0 for n_i, is given, then we let $n_i = n_0, (i = 1, \ldots, a)$ and calculate a iteratively using Procedure II from

$$a = \left\lceil \frac{\left(\hat{\theta} - \frac{1-\hat{\theta}}{n_0}\right) t^2\left(a-1;\ 1-\frac{\alpha}{2}\right)}{c^2} \right\rceil \tag{2.13}$$

where c is as in (2.12) and $\hat{\theta} = s_a^2/(s_a^2 + s_R^2)$.

If a is fixed in advance (for example, as an upper limit), we must first check the condition

$$c^2 > \frac{\hat{\theta}}{a} \cdot t^2\left(a-1;\ 1-\frac{\alpha}{2}\right) \tag{2.14}$$

If this is fulfilled, then the value n^* is calculated from

$$n^* = \left\lceil \frac{(1-\hat{\theta})t^2\left(a-1;\ 1-\frac{\alpha}{2}\right)}{c^2 a - \hat{\theta} t^2\left(a-1;\ 1-\frac{\alpha}{2}\right)} \right\rceil \tag{2.15}$$

If (2.14) is not fulfilled, the problem cannot be solved. A cost-optimal design can be found in Herrendörfer and Schmidt (1978).

An exact confidence interval can be calculated if the number n of journeys made by each lorry is constant. Let

$$\overline{\boldsymbol{y}}_{..} = \frac{1}{an}\sum_{i=1}^{a}\sum_{j=1}^{n}\boldsymbol{y}_{ij} \tag{2.16}$$

be the average of all $N = an$ observations. We obtain the $(1-\alpha)$-confidence interval

$$\left[\overline{\boldsymbol{y}}_{..} - t\left(a-1;\ 1-\frac{\alpha}{2}\right)\sqrt{\frac{\boldsymbol{MS}_A}{N}};\overline{\boldsymbol{y}}_{..} + t\left(a-1;\ 1-\frac{\alpha}{2}\right)\sqrt{\frac{\boldsymbol{MS}_A}{N}}\right] \tag{2.17}$$

with $\boldsymbol{MS}_A$ from a one-way ANOVA-table (see Table 3.2) (independent of the model).

Example 2.3 We consider the situation of Example 1.1 for the following precision requirements: we want to find a 95% confidence interval for μ, having an half expected width no greater than 0.3 (y_{ij} is the fuel consumption in litres per 100km).

If we know beforehand that the sum of the variance components is 0.36 = 0.3 + 0.06, then using equation (2.12) we find that $c = 0.5$. We want to find the design $(a, n_1, \ldots, n_a)$ which minimises $N = \sum n_i$. As we know, we must use each selected vehicle for one journey only ($n_i = 1$ for all i) and use equation (2.13) to determine the number of vehicles. We start with $a = \infty$ and find that $t(\infty; 0.975) = 1.96$, giving $a = 16$. Then $t(15; 0.975) = 2.1314$ gives the value $a = 19$, and $t(18; 0.975) = 2.1009$ leads to $a = 18$. Since $t(17; 0.975) = 2.1098$, the final value of a is 18.

Each of the 18 lorries has to be used once in the experiment; this gives a total number of trips equal to 18. If we look for a cost-optimal design and assume that a car costs \$40000 and driving with such a car costs \$50; then we have to use 16 lorries driving 3 times each; this gives a total number of trips equal to 48.

The first design is size-optimal, the second is cost-optimal. The cost of the first design is $18 * (40000 + 50) = \$720900$ and the cost of the cost-optimal design is $16 * 40000 + 48 * 50 = \$642400$.

2.2.1.4 One-sample case, σ^2 unknown, paired observations

We consider a two-dimensional random sample
$\left\{ \begin{array}{l} \boldsymbol{x}_1, \boldsymbol{x}_2, \ldots, \boldsymbol{x}_n \\ \boldsymbol{y}_1, \boldsymbol{y}_2, \ldots, \boldsymbol{y}_n \end{array} \right\}$ with expectation (of each component) $\begin{pmatrix} \mu_x \\ \mu_y \end{pmatrix}$
and unknown covariance matrix

$$\sum = \begin{pmatrix} \sigma_x^2 & \rho\sigma_x\sigma_y \\ \rho\sigma_x\sigma_y & \sigma_y^2 \end{pmatrix}$$

Again the precision requirement is given by specifying the half expected width δ of a confidence interval (CI) for the difference $\mu_x - \mu_y$ and the confidence coefficient $1-\alpha$. We consider in the sequel the case of unknown variance only; known variances are more or less academic.

For the determination of the minimal sample size, an estimate $s_d^2 = \frac{1}{n-1} \sum_{i=1}^{n} [(\boldsymbol{x}_i - \boldsymbol{y}_i) - (\overline{\boldsymbol{x}} - \overline{\boldsymbol{y}})]^2$ for the variance σ_Δ^2 of the difference $\boldsymbol{d} = \boldsymbol{x} - \boldsymbol{y}$ with $\sigma_\Delta^2 = \sigma_1^2 + \sigma_2^2 - 2\rho \cdot \sigma_2 \cdot \sigma_2$ is needed.

The two-sided $(1-\alpha)$-CI with $\overline{\boldsymbol{d}} = \frac{\sum_{i=1}^{n} (\boldsymbol{x}_i - \boldsymbol{y}_i)}{n}$ is

$$\left[\overline{\boldsymbol{d}} - t\left(n-1; 1-\frac{\alpha}{2}\right) \frac{\boldsymbol{s}_d}{\sqrt{n}}, \ \overline{\boldsymbol{d}} + t\left(n-1; 1-\frac{\alpha}{2}\right) \frac{\boldsymbol{s}_d}{\sqrt{n}} \right]$$

Depending on δ and α, the sample size n is determined for the two-sided interval analogously to (2.8) with σ_Δ^2 in place of σ^2:

$$n = \left\lceil t^2\left(n-1; 1-\frac{\alpha}{2}\right) \frac{2 \cdot \Gamma^2\left(\frac{n}{2}\right) \cdot \sigma_\Delta^2}{\Gamma^2\left(\frac{n-1}{2}\right)(n-1)\delta^2} \right\rceil \tag{2.18}$$

The one-sided intervals and the intervals for Type B can be obtained analogously to (2.9).

The calculation can be done by the R-built-in function `power.t.test`.

Example 2.4 If we want to determine the size of an experiment for paired observations so that the half expected width of the interval must not be greater than $\delta = 1.2$, we use as prior information the estimate of the standard deviation $s_d = 3.653$ of the $\boldsymbol{d}_i = \boldsymbol{x}_i - \boldsymbol{y}_i$. Performing the calculation by hand, we start with $n_0 = \infty, t(\infty;\ 0.95) = 1.6449$. Then we find $n_1 = 26$, $t(25; 0.95) = 1.7081, n_2 = 28, t(27; 0.95) = 1.7033, n_3 = 27,\ t(26; 0.95) = 1.7056, n_4 = 27$, so that 27 data pairs will be required.

With our call wrapper size.t.test in OPDOE we get

```
> size.t.test(type = "paired", power = 0.5, delta = 1.2,
+ sd = 3.653, sig.level = 0.05, alternative = "one.sided")

[1] 27
```

Hence we choose $n = 27$.

2.2.1.5 Two-sample case, σ^2 unknown, independent samples—equal variances

If in the two-sample case the two variances are equal, usually from the two samples $(\boldsymbol{x}_1, \boldsymbol{x}_2, \ldots, \boldsymbol{x}_{n_x})$ and $(\boldsymbol{y}_1, \boldsymbol{y}_2, \ldots, \boldsymbol{y}_{n_y})$ with sample means and sample variances $\boldsymbol{s}_x^2; \boldsymbol{s}_y^2$ a pooled estimator $\boldsymbol{s}^2 = \frac{(n_x-1)\boldsymbol{s}_x^2+(n_y-1)\boldsymbol{s}_y^2}{n_x+n_y-2}$ of the common variance is calculated.
The two-sided $(1-\alpha)$-CI is

$$\left[\overline{\boldsymbol{x}} - \overline{\boldsymbol{y}} - t\left(n_x + n_y - 2;\ 1 - \frac{\alpha}{2}\right) \boldsymbol{s}\sqrt{\frac{n_x + n_y}{n_x n_y}}\right.,$$

$$\left.\overline{\boldsymbol{x}} - \overline{\boldsymbol{y}} + t\left(n_x + n_y - 2;\ 1 - \frac{\alpha}{2}\right) \boldsymbol{s}\sqrt{\frac{n_x + n_y}{n_x n_y}}\right]$$

The lower $(1-\alpha)$-confidence interval is given by

$$\left[\overline{\boldsymbol{x}} - \overline{\boldsymbol{y}} - t\left(n_x + n_y - 2;\ 1 - \alpha\right) \boldsymbol{s}\sqrt{\frac{n_x + n_y}{n_x n_y}};\ \infty\right]$$

and the upper one by

$$\left[-\infty,\ \overline{\boldsymbol{x}} - \overline{\boldsymbol{y}} + t\left(n_x + n_y - 2;\ 1 - \alpha\right) \boldsymbol{s}\sqrt{\frac{n_x + n_y}{n_x n_y}}\right]$$

In our case of equal variances it can be shown that optimal plans require the two sample sizes n_x and n_y to be equal. Thus $n_x = n_y = n$, and in the case where the half expected width must be less than δ, we find n iteratively by Procedure II from

$$n = \left\lceil 2\sigma^2 \frac{t^2\left(2n-2;\ 1-\frac{\alpha}{2}\right)}{\delta^2(2n-2)} \frac{2\Gamma^2\left(\frac{2n-1}{2}\right)}{\Gamma^2(n-1)} \right\rceil \tag{2.19}$$

For the type B intervals, precision demands that the bounds of a $(1-\alpha)$-confidence interval differ at most with probability β from $\mu_x - \mu_y$. The analogue sample size formula is

$$n = \left\lceil 2\sigma^2 \frac{\left[t\left(2n-2;\, 1-\frac{\alpha}{2}\right) + t\left(2n-2;\, 1-\frac{\beta}{2}\right)\right]^2}{\delta^2(2n-2)} \frac{2\Gamma^2\left(\frac{2n-1}{2}\right)}{\Gamma^2\left(n-1\right)} \right\rceil \tag{2.20}$$

The calculation can be done by the R-built-in function `power.t.test` and the call wrapper `size.t.test`, respectively.

Example 2.5 We want to find the minimum size of an experiment to construct a two-sided 99% confidence interval for the difference of the expectations of two normal distributions. We assume equal variances and take independent samples from each population and define the precision by using (2.19) and for precision type B additionally by $\beta = 0.03$ using (2.20). For (2.19):

```
> size.t.test(power = 0.5, sig.level = 0.01, delta = 0.5,
+ sd = 1, type = "two.sample")

[1] 55
```

For (2.20):

```
> size.t.test(power = 0.97, sig.level = 0.01, delta = 0.5,
+ sd = 1, type = "two.sample")

[1] 161
```

Remark: If one is not absolutely sure that the two variances are equal, one should use the confidence interval for unequal variances described below, as it was recommended by Rasch et al. (2009).

2.2.1.6 Two-sample case, σ^2 unknown, independent samples — unequal variances

If the variances are unequal, but also if we are not sure that they are equal, the confidence interval is based on the usual estimators $\boldsymbol{s}_x^2$ and $\boldsymbol{s}_y^2$ of σ_x^2 and σ_y^2 respectively. The confidence interval (2.21) is only approximately a (1-α)-confidence interval (see Welch, 1947). It is given by

$$\left[\boldsymbol{\overline{x}} - \boldsymbol{\overline{y}} - t\left(f^*;\, 1-\frac{\alpha}{2}\right)\sqrt{\frac{\boldsymbol{s}_x^2}{n_x} + \frac{\boldsymbol{s}_y^2}{n_y}};\ \boldsymbol{\overline{x}} - \boldsymbol{\overline{y}} + t\left(f^*;\, 1-\frac{\alpha}{2}\right)\sqrt{\frac{\boldsymbol{s}_x^2}{n_x} + \frac{\boldsymbol{s}_y^2}{n_y}}\right] \tag{2.21}$$

with

$$f^* = \frac{\left(\frac{s_x^2}{n_x} + \frac{s_y^2}{n_y}\right)^2}{\frac{s_x^4}{(n_x-1)n_x^2} + \frac{s_y^4}{(n_y-1)n_y^2}} \tag{2.22}$$

To determine the necessary sample sizes n_x and n_y, besides an upper bound δ, for the half expected width we will need about the two variances. Suppose that a priori information s_x^2 and s_y^2 are available for the variances σ_x^2 and σ_y^2 which possibly may be (or are known to be) unequal. For a two-sided confidence interval, we can calculate n_x and n_y approximately (by replacing the σ's with the s's) and iteratively using Procedure II from

$$n_x = \left\lceil \frac{\sigma_x(\sigma_x + \sigma_y)}{\delta^2} t^2 \left(f^*;\ 1 - \frac{\alpha}{2}\right) \right\rceil \tag{2.23}$$

and

$$n_y = \left\lceil \frac{\sigma_y}{\sigma_x} n_x \right\rceil \tag{2.24}$$

The calculation can be done by our OPDOE-function `sizes.confint.welch`.

Example 2.6 Given the minimum size of an experiment, we like to find a two-sided 99% confidence interval for the difference of the expectations of two normal distributions with unequal variances using independent samples from each population and define the precision by $\delta = 0.4\sigma_x$ using (2.23) for n_x. If we know that $\frac{\sigma_x^2}{\sigma_y^2} = 4$, we receive $n_y = \left\lceil \frac{1}{2} n_x \right\rceil$. The R output shows that we need knowledge about 98 observations—65 from the first distribution and 33 from the second distribution.

2.2.2 Confidence intervals for probabilities

Let us presume n independent trials in each of which a certain event A occurs with the same probability p. A $(1-\alpha)$-confidence interval for p can be calculated from the number of observations $\boldsymbol{y}$ of this event as $[\boldsymbol{l}(n, \boldsymbol{y}, \alpha); \boldsymbol{u}(n, \boldsymbol{y}, \alpha)]$, with the lower bound $\boldsymbol{l}(n, \boldsymbol{y}, \alpha)$ and the upper bound, $\boldsymbol{u}(n, \boldsymbol{y}, \alpha)$, respectively, given by

$$\boldsymbol{l}(n, \boldsymbol{y}, \alpha) = \frac{\boldsymbol{y}}{\boldsymbol{y} + (n - \boldsymbol{y} + 1)F\left[2(n - \boldsymbol{y} + 1);\ 2\boldsymbol{y};\ 1 - \frac{\alpha}{2}\right]} \tag{2.25}$$

$$\boldsymbol{u}(n, \boldsymbol{y}, \alpha) = \frac{(\boldsymbol{y} + 1)F\left[2(\boldsymbol{y} + 1);\ 2(n - \boldsymbol{y});\ 1 - \frac{\alpha}{2}\right]}{n - \boldsymbol{y} + (\boldsymbol{y} + 1)F\left[2(\boldsymbol{y} + 1);\ 2(n - \boldsymbol{y});\ 1 - \frac{\alpha}{2}\right]} \tag{2.26}$$

$F(f_1, f_2, P)$ is the P-quantile of an F-distribution with f_1 and f_2 degrees of freedom. When we use the error propagation law in deriving the half expected width, only an approximation for the sample size needed is obtained.
To determine the minimum sample size approximately, it seems better to use the half expected width of an approximate confidence interval for p. Such an interval is given by

$$\left[\frac{\boldsymbol{y}}{n} - u_{1-\frac{\alpha}{2}}\sqrt{\frac{1}{n}\frac{\boldsymbol{y}}{n}\left(1-\frac{\boldsymbol{y}}{n}\right)};\ \frac{\boldsymbol{y}}{n} + u_{1-\frac{\alpha}{2}}\sqrt{\frac{1}{n}\frac{\boldsymbol{y}}{n}\left(1-\frac{\boldsymbol{y}}{n}\right)}\right]$$

This interval has a half expected width of approximately $u_{1-\frac{\alpha}{2}}\sqrt{\frac{p(1-p)}{n}}$. The requirement that this is smaller than δ leads to

$$n = \left\lceil \frac{p(1-p)}{\delta^2} u^2\left(1-\frac{\alpha}{2}\right)\right\rceil. \tag{2.27}$$

If nothing is known about p, we must take the least favorable case $p = 0.5$ which gives the maximum of the minimal sample size. Here, as well as later in ANOVA (Chapter three), this is termed the *maximin size*.

The calculation can be done by the OPDOE-function `size.prop.confint`.

Example 2.7 We require a confidence interval for the probability $p = P(A)$ of the event A: "An iron casting is faulty." The confidence coefficient is specified as 0.90. How many castings would need to be tested if the expected width of the interval shall be

a) not greater than $2\delta = 0.1$, and nothing is known about p?

b) not greater than $2\delta = 0.1$, when we know that at most 10% of castings are faulty?

For case a the R-program gives the required sample size as $n = 271$. If from prior knowledge we can put an upper limit of $p = 0.1$ (case b), then we get a smaller minimum sample size of $n = 98$:

```
> size.prop.confint(p = NULL, delta = 0.05, alpha = 0.1)

[1] 271

> size.prop.confint(p = 0.1, delta = 0.05, alpha = 0.1)

[1] 98
```

2.2.3 Confidence interval for the variance of the normal distribution

Variances or, better, standard deviations are scale parameters; therefore, we replace distances by ratios. A confidence interval for the variance σ^2 of a normal distribution with confidence coefficient $1-\alpha$ is given by

$$\left[\frac{(n-1)\boldsymbol{s}^2}{\chi^2\left(n-1;\ 1-\frac{\alpha}{2}\right)};\ \frac{(n-1)\boldsymbol{s}^2}{\chi^2\left(n-1;\ \frac{\alpha}{2}\right)}\right]$$

$\chi^2(n-1;\ 1-\alpha/2)$ and $\chi^2(n-1;\ \alpha/2)$ are the quantiles of the χ^2-distribution with $n-1$ degrees of freedom. The half expected width is

$$\delta = \frac{1}{2}\sigma^2(n-1)\left[\frac{1}{\chi^2\left(n-1;\ \frac{\alpha}{2}\right)} - \frac{1}{\chi^2\left(n-1;\ \frac{1-\alpha}{2}\right)}\right]$$

and from this we obtain

$$n = \left\lceil \frac{2\delta}{\sigma^2}\left[\frac{1}{\chi^2(n-1,\frac{\alpha}{2})} - \frac{1}{\chi^2(n-1,1-\frac{\alpha}{2})}\right]^{-1}\right\rceil + 1 \tag{2.28}$$

If we define the precision by the relative half expected width of the confidence interval $\delta_{rel} = \frac{\delta}{\tau}$ with τ defined as the expected mid-point of the interval, then

$$\delta_{rel} = \frac{\chi^2(n-1,1-\frac{\alpha}{2}) - \chi^2(n-1,\frac{\alpha}{2})}{\chi^2(n-1,1-\frac{\alpha}{2}) + \chi^2(n-1,\frac{\alpha}{2})}$$

The sample size has to be calculated so that the ratio has a value given as part of the precision requirement.

The calculation can be done by our OPDOE-function `size.variance.confint`.

Example 2.8 If δ, the half expected width of a confidence interval for the variance is fixed as $\delta = 0.25\sigma^2$ and $\alpha = 0.05$, then we need $n = 132$ observations. If we demand that $\delta_{rel} = 0.25$ and $\alpha = 0.05$, then we need $n = 120$ observations.

```
size.variance.confint(alpha = 0.05, delta = 0.25)
size.variance.confint(alpha = 0.05, deltarel = 0.25)
```

2.3 Selection procedures

The aim of research in empirical science is very often to select the best one out of a given set of possibilities. There are many examples from different areas where the aim of an investigation or experiment is to find out which is "best" out of a number of possibilities. In each particular case we have to define exactly what we mean by "best", as part of the problem formulation. If we model each of the possibilities by a population, and the characteristic of interest by a random variable whose distributions in the various populations differ only in the value of one of the parameters, then we can define "best" in terms of this parameter. The best case can be the one with the largest probability (for instance, of leading to a cure), or the one with the largest expectation (for instance, for the harvest of a particular fruit), or the one with the smallest variance (for instance, for the udder size of cows in connection with the use of a special milking technology). Although there are very many practical examples which are essentially the same as the examples above, the statistical selection methods developed during the last 45 years (see Miescke and Rasch, 1996) are almost never applied. Indeed, even in most statistical text books, neither theoretical nor practical aspects of statistical selection procedures are discussed. Another handicap is the small number of good computer programs for determining the sample sizes in this field (Rasch, 1996). The selection theory goes back to Bechhofer (1954) founding the indifference zone formulation and to Gupta (1956) founding the subset approach.

Statistical selection procedures can be divided into two groups: the ones based on the formulation of an indifference zone and subset selection methods. We restrict ourselves in this book to the indifference zone formulation of selection problems. For subset selection procedures we recommend Gupta and Panchapakesan (1972, 1979).

The basis for the indifference zone formulation of selection problems can be found in a paper by Bechhofer (1954). The formulation of the problem is as follows. We are given a "independent" treatments, procedures or populations which will be modelled by a finite set $S = \{P_{\theta 1}, \ldots, P_{\theta a}\}$ of size a with elements from a family of distributions. "Independent" means that the samples to be drawn from these distributions are assumed to be mutually independent. From this set we have to select the t "best" elements. To define what we understand by "best", we use the parameter transformation $\Psi = g(\theta)$ as a selection criterion (evaluation function). Restricting ourselves to real functions g, we say that $P_{\theta i}$ is better than $P_{\theta j}$ when $\Psi_i = g(\theta_i) > \Psi_j = g(\theta_j)$.

Since the parameters are initially unknown, the aim of an experiment is to select the t best distributions from the a distributions being investigated, with $1 \leq t < a$. In what follows, we will use the term *population* instead of *distribution*, although the terms may be used synonymously. An absolutely

correct selection is only possible if the θ_i and thus also the Ψ_i are known, but in this case we do not need an experiment. Therefore, we assume that the parameters are unknown.
The probability of a correct selection depends on the sample size, but also on the "distance" between the t best populations and the remaining $a - t$ inferior populations. This fact is the motivation for Bechhofer's introduction of the concept of *indifference zones*. He demanded that the probability that a population is erroneously denoted as one of the t best populations is not larger than β, as long as the difference between the smallest of the t best populations and the largest one of the $a - t$ remaining populations is larger than a specified value δ. This δ is the smallest difference of practical interest to the experimenter; it characterises the indifference region as the smallest distance between the t best and the $a - t$ "non-best" populations.
This can be better demonstrated if we assume for the moment that the Ψ values are known. In that case we denote by $\Psi_{[1]} < \Psi_{[2]} < \dots \Psi_{[a]}$ the ordered values of Ψ. A population will be called better than another one if it has the larger parameter value. The value $\Psi_{[a-t+1]}$ belongs to the worst of the t best populations, whereas $\Psi_{[a-t]}$ belongs to the best one of the remaining populations. When $\Psi_{[a-t+1]} - \Psi_{[a-t]} < \delta$, then the distance between both groups lies in the indifference zone in which Bechhofer states that we do not need to restrict the probability of an incorrect selection. (However Guiard, 1996, has made another proposal for a statement to be made in this case.)

The precision requirement in Bechhofer 's indifference zone formulation is: In the case $\Psi_{[a-t+1]} - \Psi_{[a-t]} \geq \delta$ the probability of an incorrect selection is at most β.

The precision requirement for the determination of the necessary size of an experiment is given by the four values (a, t, δ, β), and they suffice to determine the minimum sizes for all a samples.

The following formulae have been derived under normal assumptions. But the selection procedures are robust against non-normality (see Domröse and Rasch, 1987; Listing and Rasch, 1996).

The Bechhofer selection rule means to select that population which showed the largest sample mean $\overline{y}_{[a]}$. The corresponding expectation is called μ_s, the expectation of the population selected.

To select the one with the largest expectation, here $t = 1$, from $a > 1$ normal distributions (all have the same known variance σ^2), we choose $n_i = n$. We calculate n from

$$n = \left\lceil \frac{2\sigma^2 \cdot u^2_{(a-1)}(1-\beta)}{\delta^2} \right\rceil \tag{2.29}$$

In (2.29) $u_{(a-1)}(1 - \beta)$ is the $(1 - \beta)$-quantile of the $(a - 1)$-dimensional normal distribution. The correlation matrix of this distribution equals

$\frac{1}{2}(I_{a-1} + J_{a-1,a-1})$, with the identity matrix I_{a-1} of order $a-1$ and an $(a-1) \times (a-1)$ matrix $J_{a-1,a-1}$ with all its elements equal to 1.

If the variances are equal but unknown, the common sample size in the a populations is calculated from

$$n = \left\lceil \frac{2\sigma^2 \cdot t^2_{a-1}(a(n-1);\ 1-\beta)}{\delta^2} \right\rceil \tag{2.30}$$

In (2.30) $t_{a-1}(a(n-1);\ 1-\beta)$ is the $(1-\beta)$-quantile of the $(a-1)$-dimensional t-distribution with $a(n-1)$ degrees of freedom and correlation matrix $\frac{1}{2}(I_{a-1}+J_{a-1,a-1})$.

If n is calculated, the following interpretation for $t = 1$ of the quadruple (a, n, δ, β) may be helpful. If the population with μ_s was selected as the best one, the interval $\lfloor \mu_s, \mu_s + \delta \rfloor$ is a $(1-\beta)$-confidence interval for $\mu_{[a]}$.

The calculation can be done by our OPDOE-function `size.selection.bechhofer`.

Example 2.9 Consider the case of selecting out of $a = 6$ normal distributions with equal variances that one with the largest expectation if the variances are unknown.

Mutually independent samples $(\boldsymbol{y}_{i1}, \boldsymbol{y}_{i2}, \ldots, \boldsymbol{y}_{in_i})$ of size $n_i = n$ $(i = 1, \ldots, a)$ will be taken from the a populations. How can we choose the sample sizes satisfying the precision requirement given by $\delta = \sigma$ and $\beta = 0.01$, if we apply Bechhofer's selection rule: "Choose as the best population the one which has the largest sample mean"? The R-program gives the following solution:

```
> size.selection.bechhofer(a = 6, beta = 0.01, delta = 1,
+ sigma = NA)

[1] 17
```

2.4 Testing hypotheses

Sample sizes for hypothesis testing depend on the risk of the first kind and on the power (or the risk of the second kind) at a specific point in the region for the alternative hypothesis.

2.4.1 Testing hypotheses about means of normal distributions

In the following sections we assume normal distributions in theory. In application the violation of the normality assumption does no harm because the tests are robust against this.

2.4.1.1 One-sample problem, univariate

A random sample $\boldsymbol{y}_1, \boldsymbol{y}_2, \ldots, \boldsymbol{y}_n$ of size n will be drawn from a normally distributed population with mean μ and variance σ^2, with the purpose of testing the null hypothesis

$$\mathrm{H}_0 : \mu = \mu_0 \ (\mu_0 \text{ is a given constant})$$

against one of the following alternative hypotheses:

a) $\mathrm{H}_A : \mu > \mu_0$ (one-sided alternative)

b) $\mathrm{H}_A : \mu < \mu_0$ (one-sided alternative)

c) $\mathrm{H}_A : \mu \neq \mu_0$ (two-sided alternative)

In the known variance case, we have to put in the sequel the degrees of freedom (d.f.) equal to infinity. The test statistic is

$$\boldsymbol{t} = \frac{\overline{\boldsymbol{y}} - \mu_0}{\boldsymbol{s}} \sqrt{n} \tag{2.31}$$

which is non-central t-distributed with $n-1$ d.f. and non-centrality parameter $\lambda = \frac{\mu-\mu_0}{\sigma}\sqrt{n}$.

Under the null hypothesis, the distribution is central t.

If the Type I error probability is α, H_0 will be rejected if

in case a), $t > t(n-1; 1-\alpha)$,
in case b), $t < -t(n-1; 1-\alpha)$,
in case c), $|t| > t(n-1; 1-\alpha/2)$.

Our precision requirement is given by α *and* the risk of the second kind β if $\mu - \mu_0 = \delta$.

In the sequel for the sample size determination we will consider only two-sided alternatives, for one-sided alternatives one has to replace $1-\alpha/2$ by $1-\alpha$. From this we have the requirement

$$t(n-1;\ 1-\alpha/2) = t(n-1;\ \lambda;\ \beta) \tag{2.32}$$

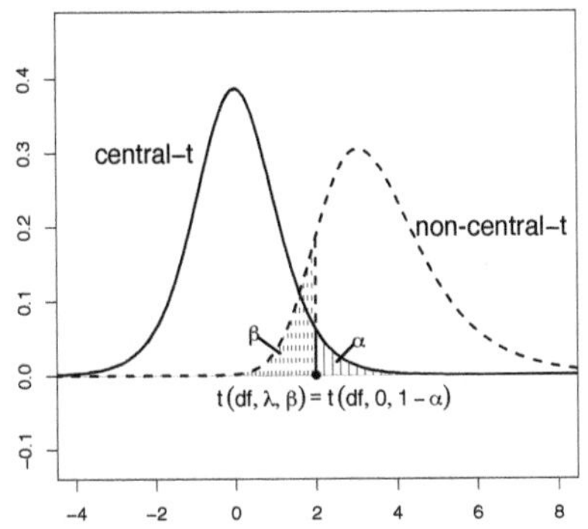

FIGURE 2.1
Graphical representation of the relation (2.32), one-sided case.

where $t(n1;\ \lambda;\ \beta)$ is the β–quantile of the non-central t-distribution with $n-1$ d.f. and non-centrality parameter $\lambda = \frac{\delta}{\sigma}\sqrt{n}$ (see Figure 2.1 for the one-sided case).

Using the approximation $t(n-1;\ \lambda;\ \beta) = t(n-1, \beta) + \lambda$ leads to the approximate formula

$$n \approx \left\lceil \left[\left\{ t\left(n-1;\ 1-\frac{\alpha}{2}\right) + t(n-1;\ 1-\beta) \right\} \frac{\sigma}{\delta} \right]^2 \right\rceil \tag{2.33}$$

Non-central distributions are defined in Appendix A.

From the requirement (2.32), the minimum sample size must be calculated iteratively from the solution of

$$t(n-1;\ 1-\frac{\alpha}{2}) = t(n-1;\ \frac{\delta}{\sigma}\sqrt{n};\ \beta) \tag{2.34}$$

Note that even in the two-sided case the value β is not to be replaced by $\beta/2$.

The requirement (2.34) is illustrated in Figure 2.1 for the one-sided case.

For calculating by hand the approximation mentioned above leads to the approximate formula (2.33).

Our R-program always gives the exact solution based on (2.34).

The calculation can be done by the R-built-in function `power.t.test` and the call wrapper `size.t.test`, respectively.

Example 2.10 Let us calculate the minimal sample size for testing the null hypothesis

$$H_0 : \mu = \mu_0$$

against one of the following alternative hypotheses:

a) $H_A : \mu > \mu_0$ (one-sided alternative)
b) $H_A : \mu \neq \mu_0$ (two-sided alternative)

The precision requirement is defined by $\delta = 0.6\sigma$; $\alpha = 0.01$ and $\beta = 0.05$. Our R-program gives for power $= 1 - \beta$ the solution $n = 47$ in the one-sided case a and $n = 53$ in the two-sided case b.

```
> size.t.test(type = "one.sample", power = 0.95, delta = 0.6, sd = 1
+      sig.level = 0.01, alternative = "one.sided")

[1] 47

> size.t.test(type = "one.sample", power = 0.95, delta = 0.6, sd = 1
+      sig.level = 0.01, alternative = "two.sided")

[1] 53
```

2.4.1.2 One-sample problem, bivariate

If we observe at the experimental units two characters at a time or the same character before a treatment and after, we can model this by a bivariate random variable. We first consider the case of quantitative variables and then the case of alternative variables.

Comparing two means

We consider a two-dimensional random sample

$(\boldsymbol{x}_1, \boldsymbol{y}_1); \ldots; (\boldsymbol{x}_n, \boldsymbol{y}_n)$ with expectation (of each component) $\begin{pmatrix} \mu_x \\ \mu_y \end{pmatrix}$

and unknown covariance matrix

$$\sum = \begin{pmatrix} \sigma_x^2 & \rho\sigma_x\sigma_y \\ \rho\sigma_x\sigma_y & \sigma_y^2 \end{pmatrix}$$

We test the null hypothesis of the equality of the two means of the vector $\begin{pmatrix} \mu_x \\ \mu_y \end{pmatrix}$, $H_0 : \mu_x = \mu_y$ against a one- or two-sided alternative. The precision requirement is given by specifying a minimum difference of interest (effect size) $\delta = \mu_x - \mu_y$ and the risks α and β. We consider in the sequel the case of unknown variance only; known variances are more or less academic.

For the determination of the minimal sample size an estimate $s_d^2 = \frac{1}{n-1}\sum_{i=1}^{n}[(x_i - y_i) - (\overline{x} - \overline{y})]^2$ for the variance σ_Δ^2 of the difference $\boldsymbol{d} = \overline{\boldsymbol{x}} - \overline{\boldsymbol{y}}$

with $\sigma_\Delta^2 = \sigma_1^2 + \sigma_2^2 - 2\sigma_{12}$ is needed. We can reduce this on the univariate case by using differences and test their mean against zero. The sample size formula is then given by

$$t(n-1;\ 1-\frac{\alpha}{2}) = t(n-1;\ \frac{\delta}{\sigma_\Delta}\sqrt{n};\beta) \tag{2.35}$$

and the approximate formula for n is analogously as (2.33) with σ_Δ instead of σ.

If without planning we took a sample of size n, say, then by the R-program in the following way the minimum difference of interest $\delta = \mu_x - \mu_y$ for given α and β can be calculated.

The calculation can be done by the R-built-in function `power.t.test`.

Example 2.11 Let us calculate the minimal sample size for testing the null hypothesis
$\mathrm{H}_0 : \mu_x = \mu_y$
against one of the following alternative hypotheses using a bivariate sample of a two-dimensional normal distribution:
a) $\mathrm{H}_0 : \mu_x > \mu_y$ (one-sided alternative),
b) $\mathrm{H}_0 : \mu_x \neq \mu_y$ (two-sided alternative).
The precision requirement is defined by $\delta = 0.8\sigma$; $\alpha = 0.03$ and $\beta = 0.10$. Our R-program for power $= 1-\beta$ gives the solution $n = 18$ in the one-sided case a and $n = 22$ in the two-sided case b. If we use a sample of size $n = 20; \alpha = 0.05$ and $\beta = 0.2$, we can detect a δ-value of 0.5769σ or larger in the one-sided case.

```
> size.t.test(type = "one.sample", power = 0.9, delta = 0.8,
+       sd = 1, sig.level = 0.03, alternative = "one.sided")

[1] 18
```

Comparing two probabilities

Let the two-dimensional random sample $\begin{pmatrix} \boldsymbol{x}_1, \boldsymbol{x}_2, \ldots, \boldsymbol{x}_n \\ \boldsymbol{y}_1, \boldsymbol{y}_2, \ldots, \boldsymbol{y}_n \end{pmatrix}$ be bivariate binomially distributed with $E(\boldsymbol{x}_i) = p_1; i = 1, \ldots, n$ and $E(\boldsymbol{y}_i) = p_2; i = 1, \ldots, n$.

We test the null hypothesis of the equality of the two probabilities $H_0 : p_1 = p_2$ against a one- or two-sided alternative. The precision requirement is given by specifying a minimum difference of interest $\delta = p_1 - p_2$ and the risks α and β.

It is well known that the analysis is done by the McNemar test. To explain this and the sample size formula let us assume that the two random variables $\boldsymbol{x}$ and $\boldsymbol{y}$ can take the values $(A, \overline{A}); \overline{A} \equiv$ not A and $(B, \overline{B})$, respectively. In the bivariate sample four combinations of these values are possible. The probabilities for these combinations are given in the following scheme:

		y		
		B	$\overline{B}$	$\sum$
x	A	p_{11}	p_{12}	$p_{1.}$
	$\overline{A}$	p_{21}	p_{22}	$p_{2.}$
	$\sum$	$p_{.1}$	$p_{.2}$	1

Because $p_{1.} = p_{11} + p_{12}$ and $p_{.1} = p_{11} + p_{21}$ we have $p_1 = p_{1.}$ and $p_2 = p_{.1}$ and the hypothesis "$p_{1.} = p_{.1}$" can be written as "$p_{12} = p_{21}$". If we have prior information $\hat{p}_1$ and $\hat{p}_2$ for p_1 and p_2, we can determine the sample size from

$$n = \frac{u(p)h + u(1-\beta)[h^2 - \frac{1}{4}\delta^2(3+h)]^{\frac{1}{2}}}{h\delta^2}$$

using $h = \hat{p}_1 + \hat{p}_2 - 2\hat{p}_1\hat{p}_2$.

If no prior information is available, we use the least favourable case $\hat{p}_1 = \hat{p}_2 = 0.5$ (i.e., $h = 0.5$). $u_{(P)}$ is the P-quantile of the standard normal distribution, and we put $P = 1 - \alpha$ in case of a one-sided alternative and $P = 1 - \frac{\alpha}{2}$ in case of a two-sided alternative.

For the analysis based on a sample of size n we write the number of observations in the four combinations in the contingency table:

		y		
		B	$\overline{B}$	$\sum$
x	A	n_{11}	n_{12}	$n_{1.}$
	$\overline{A}$	n_{21}	n_{22}	$n_{2.}$
	$\sum$	$n_{.1}$	$n_{.2}$	n

We calculate $u = \frac{n_{12} - n_{21}}{\sqrt{n_{12} + n_{21}}}$ and reject the null hypothesis if $|u| > u_{1-\alpha}$ and in the two-sided case if $|u| > u_{1-\frac{\alpha}{2}}$.

Example 2.12 In an experiment with perennial one flower plants which are pollenised by either bees or several other insects, it should be found out whether in the first two flowering years the relative frequency of bees is the same. It is known that in the first year about 65% of the flowers are pollenised by bees. The same n plants will be observed in the two years. If we fix the risk of the first kind by $\alpha = 0.05$ and that of the second kind by $\beta = 0.10$ and if a difference between the relative frequency of bees and other insects is 0.10, what is then the minimal n in case of a two-sided alternative?

We use $\hat{p}_1 = 0.65$ and have $\hat{p}_1 - \delta = 0.55$ and use $\hat{p}_2 = 0.55$ as well as $h = 0.65 + 0.55 - 2 \cdot 0.65 \cdot 0.55 = 0.485$. Then we obtain

$$n = \frac{1.96 \cdot 0.485 + 1.282 \cdot (0.485^2 - 0.0087125)^{\frac{1}{2}}}{0.00485}$$

and hence $n = 322$.

With our OPDOE-program `size.comparing_probs.mcnemar` we obtain:

```
> size.comparing_probs.mcnemar(p1 = 0.65, p2 = 0.55, alpha = 0.05,
+      beta = 0.1, alternative = "two.sided")

[1] 322
```

2.4.1.3 Two-sample problem, equal variances

We have two normally distributed populations with means μ_x, μ_y and variances σ_x^2, σ_y^2, respectively. Our purpose is to take two independent random samples $(\boldsymbol{x}_{11}, \ldots, \boldsymbol{x}_{1n_x})$ and $(\boldsymbol{y}_{21}, \ldots, \boldsymbol{y}_{2n_y})$ of sizes n_x and n_y from the two populations in order to test the null hypothesis

$$\mathrm{H}_0 : \mu_x = \mu_y$$

against one of the following one- or two-sided alternative hypotheses

a) $\mathrm{H}_A : \mu_x > \mu_y$

b) $\mathrm{H}_A : \mu_x < \mu_y$

c) $\mathrm{H}_A : \mu_x \neq \mu_y$

The sample sizes n_x and n_y should be determined in such a way that for a given risk of the first kind α, the risk of the second kind β as a function of $(\mu_x - \mu_y)$ does not exceed a predetermined upper bound β_0 as long as, for the respective alternative hypothesis, we have either

a) $\mu_x - \mu_y \geq \delta$,

b) $\mu_x - \mu_y \leq -\delta$ or

c) $|\mu_x - \mu_y| \geq \delta$.

As before, the value of δ is called the minimal difference of practical interest between μ_x and μ_y.

The minimum sample size is to compute an integer solution iteratively from

$$t(2n-2; P) = t(2n-2;\ \lambda;\ \beta_0) \tag{2.36}$$

with $P = 1 - \alpha$ in the one-sided cases and $P = 1 - \alpha/2$ in the two-sided case and $\lambda = \frac{\mu_x - \mu_y}{\sigma}\sqrt{\frac{n}{2}}$.

Using the approximation $t(n-1;\ \lambda;\ \beta) = t(n-1, \beta) + \lambda$ leads to the approximate formula

$$n \approx \left\lceil 2\left[\{t(2n-2; P) + t(2n-2;\ 1-\beta)\}\frac{\sigma}{\delta}\right]^2\right\rceil \tag{2.37}$$

In the test, the common variance is estimated by $s^2 = \frac{(n_x-1)s_x^2+(n_y-1)s_y^2}{n_x+n_y-2}$.

Remark: If one is not absolutely sure that the two variances are equal, one should use the test described below for unequal variances, as it was recommended by Rasch et al. (2009).

The calculation can be done by the R-built-in function `power.t.test` and the call wrapper `size.t.test`, respectively.

Example 2.13 We like to know the minimal sample sizes $n_x = n_y = n$ to be drawn independently from two normal distributions to test the null hypothesis above against a two-sided alternative when the precision demanded is defined by $\delta = |\mu_x - \mu_y| = 0.9\sigma$; $\alpha = 0.05$; $\beta = 0.1$. Our R-program shows the result $n = 27$.

```
> size.t.test(delta = 0.9, sd = 1, sig.level = 0.05, power =
+ 0.9, type = "two.sample", alternative = "two.sided")

[1] 27
```

Example 2.11 – continued Assume in our example with $\alpha = 0.01$ there are only 20 measurements per group possible. How large is the smallest difference $|\mu_x - \mu_y|$ which can be detected with probability $\beta = 0.1$ and what is the maximum power achievable for a difference of 0.9?

```
> delta.t.test(n = 20, sd = 1, sig.level = 0.01, power = 0.9,
+    type = "two.sample",  alternative = "two.sided")

[1] 1.276241

> power.t.test(n = 20, sd = 1, sig.level = 0.01, delta = 0.9,
+    type = "two.sample",    alternative = "two.sided")$power

[1] 0.5580143
```

Hence the difference is $\delta = 1.276$ and the power is at most 0.558.

2.4.1.4 Two-sample problem, unequal variances

With f^* from (2.22) but using there $n_y = \frac{\sigma_y}{\sigma_x} n_x$, we calculate the sample size n_y from

$$t(f^*;\ 1 - \frac{\alpha}{2}) = t(f^*;\ \lambda;\ \beta_0)) \tag{2.38}$$

with

$$\lambda = \frac{\mu_x - \mu_y}{\sqrt{\frac{\sigma_x^2}{n_x} + \frac{\sigma_y^2}{n_y}}}$$

and

$$n_x = \left\lceil \frac{\sigma_x}{\sigma_y} n_y \right\rceil \tag{2.39}$$

Using the approximation $t(n-1;\ \lambda;\ \beta) = t(n-1, \beta) + \lambda$ leads to the approximate formula

$$n_x \approx \left\lceil \frac{\sigma_x(\sigma_x + \sigma_y)}{\delta^2} \left[\left\{t\left(f^*; 1 - \frac{\alpha}{2}\right) + t(f^*; 1 - \beta)\right\}\right]^2 \right\rceil \tag{2.40}$$

This is the optimal solution for the Welch test; see also Section 2.2.1.6. If we take without planning sample sizes $n_x = 20$; $n_y = 40$, say, then by the R-program in the following way the minimum difference of interest $\delta = \mu_x - \mu_y$ for given α β and the ratio $\frac{\sigma_x^2}{\sigma_y^2}$ can be calculated.

The calculation can be done by the R-built-in function `power.t.test`.

Example 2.14 We would like to know the minimal sample sizes $n_x; n_y$ to be drawn independently from two normal distributions with variances $\frac{\sigma_x^2}{\sigma_y^2} = 16$ to test the null hypothesis above against a two-sided alternative when the precision demanded is defined by $|\mu_x - \mu_y| = 0.9\sigma_y; \alpha = 0.05;\ \beta = 0.1$. We may conjecture that n_x is 4 times as large as n_y because $\sigma_x = 4\sigma_y$. Our R-program shows the result $n_y = 27$; $n_x = 105$. The fact that $n_x \neq 4 \cdot n_y$ is due to the integer rounding during the calculation by the $\lceil\ \rceil$ operator.

```
> power.t.test(sd = 1, sig.level = 0.05, delta = 0.9, power =
+ 0.9, type = "two.sample", alternative = "two.sided")

     Two-sample t test power calculation

              n = 26.94187
          delta = 0.9
             sd = 1
      sig.level = 0.05
          power = 0.9
```

```
    alternative = two.sided

NOTE: n is number in *each* group
```

2.4.1.5 Comparing more than two means, equal variances

The following assumptions hold: pair-wise independent random samples will be taken from a populations; the sample from the i-th population is denoted by $(\boldsymbol{y}_{i1}, \boldsymbol{y}_{i2}, \ldots, \boldsymbol{y}_{in_i})$. The $\boldsymbol{y}_{ij}(i = 1, \ldots, a > 2;\ j = 1, \ldots, n_i)$ are distributed independently of each other as $N(\mu_i; \sigma^2)$ with the same variance σ^2. $N = \sum_{i=1}^{a} n_i$ is the total number of observations thus the size of the experiment. We would like to test the hypotheses

$$\mathrm{H}_{0ij} : \mu_i = \mu_j;\ i \neq j = 1, \ldots, a$$

against the alternatives

$$\mathrm{H}_{Aij} : \mu_i \neq \mu_j;\ i \neq j = 1, \ldots, a.$$

In this connection, we distinguish between two types of risks of the first and second kind (error probabilities).
Individual Risk: The individual risk applies to every single comparison. The greater the number of pair-wise comparisons in an experiment, the greater is the chance (assuming a fixed pair-wise risk) of a false decision in at least one of these comparisons. The individual risk is also called the *per comparison*, or *comparison-wise risk.*

Simultaneous Risk: The simultaneous risk applies overall to several comparisons simultaneously; it is also called the *family-wise risk*. Frequently we consider all the comparisons in the experiment, in which case the risk is called the *experiment-wise risk*. Under the null hypothesis $\mathrm{H}_0 : \mu_1 = \mu_2 = \ldots = \mu_a$ the experiment-wise risk of the first kind is the probability that at least one of the alternative hypotheses $\mathrm{H}_{Aij} : \mu_i \neq \mu_j$ will be falsely accepted. We consider this case in Chapter three.

If the simultaneous risk of the first kind α^* applies only to a subset of size $m > 1$ of all $\binom{a}{2}$ pair-wise comparisons, its size can be assessed in terms of the per comparison risk of the first kind α with the aid of the *Bonferroni Inequality* as follows:

$$\alpha < \alpha^* < m\alpha \tag{2.41}$$

Before an experiment is planned or analysed, a choice must be made between fixing the experiment-wise or the per comparison risks of the first and second kind. To fix the experiment-wise risk at a certain level is a stricter condition, and will require a larger experiment than if the same level is fixed for per comparison risk. For the F-test in the analysis of variance the risks of both kinds are defined experiment-wise. In what follows, the risk of the second kind β is chosen per comparison. We consider the following cases.

Case 1. Comparison of the means of the $a > 2$ populations pair-wise with each other.

Methods: For sample size determination we consider only some of a large set of different multiple comparison methods.

Multiple t-Test: pair-wise comparisons with per comparison risk of the first kind. We use equal sample sizes n in each population and estimate the common variance by the residual MS with $a(n-1) = N - a$ degrees of freedom (cf. Table 3.2). We solve

$$t\left[a(n-1); 1-\frac{\alpha}{2}\right] = t\left[a(n-1); \frac{\delta}{\sigma}; \beta_0\right] \tag{2.42}$$

We can use the R-program of Section 2.2.1.5 and denote the output not by n but by ν. Then $\nu - 1 = a(n-1)$ or $n = \lceil \frac{\nu+a-1}{a} \rceil$.

Example 2.15 We wish to test with a pair-wise risk of the first kind of $\alpha = 0.05$ the equality of the expectations of 9 normal distributions. We take $\beta_0 = 0.20, \delta = 4$ and assume that we have an estimate of σ^2, namely $\tilde{\sigma}^2 = 16$. From the R-program we obtain $n = 17$ as the size of each of the 9 samples and $N = 9 \cdot 17 = 153$ as the size of the experiment.

```
> size.multiple_t.test(a = 9, alpha = 0.05, beta = 0.2,
+ delta = 4, sd = 4)

$size
[1] 17

$total
[1] 153
```

Tukey-Test: pair-wise comparisons with experiment-wise risk of the first kind. Let $q(a; f; P)$ be the P-quantile of the studentised range of a means with f degrees of freedom. The common sample size for the a samples is calculated from

$$\frac{q[a; a(n-1); 1-\alpha]}{\sqrt{2}} = t\left[a(n-1); \frac{\delta}{\sigma}; \beta_0\right] \tag{2.43}$$

Example 2.16 We wish to use an experiment-wise risk of the first kind $\alpha = 0.05$ and to test the equality of the expectations of 9 normal distributions. We take $\beta_0 = 0.20, \delta = 4$ and assume that we have an estimate of σ^2, namely $\tilde{\sigma}^2 = 16$.

From our OPDOE-program we obtain $n = 32$ as the size of each of the 9 samples and $N = 9 \cdot 32 = 288$ as the size of the experiment.

Case 2. Comparison of each of the means of $a - 1$ populations with that of a standard population, all variances are equal.

We renumber the populations such that population a is the standard. It is now no longer optimal to use equal sample sizes, but the standard has a higher size n_0 than the other samples with equal sizes n for $i = 1, \ldots, a-1$. We use approximately $n_0 = n\sqrt{a-1}$.

Multiple t-Test: Comparison with per comparison risk of the first kind.

We first determine the total (integer) size N of the experiment with $f = N - a$ from

$$t(f; 1 - \frac{\alpha}{2}) = t(f; \frac{\delta}{\sigma}; \beta_0) \tag{2.44}$$

From N we next calculate

$$n^* = \left\lceil \frac{N}{a - 1 + \sqrt{a-1}} \right\rceil, \tag{2.45}$$

$$n_0^* = \sqrt{a-1} n^* \tag{2.46}$$

and

$$N^* = n_0^* + (a-1) \cdot n^* \tag{2.47}$$

For (n, n_0) we choose either (n^*, n_0^*) or an admissible (n, n_0) combination in the neighbourhood.

To check admissibility of a combination (n, n_0), calculate N_{new} from (2.44) with the d.f. from the new candidate (n, n_0). If N_{new} does not exceed N, it is admissible. The neighbourhood to check may be determined as follows: Select for n consecutively $n^* - 1, n^*$ and $n^* + 1$. For each of these values choose $n_0 = N - (a-1) \cdot n$ and decrease n_0 until the result is an admissible combination.

By R we solve the problem as follows: First we use the method of Case 1 with $N = \nu$ and then proceed as described above.

The calculation can be done by our OPDOE-function:
`sizes.multiple_t.comp_standard`.

Example 2.17 We wish to test with a pair-wise risk of the first kind $\alpha = 0.05$ the equality of the expectations of 8 normal distributions with that of a control distribution (number 9). We take $\beta_0 = 0.20, \delta = 4$ and assume that we have an estimate of σ^2, namely $\tilde{\sigma}^2 = 16$.

From our OPDOE-program we obtain $n_1 = \ldots = n_8 = 11; n_9 = 30$ as the size of the 9 samples and $N = 118$ as the size of the experiment.

Dunnett-Test: Comparison with experiment-wise risk of the first kind.

Let $d\left(a-1; f; \rho; 1-\frac{\alpha}{2}\right)$ be the $\left(1-\frac{\alpha}{2}\right)$-quantile of the $(a-1)$-dimensional t-distribution with correlation coefficients all equal to $\rho = \frac{1}{1+\sqrt{a-1}}$.

Search an integer solution for $f = N - a$ from

$$d\left(a-1; f; \rho; 1-\frac{\alpha}{2}\right) = t\left(f; \frac{\delta}{\sigma}; \beta_0\right) \tag{2.48}$$

and then continue as in (2.45)–(2.47) and below.

By R we solve the problem using our OPDOE-function `sizes.dunnett.exp_wise`.

Example 2.18 We wish to test the equality of the expectations of 8 normal distributions with that of a standard distribution (number 9) by using an experiment-wise risk of the first kind $\alpha = 0.05$. We take $\beta_0 = 0.20, \delta = 4$ and assume that we have an estimate of σ^2, namely $\tilde{\sigma}^2 = 16$.

From our OPDOE-program `sizes.dunnett.exp_wise` we obtain $n_1 = \ldots = n_8 = 18; n_9 = 45$ as the size of the 9 samples and $N = 189$ as the size of the experiment.

Further multiple comparison procedures for means are described by Hochberg and Tamhane (1987) and in Rasch et al. (2008; 3/26/0000 – 3/26/9104).

2.4.2 Testing hypotheses about probabilities

2.4.2.1 One-sample problem

When we test the pair of hypotheses

$$\mathrm{H}_0: p \leq p_0; \ \mathrm{H}_A: p > p_0$$

with a risk of the first kind α and want the risk of the second kind to be no larger than β as long as $p > p_1 > p_0$, we can approximately determine the minimum sample size from

$$n = \left\lceil \frac{\left[u_{1-\alpha}\sqrt{p_0(1-p_0)} + u_{1-\beta}\sqrt{p_1(1-p_1)}\right]^2}{(p_1 - p_0)^2} \right\rceil \tag{2.49}$$

The same size is needed for the other one-sided test. In the two-sided case

$$\mathrm{H}_0 : p = p_0; \ \mathrm{H}_A : p \neq p_0$$

we replace α by $\alpha/2$ and calculate (2.49) twice for $p_1 = p_0 - \delta > 0$ and for $p_1 = p_0 + \delta < 1$ if δ is the difference from p_0 which should not be overseen with a probability of at least $1 - \beta$. From the two n-values, we then take the maximum.

The calculation can be done by the R-built-in function `power.prop.test`.

Example 2.19 We would like to test that the probability $p = P(A)$ of the event A: "an iron casting is faulty" equals $p_0 = 0.1$ against the alternative that it is larger. The risk of the first kind is planned to be $\alpha = 0.05$ and the power should be at least 0.9 as long as $p = P(A) \geq 0.2$. How many castings should be tested?

The R-program gives the required sample size as $n = 102$.

2.4.2.2 Two-sample problem

Using independent samples from two populations, we wish to test the null hypothesis

$$\mathrm{H}_0 : p_1 = p_2$$

against either one of the alternative hypotheses:

a) $\mathrm{H}_A : p_1 > p_2$

b) $\mathrm{H}_A : p_1 < p_2$

c) $\mathrm{H}_A : p_1 \neq p_2$.

In each population it is also assumed that, out of n_i observations, the number $\boldsymbol{y}_i$ of those which result in a certain event A follows a binomial distribution with parameter p_i $(i = 1, 2)$.

To design an experiment, we would have to determine n_1 and n_2 so that pre-assigned precision requirements will be satisfied.

In contrast to the other parts of this chapter, in the case of two independent proportions it is not possible to give a generally applicable recommendation for the test to be used and for the best formula for calculating the sample size. Most often, one uses one of the numerous approximate formulae discussed in, for example, Sahai and Khurshid (1996).

Also the test itself is frequently carried out in an approximate form (apart from Fisher's exact test, which is conservative). We give here the formulae which, following Bock (1998), are recommended for the test and also for the sample size calculation, when the conditions

$$0.5 \leq p_1, p_2 \leq 0.95$$

$$n_1 = n_2 = n$$

are satisfied.
As in the previous sections, the precision requirements stipulate the values of both risks α and β as well as a minimal difference of practical interest

$$\begin{array}{ll} \delta = p_1 - p_2 & \text{in case a} \\ \delta = p_2 - p_1 & \text{in case b} \\ \delta = |p_1 - p_2| & \text{in case c.} \end{array}$$

In addition, $n = n_1 = n_2$ depends on prior information about the values of p_1 and p_2.
Following Fleiss (1981) we have

$$n = \left\lceil \frac{1}{\delta^2} \left[u_{(P)} \sqrt{(p_1 + p_2)\left(1 - \frac{1}{2}(p_1 + p_2)\right)} \right. \right.$$
$$\left. \left. + u(1-\beta)\sqrt{p_1(1-p_1) + p_2(1-p_2)} \right]^2 \right\rceil \quad (2.50)$$

In cases a and b we use $P = 1 - \alpha$ in (2.50), and in case c we use $P = 1 - \frac{\alpha}{2}$.
After this we make the following modification, following Casagrande, Pike and Smith (1978), which takes the form

$$n_{corr} = \left\lceil \frac{n}{4} \left[1 + \sqrt{1 + \frac{4}{n\delta}} \right]^2 \right\rceil \quad (2.51)$$

where n is taken from (2.50).
By R we solve the problem as follows:

The calculation of n according to (2.50) can be done by the R-built-in function `power.prop.test`; the corrected value according to (2.51) can be obtained with our OPDOE-function `size.prop_test.two_sample`.

Example 2.20 In an experiment to compare two breeds (B1, B2) of carrier pigeons, the proportions of the two breeds that successfully returned home will be observed in independent samples.
How large must the common sample size be, if we want to test the above hypothesis and give $\alpha = 0.05, \beta = 0.20, \delta = 0.1$? We may have $p_1 = 0.95, p_2 = 0.85$ given as prior estimates for the return probabilities of the two breeds.
Using (2.50) by hand we obtain

$$n = \lceil 100[1.96\sqrt{1.8 \cdot 0.1} + 0.8416\sqrt{0.95 \cdot 0.05 + 0.85 \cdot 0.15}]^2 \rceil = 141$$

and using (2.51)

$$n_{corr} = \left\lceil \frac{141}{4}\left(1+\sqrt{1+\frac{4}{14.1}}\right)^2 \right\rceil = 161$$

With our OPDOE-program `size.prop_test.two_sample` we obtain:

```
> size.prop_test.two_sample(p1 = 0.95, p2 = 0.85, alpha = 0.05,
+     power = 0.8, alt = "two.sided")

[1] 161
```

2.4.3 Comparing two variances

We wish to compare the variances σ_1^2 and σ_2^2 of two normal distributions and for this purpose we will take a random sample from each distribution (independent from each other). Let the sample sizes be n_1 and n_2. If n_1 and n_2 can be freely chosen, then we choose $n_1 = n_2 = n$. In most textbooks the F-test is recommended for the analysis.

However, simulation experiments have shown (Rasch and Guiard, 2004; Rasch, Teuscher and Guiard, 2007) that this test is very sensitive to even small deviations from the assumed normality of the underlying distributions of the variables. The F-test is the best (optimal) test if the two variables really are normally distributed, but in practice we can never be sure that the distributions do not deviate slightly from normality. This does not matter in the case of confidence intervals or tests about means, because the procedures based on the t-test have shown themselves to be extremely robust. However for the comparison of two variances we only use the F-test for the calculation of the sample sizes n; for the analysis we recommend Levene's test. For this test there exists an R-built-in function `levene.test` in the package CAR.

Independent samples of size n will be drawn from two populations with variances σ_1^2 and σ_2^2 in order to test the hypothesis

$$\mathrm{H}_0 : \sigma_1^2 = \sigma_2^2$$

against the alternative hypothesis

$$\mathrm{H}_A : \sigma_1^2 \neq \sigma_2^2$$

How do we choose n if, for a significance level of α and for a given $\tau > 1$, the Type II error probability when $\frac{\sigma_{\max}^2}{\sigma_{\min}^2} \geq \tau$ must not be greater than β?

Here $\sigma_{\max}^2$ is the larger and $\sigma_{\min}^2$ the smaller of the two variances.

The required n is the solution of

$$\tau = F(n-1; n-1; 1-\frac{\alpha}{2})F(n-1;\; n-1;\; 1-\beta) \tag{2.52}$$

involving the quantiles of the F-distribution with $n-1$ and $n-1$ degrees of freedom.

The calculation can be done by our R-function `size.comparing.variances`.

Example 2.21 For $\alpha = 0.05, \beta = 0.2$ and $\tau = 5.5$ we obtain the value of $n = 13$:

```
> size.comparing.variances(ratio = 5.5, alpha = 0.05, power = 0.8)

[1] 13
```

2.5 Summary of sample size formulae

Summarising the different sample size formulae of this chapter we give two tables, Table 2.1 shows formulae for determining the sample size for confidence estimation and Table 2.2 for hypothesis testing.

But all the R-programs for calculating sample sizes can also be used to calculate each of the other parameters in the formula (besides α; this is not recommended) for given sample sizes.

TABLE 2.1
Sample Size Formulae for Constructing Two-Sided Confidence Intervals with Expected Length δ (Procedure 1) and for Selection.

Parameter(s)	Sample size	Formula
μ	$n = \left\lceil t^2(n-1; 1-\frac{\alpha}{2}) \frac{2 \cdot \Gamma^2\left(\frac{n}{2}\right)}{\Gamma^2\left(\frac{n-1}{2}\right)(n-1)} \frac{\sigma^2}{\delta^2} \right\rceil$	(2.8)
$\mu_x - \mu_y$ paired observations	$n = \left\lceil t^2(n-1; 1-\frac{\alpha}{2}) \frac{2 \cdot \Gamma^2\left(\frac{n}{2}\right)}{\Gamma^2\left(\frac{n-1}{2}\right)(n-1)} \frac{\sigma_\Delta^2}{\delta^2} \right\rceil$	(2.18)
$\mu_x - \mu_y$ ind. samples, equal variances	$n = \left\lceil 2\sigma^2 \frac{t^2(2n-2; 1-\frac{\alpha}{2})}{\delta^2(2n-2)} \frac{2\Gamma^2\left(\frac{2n-1}{2}\right)}{\Gamma^2(n-1)} \right\rceil$	(2.19)
$\mu_x - \mu_y$ ind. samples, unequal variances	$n_x = \left\lceil \frac{\sigma_x(\sigma_x+\sigma_y)}{\delta^2} t^2\left(f^*; 1-\frac{\alpha}{2}\right) \right\rceil ; n_y = \left\lceil \frac{\sigma_y}{\sigma_x} n_x \right\rceil$	(2.23)
Probability p	$n = \left\lceil \frac{p(1-p)}{\delta^2} u^2(1-\frac{\alpha}{2}) \right\rceil$	(2.27)
σ^2	$n = \left\lceil \frac{2\delta}{\sigma^2} \left[\frac{1}{\chi^2\left(n-1; \frac{\alpha}{2}\right)} - \frac{1}{\chi^2\left(n-1; 1-\frac{\alpha}{2}\right)} \right]^{-1} \right\rceil + 1$	(2.28)
μ, variances equal and known	$n = \left\lceil \frac{2\sigma^2 \cdot u_{a-1}^2(1-\beta)}{\delta^2} \right\rceil$	(2.29)
μ, variances equal and unknown	$n = \left\lceil \frac{2\sigma^2 \cdot t_{a-1}^2(a(n-1); 1-\beta)}{\delta^2} \right\rceil$	(2.30)

TABLE 2.2
(Approximate) Sample Size Formulae for Testing Hypotheses for Given Risks α and β for a Given Distance δ (or Ratio τ of Variances).

Parameter(s)	Sample size	Formula
μ	$n \approx \left\lceil \left[\left\{ t(n-1; 1-\frac{\alpha}{2}) + t(n-1; 1-\beta) \right\} \frac{\sigma}{\delta} \right]^2 \right\rceil$	(2.33)
$\mu_x - \mu_y$ paired observations	$n \approx \left\lceil \left[\left\{ t(n-1; 1-\frac{\alpha}{2}) + t(n-1; 1-\beta) \right\} \frac{\sigma_\Delta}{\delta} \right]^2 \right\rceil$	(2.35)
$\mu_x - \mu_y$ ind. samples, equal variances	$n \approx \left\lceil 2 \left[\left\{ t(2n-2; 1-\frac{\alpha}{2}) + t(2n-2;\ 1-\beta) \right\} \frac{\sigma}{\delta} \right]^2 \right\rceil$	(2.37)
$\mu_x - \mu_y$ ind. samples, unequal variances	$n_x \approx \left\lceil \frac{\sigma_x(\sigma_x+\sigma_y)}{\delta^2} \left[\left\{ t(f^*; 1-\frac{\alpha}{2}) + t(f^*; 1-\beta) \right\} \right]^2 \right\rceil$	(2.40)
Probability p	$n = \left\lceil \frac{\left[u_{1-\alpha}\sqrt{p_0(1-p_0)} + u_{1-\beta}\sqrt{p_1(1-p_1)} \right]^2}{(p_1-p_0)^2} \right\rceil$	(2.49)
Probabilities $p_1 - p_2$	$n = \left\lceil \frac{1}{\delta^2} \left[u(P)\sqrt{(p_1+p_2)\left(1-\frac{1}{2}(p_1+p_2)\right)} + u(1-\beta)\sqrt{p_1(1-p_1)+p_2(1-p_2)} \right]^2 \right\rceil$	(2.50)
Variances σ_1^2/σ_2^2	$\tau = F(n-1; n-1; 1-\frac{\alpha}{2}) F(n-1; n-1; 1-\beta)$	(2.52)

3

Size of Experiments in Analysis of Variance Models

The analysis of variance (ANOVA) comprises several linear statistical models for the description of the influence of one or more qualitative factor(s) on a quantitative character y.

3.1 Introduction

For all models in the analysis of variance, the linear model equation has the form

$$\boldsymbol{y} = E(\boldsymbol{y}) + \boldsymbol{e}$$

In this equation the random variable $\boldsymbol{y}$ models the observed character. The observation $\boldsymbol{y}$ is the sum of the expectation (mean) $E(\boldsymbol{y})$ of $\boldsymbol{y}$ and an error term $\boldsymbol{e}$, containing observational errors with $E(\boldsymbol{e}) = 0, var(\boldsymbol{e}) = \sigma^2$. The variability in $E(\boldsymbol{y})$ between experimental units depends linearly on model parameters. The models for the analysis of variance differ in the number and the nature of these parameters.

The observations in an analysis of variance are allocated to at least two classes which are determined by the levels of the factors.

Each of the models of the analysis of variance contains the general mean μ; i.e., we write $E(\boldsymbol{y})$ in the form

$$E(\boldsymbol{y}) = \mu + EC(\boldsymbol{y});\ var(\boldsymbol{y}) = \sigma_y^2 \tag{3.1}$$

where $EC(\boldsymbol{y})$ is the mean deviation from μ within the corresponding class. In the case of p factors the analysis of variance is called p-way.

It follows that the total set of the $\boldsymbol{y}$ does not constitute a random sample because not all the $\boldsymbol{y}$ have the same expectation. Furthermore, in models with random factors, the $\boldsymbol{y}$-values within a class are not independent.

We need to consider two basic situations, depending on how we select the levels of a factor (following Eisenhart, 1947).

Definition 3.1 We define fixed and random factors as follows:

Situation 1: There are exactly a levels, all included in the experiment. We call this factor a fixed factor. If all factors in the model are fixed, we speak about Model I analysis of variance or the fixed effects model.

Situation 2: There are many levels, whose number is in theory considered to be infinite. The levels to be included in the experiment have to be selected randomly from the universe of levels; we call this factor a random factor. If all factors in the model are random, we speak about Model II analysis of variance or the random effects model.

In higher classifications like the two- or three-way classification, both fixed as well as random factors may occur; we then speak about mixed models.

The overall mean μ and the components of $EC(\boldsymbol{y})$ are estimated by the method of (ordinary) least squares (OLS). In the case of correlated errors or incomplete block designs, OLS will be generalised and becomes the generalised least squares method (GLS).

The theoretical background can be found in many statistical textbooks and special books on ANOVA (for instance, Scheffé, 1959).

For all the models we assume that the variance σ^2 of the error terms in the equations is the same in all subclasses and that all the random variables in the r.h.s. of the model equations are mutually independent and have expectation zero.

If $\boldsymbol{y}$ has a normal distribution, we can test the following null hypothesis in all models with a fixed factor A having effects $a_1, \ldots, a_a$.

> H_0: "The factor A has no effect on the dependent variable $\boldsymbol{y}$." In other words: "All the a_i are equal." If it is assumed that the sum of the a_i is zero, this is the same as "All the a_i are equal to zero."

The alternative hypothesis is:

H_A: "At least two of the a_i are different."

The test statistic for this test is a variate $\boldsymbol{F}$ which (if the null hypothesis is true) follows a (central) F-distribution with f_1 and f_2 degrees of freedom. The $(1-\beta)$-quantile of the distribution of $F(f_1; f_2)$ is denoted by $F(f_1; f_2;\ 1-\beta)$.

This test statistic is generally calculated by following the next 8 steps—here *generally* means that these steps should be used for all situations and models in this section but also for any other ANOVA situation.

1. Define the null hypothesis.
2. Choose the appropriate model (I, II, or mixed).
3. Find the $E(\boldsymbol{MS})$ column in the appropriate ANOVA table (if there are several such columns, find the one that corresponds to your model).
4. In the same table find the row for the factor that appears in your null hypothesis.
5. Change the $E(\boldsymbol{MS})$ in this row to what it would be if the null hypothesis were true.
6. Search in the same table (in the same column) for the row which now has the same $E(\boldsymbol{MS})$ as you found in the fifth step.
7. The F-value is now the value of the $\boldsymbol{MS}$ of the row you found in the fourth step divided by the value of the $\boldsymbol{MS}$ of the row you found in the sixth step.
8. Note: in ANOVA with higher classifications, the sixth step may not be successful, in which case one can use the so-called Satterthwaite approximation described in Section 3.4.1.3.

The minimum size of the experiment should be determined so that the precision requirements are fulfilled. The size of the experiment depends on the degrees of freedom (d.f.) of the numerator f_1 and the denominator f_2 of the F-statistic. f_2 does not always depend on the subclass number n. If we sample factor levels of random factors, the size of those samples also determines the size of the experiment.

The minimal size is determined in dependence on a lower bound δ of the difference between the maximum and the minimum of the effects to be tested for equality by an F-test, further on the risks α and β of the first and second kind of the test, respectively, and on a presumable value of the variance σ_y^2.

The problem of the determination of the size of an experiment for the analysis of variance has been investigated by, among others, Tang (1938), Lehmer (1944), Fox (1956), Tiku (1967, 1972), Das Gupta (1968), Bratcher et al. (1970), Kastenbaum et al. (1970a, b), Bowman (1972), Bowman and Kastenbaum (1975), Rasch et al. (1997b), Herrendörfer et al. (1997) and Rasch (1998).

For a hypothesis testing problem that leads to an exact F-test, we can determine the size of the experiment which depends on the denominator degrees of freedom f_2 for a given power $1-\beta$ of the F-test at given significance level α and the non-centrality parameter λ of the non-central F-distribution by solving the equation

$$F(f_1 f_2, 0, 1-\alpha) = F(f_1, f_2, \lambda, \beta) \tag{3.2}$$

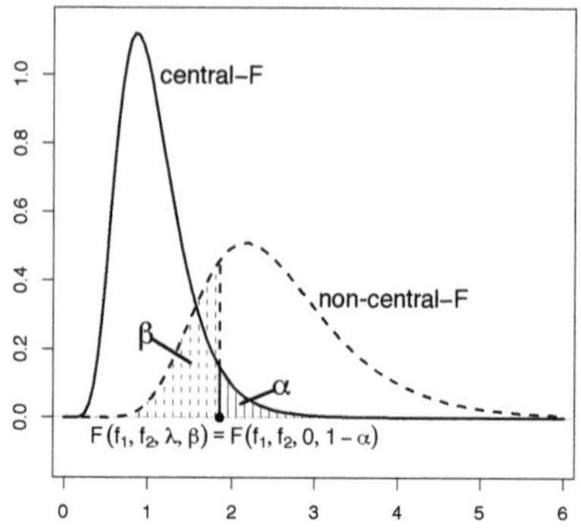

FIGURE 3.1
Graphical representation of the relation (3.2)

Here $F(f_1, f_2, 0, 1-\alpha) = F(f_1, f_2, 1-\alpha)$ is the $(1-\alpha)$-quantile of the (central) F-distribution with degrees of freedom f_1 and f_2 and non-centrality parameter 0; $F(f_1, f_2, \lambda, \beta)$ is the β-quantile of the F-distribution with degrees of freedom f_1 and f_2 and non-centrality parameter λ (see Figure 3.1).

Let $E_{\min}, E_{\max}$ denote, respectively, the minimum and the maximum of a set of q main effects $E_1, E_2, \ldots, E_q$ of the levels of some fixed factor E. The precision of the test can be characterised by the following:

α: the risk of the first kind
$1-\beta$: power of the F-test for a given δ; thus β is the risk of the second kind
σ_y^2: the population variance of the model $\boldsymbol{y}$
δ: the minimum difference to be detected between the largest and the smallest treatment effect $E_{\max} - E_{\min}$; it is a common practice to standardize this difference using $\tau = (E_{\max} - E_{\min})/\sigma_y$

We write $\lambda = C \cdot \psi^2$ with $\psi^2 = \frac{1}{\sigma_y^2} \sum_{i=1}^{q} (E_i - \overline{E})^2$.

We determine the minimum size of an experiment for testing the effects of the fixed factor E for equality for the least favourable and the most favourable case, which we refer to as maxi-min ($n_{\max}$) and mini-min ($n_{\min}$) size, respectively. This means the following:
When $E_{\max} - E_{\min} \geq \delta$, the non-centrality parameter of the F-distribution

satisfies (for even q) $\lambda = C \cdot \sum_{i=1}^{q} (E_i - \overline{E})^2/\sigma_y^2$, where C is some factor depending on the model. See Table 3.2 for C in several ANOVA models.

The least favourable case is leading to the minimal non-centrality parameter and by this to the maxi-min size. The most favourable case is leading to the maximal non-centrality parameter and by this to the mini-min size. Our objective is to determine the minimal size of an experiment for the least as well as for the most favourable case.

The Algorithm and the Program

The solution of (3.2) is a known function $\lambda(\alpha, \beta, f_1, f_2)$ of the risk of the first kind α, the risk of the second kind β, and the degrees of freedom f_1 and f_2 of the numerator and denominator, respectively. We assume that σ_y^2 (or a guess of σ_y^2) is given in advance. The solution is found by iteration. The algorithm for this iteration is given by Lenth (1986) and Rasch et al. (1997b).

In cases when the iteration does not converge, we use a systematic search.

The R-programs are based on the basic programs `ncp()` and `beta()`.

3.2 One-way layout

If the influence of only one factor is considered in the analysis of variance, we call this a *one-way analysis of variance*. The number of levels of the treatment factor will be denoted by a; the factor itself is A, and its a levels are written as $A_1, \ldots, A_a$. We here consider only the following situation:
There are exactly a levels, all included in the experiment. We call this *Model I analysis of variance* or the *fixed effects model*. The observations y_{ij} in the factor levels are as follows for a *balanced* one-way layout where each level has n observations.

Factor Levels	A_1	A_2	$\ldots$	A_a
Observations	y_{11}	y_{21}	$\ldots$	y_{a1}
	y_{12}	y_{22}		y_{a2}
	$\ldots$	$\ldots$		$\ldots$
	y_{1n}	y_{2n}		y_{an}

The model equation is written in the form

$$\boldsymbol{y}_{ij} = E(\boldsymbol{y}_{ij}) + \boldsymbol{e}_{ij} = \mu + a_i + \boldsymbol{e}_{ij} \quad (i = 1, \ldots, a; j = 1, \ldots, n) \tag{3.3}$$

The a_i are called the main effects of the factor levels A_i; they are real numbers, i.e., not random. The model is completed by the following constraints (sometimes called side conditions): the $\boldsymbol{e}_{ij}$ are mutually independent with $E(\boldsymbol{e}_{ij}) = 0$ and $var(\boldsymbol{e}_{ij}) = \sigma^2$. We assume either that the sum of the a_i is

TABLE 3.1
ANOVA Table: One-Way Classification, Model $1(n_i = n)$.

Source of Variation	SS	df
Main Effect A	$SS_A = \frac{1}{n}\sum_i Y_{i.}^2 - \frac{Y_{..}^2}{N}$	$a-1$
Residual	$SS_R = \sum_{i,j} y_{ij}^2 - \frac{1}{n}\sum_i Y_{i.}^2$	$N-a$
Total	$SS_T = \sum_{i,j} y_{ij}^2 - \frac{Y_{..}^2}{N}$	$N-1$

Source of Variation	MS	$E(MS)$
Main Effect A	$MS_A = \frac{SS_A}{a-1}$	$\sigma^2 + \frac{n}{a-1}\sum a_i^2$
Residual	$MS_R = \frac{SS_R}{N-a}$	σ^2
Total		

Note: A dot means summation over the suffix replaced by a dot. SS = Sum of Squares, MS = Mean Squares, d.f. = degrees of freedom.
We use the convention for writing sums by capitalization of the corresponding letter and replacing suffixes by a dot if summation over them has taken place.
For example, $Y_{i.} = \sum_{j=1}^{n} y_{ij}$, $Y_{..} = \sum_{i=1}^{a}\sum_{j=1}^{n} y_{ij}$.

zero or in the case of an unbalanced design that the sum of the products $n_i a_i$ equals zero (which are equivalent if the n_i are all equal).
Table 3.1 gives the analysis of variance (ANOVA) table for the model, with $N = \sum_{i=1}^{a} n_i = an$, since we shall restrict ourselves to the optimal case where $n_i = n$ and $\sum_{i=1}^{a} a_i = 0$.
The R-program in this book tries to generalise the procedures that are used to determine the sample sizes. Meanwhile, there are differences of the classifications in Table 3.2 which require that different functions be used during the calculation of the sample sizes. It is efficient to group the classifications in Table 3.2 into 3 categories:

- Category 1: One way classification. In this category, the number of levels of the factor A is fixed; the subclass number n is determined.
- Category 2: Two way classifications.
- Category 3: Three way classifications.

If $\boldsymbol{y}$ has a normal distribution, we can test the following null hypothesis:

TABLE 3.2
Parameters of C, f_1 and f_2 for Some of the Classifications in ANOVA Models for Testing $H_0 : E_1 = E_2 = \ldots = E_q$ of Some Fixed Factor.

Classification	Effects		C	f_1	f_2
	Fixed	Random			
One Way Classification	A		n	$a-1$	$a(n-1)$
Two Way Cross	$A,\ B$		bn	$a-1$	$ab(n-1)$
BIBD, PBIBD (Two Way Cross) with block size k	A, B		b	$a-1$	$bk-a-b+1$
Two-Way Cross $A \times B$	A	B	bn	$a-1$	$(a-1)(b-1)$
Two-Way Nested $A \succ B$	A, B		bn	$a-1$	$ab(n-1)$
Two-Way Nested $A \succ B$	A	B	bn	$a-1$	$a(b-1)$
Two-Way Nested $A \succ B$	B	A	n	$a(b-1)$	$ab(n-1)$
Three Way Cross $A \times B \times C$	A, B, C		bcn	$a-1$	$abc(n-1)$
Three-Way Cross $A \times B \times C$	A, B	C	bcn	$a-1$	$(a-1)(c-1)$
Three Way Nested $A \succ B \succ C$	A, B, C		bcn	$a-1$	$abc(n-1)$
Three-Way Nested $A \succ B \succ C$	A	B, C	bcn	$a-1$	$a(b-1)$
Three-Way Nested $A \succ B \succ C$	B	A, C	cn	$a(b-1)$	$ab(c-1)$
Three-Way Nested $A \succ B \succ C$	C	A, B	n	$ab(c-1)$	$abc(n-1)$
Three-Way Nested $A \succ B \succ C$	A, B	C	bcn	$a-1$	$ab(c-1)$
Three-Way Nested $A \succ B \succ C$	A, C	B	bcn	$a-1$	$a(b-1)$
Three-Way Nested $A \succ B \succ C$	B, C	A	cn	$a(b-1)$	$abc(n-1)$
Three-Way Mixed $(A \times B) \succ C$	A, B, C		bcn	$a-1$	$abc(n-1)$
Three-Way Mixed $(A \times B) \succ C$	A, C	B	bcn	$a-1$	$(a-1)(b-1)$
Three-Way Mixed $(A \times B) \succ C$	A, B	C	bcn	$a-1$	$ab(c-1)$
Three-Way Mixed $(A \times B) \succ C$	A	B, C	bcn	$a-1$	$(a-1)(b-1)$
Three-Way Mixed $(A \succ B) \times C$	A, B, C		bcn	$a-1$	$abc(n-1)$
Three-Way Mixed $(A \succ B) \times C$	B, C	A	cn	$a(b-1)$	$abc(n-1)$
Three-Way Mixed $(A \succ B) \times C$	A, C	B	bcn	$a-1$	$a(b-1)$
Three-Way Mixed $(A \succ B) \times C$	A, B	C	n	$a-1$	$(a-1)(c-1)$
Three-Way Mixed $(A \succ B \times C)$	A	B, C	bcn	$a-1$	f in 3.4.4.5
Three-Way Mixed $(A \succ B) \times C$	B	A, C	cn	$a(b-1)$	$a(b-1)(c-1)$
Three-Way Mixed $(A \succ B) \times C$	C	A, B	abn	$c-1$	$(a-1)(c-1)$

H_0: "The factor A has no effect on the dependent variable $\boldsymbol{y}$."

In other words: "All the a_i are equal." If it is assumed that the sum of the a_i is zero, this is the same as "All the a_i are zero."
The alternative hypothesis is:

H_A: "The factor A has an effect on the dependent variable $\boldsymbol{y}$" or "At least two of the a_i are different."

The general rules in steps 1 through 8 from Section 3.1 will now be applied to our very simple case. The null hypothesis (step 1) has already been defined above; our model is a Model I (step 2). Step 3 is unnecessary, as there is only one $E(\boldsymbol{MS})$ column in Table 3.1. Step 4 leads us to the "Main effect A" row in Table 3.1. If the null hypothesis that all the a_i are equal is true, then the $E(\boldsymbol{MS})$ for this row reduces to σ^2 (step 5), and this is the $E(\boldsymbol{MS})$ for the residual (step 6). Therefore

$$\boldsymbol{F} = \frac{\boldsymbol{MS}_A}{\boldsymbol{MS}_R}$$

is the required test statistic (step 7). It is under H_0 distributed as $F(a-1; N-a)$, i.e., it follows a central F-distribution with $f_1 = a - 1$ and $f_2 = N - a$ degrees of freedom.
To determine the minimum size to test the above null hypothesis H_0, we have seen in Section 3.1 that we have to specify α and β, and we also need an estimate of σ^2. If we had $a = 2$ in the present case, then the F-test would be identical with the two-sided two-sample t-test for independent samples in Section 2.4.1.3 and $n_{\min} = n_{\max}$.
The power of the F-test depends on the non-centrality parameter λ. For the one-way Model I analysis of variance in this section it is proportional to

$$\sum_{i=1}^{a}(a_i - \overline{a})^2 \text{ where } \overline{a} = \frac{1}{a}\sum a_i \tag{3.4}$$

If we use the side condition $\sum\limits_{i=1}^{a} a_i = 0$, we have $\sum\limits_{i=1}^{a}(a_i - \overline{a})^2 = \sum\limits_{i=1}^{a} a_i^2$. Of course when H_0 is true, λ equals zero.
Let us repeat the general discussion in Section 3.1 for the special case of one-way ANOVA. If the a_i are not all equal, the value of λ depends on their values. The least favourable case from the point of view of the size required (leading to the maximum number n for each factor level) is the case with the smallest possible λ-value if at least two of the a_i are different. Let $a_{\max} = \max(a_i)$ be the largest and $a_{\min} = \min(a_i)$ be the smallest of the a effects. If the $a - 2$ remaining effects are equal to

$$\frac{1}{2}(a_{\min} + a_{\max}) \tag{3.5}$$

we have the least favourable case.

Lemma 3.1 Without loss of generality (w.l.o.g.) we assume: $\sum_{i=1}^{a} a_i = 0,\ a_1 \leq a_2 \leq \ ... \leq a_a,\ a_{\min} = -\frac{\delta}{2},\ a_{\max} = \frac{\delta}{2}$, and further w.l.o.g. $\delta = \sigma$. We consider a Model I of ANOVA (all the factors are fixed) with a cross classification and equal sub-class numbers.

a) Under the conditions above, the minimin size $n_{\min}$ (the most favourable case) occurs if we split the a_i into two groups of size a_I and a_{II}, respectively, with $a = a_I + a_{II}$ and $|a_I - a_{II}| \leq 1$ and the a_I elements of the first group equal $-\frac{a_{II}}{a}\delta$ and the a_{II} remaining equal $\frac{a_I}{a}\delta$. Thus there are two solutions for odd a and for even a half of the effects are equal to $-\frac{\delta}{2}$ and half of them are equal to $\frac{\delta}{2}$. Then $\sum_{i=1}^{a} a_i^2 = \frac{\delta^2}{a}(a_I a_{II})$ and is a maximum.

b) Under the conditions above, the maximin size $n_{\max}$ (the least favourable case) occurs if $a_1 = -\frac{\delta}{2}$, and $a_a = \frac{\delta}{2}$ and all the other effects are zero. Then $\sum_{i=1}^{a} a_i^2 = \frac{\delta^2}{2}$ and is a minimum.

c) In the singular case (two-sample problem) $a = 2$; both sizes are identical.

Proof: It is easy to see that the condition $\sum_{i=1}^{a} a_i = 0,\ a_1 \leq a_2 \leq \cdots \leq a_a$ is fulfilled as well as in case a) and in case b).

a) For case a) and even a the statement is evident. In general we know that with $a = a_I + a_{II}$ the product $a_I \cdot a_{II}$ is maximum if a_I and a_{II} are as equal as possible. That makes $\sum_{i=1}^{a} a_i^2 = a_I \frac{a_{II}^2}{a^2} \cdot \delta^2 + a_{II} \frac{a_I^2}{a^2} \cdot \delta^2 = \frac{a}{a^2} a_I \cdot a_{II} \delta^2 = \frac{a_I \cdot a_{II}}{a} \delta^2$ a maximum if a_I and a_{II} differ at most by 1.

b) For even a the result follows from the theory of D-optimal designs in regression. For odd a we obtain the result by equating the partial derivatives with respect to the effects and κ of $w = x - \kappa z$ to zero. Hereby is $x = \sum_{i=1}^{a} a_i^2 = a_1^2 + ... + a_{a-1}^2 + (\delta + a_1)^2$ and $z = \sum_{i=1}^{a} a_i = a_1 + ... + a_{a-1} + (\delta + a_1)$. This completes the proof because w is a convex function.

In the most favourable case we have as examples the following values for $\sum_{i=1}^{a} a_i^2$

a	2	3	4	5	6	7	8	9	10	11
$\sum_{i=1}^{a} a_i^2$	0.5	0.667	1	1.2	1.5	1.7143	2	2.222	2.5	2.7272

The total size of the experiment is minimised if all n_i are equal; this means that we design the experiment so that

$$n_i = n$$

We already called the mini-min size $n_{\min}$ and the maxi-min size $n_{\max}$. The experimenter now has to choose the number of observations n per factor level (class) between the lower bound $n_{\min}$ and the upper bound $n_{\max}$:

$$n_{\min} \leq n \leq n_{\max} \tag{3.6}$$

All that remains to be done is to calculate the bounds $n_{\min}$ and $n_{\max}$.

Example 3.1 We plan to perform an experiment with four levels of a fixed factor A and measure the yield of a crop in dt per ha. The four levels are four varieties of a cereal crop. We formulate the null hypothesis:

H_0: All the varieties have the same mean yield, i.e.,

$$a_1 = a_2 = a_3 = a_4$$

and the alternative hypothesis:

H_A: At least two varieties differ in their average yield, that is to say,

$$a_i \neq a_j \text{ for at least one pair } i \neq j.$$

Suppose the number n of plots per variety has to be determined to satisfy the following conditions: Type I error probability $\alpha = 0.05$ and Type II error probability $\beta = 0.1$ if $a_{\max} - a_{\min} = 2\sigma$.

The R-programs in this chapter must be understood as follows. The names start with **size**, after this is written what is to be determined; this could be n but also the number (a, say) of levels of a random factor in mixed models. Next position is the classification (one-, two- or three-way and the type of classification (cross, nested or mixed); the next entry is the number of the model (numbers in the text of the corresponding section). The next entry is the factor or factor level combination under test; for example, a, b or c means factor A, B, or C, respectively, and $a \times b$ means interactions $A \times B$. Then follow the precision requirements including the number of levels of the fixed factor(s) and at the end it is stated whether we are to calculate the minimin or the maximin size.

For example > `size_b.three_way_mixed_cxbina.model_4_a(`0.05, 0.1, 0.5,

4, "maximin") means that we calculate the maximin size b of the number of levels of the random factor B in a three-way mixed classification $C \times (B \prec A)$, model 4 when the effects of the factor A are tested and C has 4 levels.

The R-program to calculate the sizes is `size_n.one_way.model_1()`.

```
> size_n.one_way.model_1(0.05, 0.1, 2, 4, "maximin")

[1] 9

> size_n.one_way.model_1(0.05, 0.1, 2, 4, "minimin")

[1] 5
```

Thus the output of R gives the mini-min size $n_{\min} = 5$ and the maxi-min size $n_{\max} = 9$.

If the a_i in (3.3) are random (Model II), we obtain $\boldsymbol{y}_{ij} = \mu + \boldsymbol{a}_i + \boldsymbol{e}_{ij}$ $(i = 1, \ldots, a; j = 1, \ldots, n)$ (the so-called model II of the one-way ANOVA), while model (3.3) is called Model I. Sample size problems for Model II have already been discussed in Chapter two.

3.3 Two-way layout

The two factors are denoted by A and B, respectively. We distinguish between two kinds of combinations of the levels of the two factors:

Cross-classification ($A \times B$):

Observations can (but do not have to) be made in all combinations of A- and B-levels. If observations are available for all of the $a \cdot b$ factor level combinations of the a levels A_i of the factor A with the b levels B_j of B, then the cross-classification is called complete (e.g., a complete block design); otherwise it is called incomplete (for instance, an incomplete block design such as a BIBD as in Chapter six). We always assume that incomplete cross-classifications are connected (see Chapter six). A typical example of a two-way classification occurs if the effect of a treatment factor A is investigated in a block experiment. Then the blocks are the levels of a further (noise) factor B.

Nested (or hierarchical) classification ($A \succ B$):

The levels of B can only occur jointly with exactly one level of A. We write $(B \prec A)$ which means B is inferior to A or we can write $A \succ B$ which means A is superior to B.

Such a case occurs if we consider the milk yield of daughters of several dams (levels of factor B) artificially inseminated by several sires (levels of the superior factor A). Another example is member states of the EU as levels of a factor A and administrative subunits of the states such as provinces (The Netherlands), departments (France) or Bundesländer (Austria and Germany) as levels of a factor B.

3.3.1 Two-way analysis of variance—cross-classification

In the two-way cross-classification the classes are defined by the factor level combinations (i, j) of the levels A_i of A and the levels B_j of B.

The observations y_{ijk} in the combinations of factor levels are as follows.

Factor Levels	B_1	B_2	...	B_b
A_1	y_{111}	y_{121}	...	y_{1b1}
	y_{112}	y_{122}		y_{1b2}
	...	...		...
	y_{11n}	y_{12n}		y_{1bn}
A_2	y_{211}	y_{221}	...	y_{2b1}
	y_{212}	y_{222}		y_{2b2}
	...	...		...
	y_{21n}	y_{22n}		y_{2bn}
⋮	⋮	⋮	⋮	⋮
A_a	y_{a11}	y_{a21}	...	y_{ab1}
	y_{a12}	y_{a22}		y_{ab2}
	...	...		...
	y_{a1n}	y_{a2n}		y_{abn}

Let $\boldsymbol{y}_{ijk}$ be the k-th observation in the class (i, j). Then we get

$$\boldsymbol{y}_{ijk} = \mu + a_i + b_j + (ab)_{ij} + \boldsymbol{e}_{ijk} (i = 1, \ldots, a; j = 1, \ldots, b; k = 1, ..., n_{ij})$$

In this equation μ is the general mean; a_i are the main effects of levels A_i of $A; b_j$ the main effects of levels B_j of B; $(ab)_{ij}$ the interaction effects between A_i and B_j, (they are defined only if more than one observation is available for the class (i, j)). We have runs in class (i, j) from 1 to n_{ij}. If $n_{ij} = n$, we have the case of equal class numbers (replications); we call this a balanced ANOVA. Optimal experimental designs in the two-way analysis of variance often require equal class numbers. The main effects and interaction effects are not at the moment defined as being fixed or random—these differences will be discussed in the following sections. For all the subsequent models, we assume that the $\boldsymbol{e}_{ijk}$ are mutually independent with expectation zero and have a common variance σ^2 and that $n_{ij} = n$.

The ANOVA Table for all models (without the column $E(\boldsymbol{MS})$) is given for this balanced case in Table 3.3. Because $n_{ij} = n$ for all i, j, we have $N = abn$. The $E(\boldsymbol{MS})$ are given in Table 3.4.

TABLE 3.3
ANOVA Table for the Two-Way Cross-Classification ($n_{ij} = n > 1$).

Sources of Variation	$\boldsymbol{SS}$	*df*	$\boldsymbol{MS}$
Main Effect A	$\boldsymbol{SS}_A = \frac{1}{bn}\sum_i \boldsymbol{Y}^2_{i..} - \frac{1}{N}\boldsymbol{Y}^2_{...}$	$a-1$	$\boldsymbol{MS}_A = \frac{\boldsymbol{SS}_A}{a-1}$
Main Effect B	$\boldsymbol{SS}_B = \frac{1}{an}\sum_j \boldsymbol{Y}^2_{.j.} - \frac{1}{N}\boldsymbol{Y}^2_{...}$	$b-1$	$\boldsymbol{MS}_B = \frac{\boldsymbol{SS}_B}{b-1}$
Interaction $A \times B$	$\boldsymbol{SS}_{AB} = \frac{1}{n}\sum_{i,j} \boldsymbol{Y}^2_{ij.} - \frac{1}{bn}\sum_i \boldsymbol{Y}^2_{i..}$ $- \frac{1}{an}\sum_j \boldsymbol{Y}^2_{.j.} + \frac{\boldsymbol{Y}^2_{...}}{N}$	$(a-1)\cdot(b-1)$	$\boldsymbol{MS}_{AB} = \frac{\boldsymbol{SS}_{AB}}{(a-1)(b-1)}$
Residual	$\boldsymbol{SS}_R = \sum_{i,j,k} \boldsymbol{y}^2_{ijk} - \frac{1}{n}\sum_{ij} \boldsymbol{Y}^2_{ij.}$	$ab(n-1)$	$\boldsymbol{MS}_R = \frac{\boldsymbol{SS}_R}{ab(n-1)}$
Total	$\boldsymbol{SS}_T = \sum_{i,j,k} \boldsymbol{y}^2_{ijk} - \frac{1}{N}\boldsymbol{Y}^2_{...}$	$N-1$	

Note: We use the convention for writing sums by capitalization of the corresponding letter and replacing suffixes by a dot if summation over it has taken place as has been explained in Table 3.2.

TABLE 3.4
The Expected Mean Squares for the Models of the Two-Way Cross-Classification ($n_{ij} = n > 1$).

Sources of Variation	Model I	Mixed model, (A fixed, B random)
Main Effect A	$\sigma^2 + \frac{bn}{a-1}\sum_i^a a_i^2$	$\sigma^2 + n\sigma_{ab}^2 + \frac{bn}{a-1}\sum_i^a a_i^2$
Main Effect B	$\sigma^2 + \frac{an}{b-1}\sum_j b_j^2$	$\sigma^2 + an\sigma_b^2$
Interaction $A \times B$	$\sigma^2 + \frac{n}{(a-1)(b-1)} \cdot \sum_{i,j}(ab)_{ij}^2$	$\sigma^2 + n\sigma_{ab}^2$
Residual	σ^2	σ^2

We are interested here in models with at least one fixed factor and not in a model with two random factors. We are also only interested in hypotheses about fixed factors and fixed interaction effects for determining the minimum size of the experiment.

3.3.1.1 Two-way analysis of variance—cross-classification—Model I

A Model I cross-classified two-way analysis of variance is specified as follows: The model equation with equal subclass numbers is given by

$$\boldsymbol{y}_{ijk} = \mu + a_i + b_j + (ab)_{ij} + \boldsymbol{e}_{ijk} (i = 1, \ldots, a;\ j = 1, \ldots, b;\ k = 1, \ldots, n) \quad (3.7)$$

All main effects a_i and b_j and interaction effects $(ab)_{ij}$ are real numbers; i.e., they are not random. The model is completed by the side conditions that the sums of the a_i and of the b_j and of the $(ab)_{ij}$ (separately over each index) all equal zero.
The expected mean squares can be found in Table 3.4 for the case of $n_{ij} = n > 1$.

The estimates of the fixed effects are obtained by the method of least squares and are given by the following equations: $\hat{\mu} = \overline{y}_{...}, \hat{a}_i = \overline{y}_{i..} - \overline{y}_{...}, \hat{b}_j = \overline{y}_{.j.} - \overline{y}_{...}, \widehat{(ab)}_{ij} = \overline{y}_{ij.} - \overline{y}_{i..} - \overline{y}_{.j.} + \overline{y}_{...}$

The null hypotheses that can be tested (under the assumption that the $\boldsymbol{e}_{ijk}$ are independently distributed as $N(0; \sigma^2)$) are
- H_{01}: "All a_i are zero."
- H_{02}: "All b_j are zero."
- H_{03}: "All $(ab)_{ij}$ are zero."

The alternative hypotheses are
- H_{A1}: "At least one a_i differs from zero."
- H_{A2}: "At least one b_j differs from zero."

H_{A3}: "At least one $(ab)_{ij}$ differs from zero."
With the help of the 8 steps defined in Section 3.1, and by using Tables 3.3 and 3.4 we get

$\boldsymbol{F}_A = \frac{\boldsymbol{MS}_A}{\boldsymbol{MS}_R}$ is under H_{01} $F((a-1); ab(n-1))$-distributed; $\boldsymbol{F}_B = \frac{\boldsymbol{MS}_B}{\boldsymbol{MS}_R}$ is under H_{02} $F(b-1; ab(n-1))$ distributed;

$\boldsymbol{F}_{AB} = \frac{\boldsymbol{MS}_{AB}}{\boldsymbol{MS}_R}$ is under H_{03} $F((a-1)(b-1);\ ab(n-1))$ distributed.

To find the minimum class size (number of replications) n which will satisfy given precision requirements, we can in principle proceed as in Section 3.2. Analoguous to Section 3.2, we can consider the most favourable and the least favourable situation concerning the effects of the levels of factor A. The non-centrality parameter is $\lambda = bn \sum a_i^2/\sigma^2$.

The corresponding R-program is `size_n.two_way_cross.model_1_a()`.

Example 3.2 We want to test the null hypothesis that six wheat varieties do not differ in their yields. For the experiment we can use a number of plots for each of the varieties at each of four experimental stations (farms, for short). The varieties are the levels of a fixed factor A, and the four farms are also considered as fixed levels of a (block) factor B because just these 4 experimental stations are used for such investigations. Let us take the precision demands as Type I error probability $\alpha = 0.05$, and Type II error probability $\beta \leq 0.1$ if $a_{max} - a_{min} \geq \sigma$.

```
> size_n.two_way_cross.model_1_a(0.05, 0.1, 1, 6, 4, "maximin")

[1] 9

> size_n.two_way_cross.model_1_a(0.05, 0.1, 1, 6, 4, "minimin")

[1] 4
```

Hence $n_{min} = 4$ and $n_{max} = 9$.

To test the hypothesis that no interaction effects exist, i.e., we like to test $H_0 : \forall (ab)_{ij} = 0$ against $H_A : \exists (ab)_{ij} \neq 0$, we use the F-statistic $\boldsymbol{F}_{AB} = \frac{\boldsymbol{MS}_{AB}}{\boldsymbol{MS}_R}$ which is under $H_0 : \forall (ab)_{ij} = 0$ centrally F-distributed with $(a-1)(b-1)$ and $ab(n-1)$ degrees of freedom. In Lemma 3.2 the interaction $(ab)_{ij}$ is denoted as a_{ij}.

Lemma 3.2 A balanced two-way layout as well as higher classifications with two fixed factors (A;B, say) under the side conditions $\sum_{i=1}^{a} a_{ij} = 0$; $\sum_{j=1}^{b} a_{ij} = 0$; and $max(a_{ij}) - min(a_{ij}) = \delta = 1$ and w.l.o.g. $a \leq b$ are now considered.

a) The minimum of $\sum_{i=1}^{a} \sum_{j=1}^{b} a_{ij}^2$ of the interaction effects $a_{ij}; i = 1, \dots, a; j = 1, \dots, b$ is obtained for $a_{11} = \frac{(a-b)(b-1)}{a(b-1)}$; $a_{i1} = -\frac{b-1}{a(b-1)}$; $i =$

$2, \dots, a; a_{1j} = -\frac{a-1}{a-(b-1)}; j = 2, \dots, b; a_{ij} = \frac{1}{a(b-1)}; i = 2, \dots, a; j = 2, \dots, b.$

If $a = b$, this solution is equivalent with a further non-isomorphic solution $a_{11} = a_{aa} = -0.5; a_{1a} = a_{a1} = 0.5$ and all the other a_{ij} equal zero

In both cases $\underset{max(a_{ij}) - min(a_{ij}) = 1}{min} \left\{ \sum_{i=1}^{a} \sum_{j=1}^{b} a_{ij}^2 \right\} = 1.$ If w.l.o.g. a < b, we obtain $S(a,b) = \left[\frac{(a-1)(b-1)}{a(b-1)}\right]^2 + (a-1)\left[\frac{b-1}{a(b-1)}\right]^2 + (b-1)\left[\frac{a-1}{a(b-1)}\right]^2 + (a-1)(b-1)\left[\frac{1}{a(b-1)}\right]^2 < 1.$

This is monotonously decreasing in b. A lower bound is given for $b \to \infty$. Because in our case with fixed factors A and B the number of levels a and b are known, $S(a,b)$ will be used in our R-program to calculate the corresponding size of the experiment.

Other solutions of the problem are given for all permutations of rows and/or columns; these solutions are isomorphic.

Some values of $S(a,b)$ are given below.

		b		
		3	4	5
a	2	0.75	0.667	0.625
	3	1	0.88	0.8333
	4	...	1	0.890

b) The maximum of $\sum_{i=1} \sum_{j=1} a_{ij}^2$ of the interaction effects is for even a and b given if one half of the a_{ij} equal -0.5 and the other half equal 0.5. Then

$$\underset{[-0.5, 0.5]}{max} \left\{ \sum_{i=1}^{a} \sum_{j=1}^{b} a_{ij}^2 \right\} = \frac{ab}{4}$$

Conjecture 3.1 In a complete two-way layout or in a three-way layout with two fixed factors (A, B, say) the maximum of $\sum_{i=1}^{a} \sum_{j=1}^{b} a_{ij}^2$ of the interaction effects under the side conditions $\sum_{i=1}^{a} a_{ij} = 0; \sum_{j=1}^{b} a_{ij} = 0$ and $max(a_{ij}) - min(a_{ij}) = \delta = 1$ are for

(i) even a and b odd,
(ii) odd a and even b and
(iii) a and b both odd

identical with that value for the cases where the odd number(s) is(are) reduced by one. Thus the maximum ncp is proportional to

(i) $\sum_{i=1}^{a} \sum_{j=1}^{b} a_{ij}^2 = \frac{a(b-1)}{4}$

(ii) $\sum_{i=1}^{a}\sum_{j=1}^{b} a_{ij}^2 = \frac{(a-1)b}{4}$ and

(iii) $\sum_{i=1}^{a}\sum_{j=1}^{b} a_{ij}^2 = \frac{(a-1)(b-1)}{4}$, respectively.

So we may trust in the conjecture, we prove it first for $a = 2, b = 3$ (that means also for $a = 3, b = 2$).
The effects are (due to the side conditions)

a_{11}	a_{12}	a_{13}
$-a_{11}$	$-a_{12}$	$-a_{13}$

$-a_{12} - a_{13}$	a_{12}	a_{13}
$a_{12} + a_{13}$	$-a_{12}$	$-a_{13}$

The maximum cannot be larger than that for $a = 2, b = 4$, which following Lemma 3.2 equals 2. But this is the case for $a_{12} = a_{13} = 0.5$. Therefore $a_{12} \cdot a_{13}$ must be negative which leads to $a_{12} = a_{13} = -0.5$ and a maximum of $\sum_{i=1}^{2}\sum_{j=1}^{3} a_{ij}^2$ of 1. Therefore the table above becomes

$a_{11} = 0$	$a_{12} = 0.5$	$a_{13} = -0.5$
$-a_{11} = 0$	$-a_{12} = -0.5$	$-a_{13} = 0.5$

Next we prove the conjecture for $a = b = 3$. We drop the suffixes and write the effects as in the following table.

x	u	c
y	v	d
z	w	e

From the side conditions it follows that

$v + d + w + e$	$-v - w$	$-d - e$
$-v - d$	v	d
$-w - e$	w	e

The sum of the squares of the entries becomes a maximum (in $[-0.5; 0.5]$) if $v = e = 0.5$ and $d = w = -0.5$.

The corresponding R-program is `size_n.two_way_cross.model_1_axb()`.

Example 3.2 – continued: We want to test the null hypothesis that there is no interaction of wheat varieties and experimental farms in the experiment described above. Let us take the same precision demands as Type I error probability $\alpha = 0.05$, and Type II error probability $\beta = 0.1$ if $(ab)_{\max} - (ab)_{\min} \geq \delta = \sigma$. Then the output of R gives the mini-min size $n_{\min} = 5$ and the maxi-min size $n_{\max} = 48$.

```
> size_n.two_way_cross.model_1_axb(0.05, 0.1, 1, 6, 4, "maximin")

[1] 48
```

```
> size_n.two_way_cross.model_1_axb(0.05, 0.1, 1, 6, 4, "minimin")

[1] 5
```

Hence $n_{max} = 48$ and $n_{min} = 5$.

3.3.1.2 Two-way analysis of variance—cross-classification—mixed model

The situation in a mixed model of the cross-classified two-way analysis of variance is as follows. Without loss of generality we assume that the levels of A are fixed and those of B are random. The model equation in the balanced case is given by

$$\boldsymbol{y}_{ijk} = \mu + a_i + \boldsymbol{b}_j + (\boldsymbol{ab})_{ij} + \boldsymbol{e}_{ijk} (i = 1, \ldots, a;\ j = 1, \ldots, b;\ k = 1, \ldots, n) \tag{3.8}$$

All effects except μ and the a_i are random. The model is completed by the following conditions: $E(\boldsymbol{b}_j) = E((\boldsymbol{ab})_{ij}) = E(\boldsymbol{e}_{ijk}) = 0$ and $var(\boldsymbol{e}_{ijk}) = \sigma^2, var(\boldsymbol{b}_j) = \sigma_b^2, var((\boldsymbol{ab})_{ij}) = \sigma_{ab}^2$; the mutual independence of all random variables; and that the a_i sum to zero and $\sum_{i=1}^{a}(\boldsymbol{ab})_{ij} = 0$ for each j. σ^2, σ_b^2 and σ_{ab}^2 are called variance components. Due to the condition $\sum_{i=1}^{a}(\boldsymbol{ab})_{ij} = 0$ for each j, the $E(\boldsymbol{MS}_B)$ is $\sigma^2 + an\sigma_b^2$. Note that, e.g., SAS and SPSS use independent $(\boldsymbol{ab})_{ij}$; thus without $\sum_{i=1}^{a}(\boldsymbol{ab})_{ij} = 0$ for each j, is $E(\boldsymbol{MS}_B) = \sigma^2 + n\sigma_{ab}^2 + an\sigma_b^2$.

The expected mean squares for this model are given in Table 3.4 for $n_{ij} = n$. The null hypotheses that can be tested are

H_{01}: "All a_i are zero"
H_{02}: "$\sigma_b^2 = 0$"
H_{03}: "$\sigma_{ab}^2 = 0$"

We determine the F-ratios using the 8 steps from Section 3.1. If H_{01} is true, then $E(\boldsymbol{MS}_A)$ equals $E(\boldsymbol{MS}_{A\times B})$, and we test H_{01} using

$$\boldsymbol{F}_A = \frac{\boldsymbol{MS}_A}{\boldsymbol{MS}_{A\times B}}$$

which has under H_{01} an F-distribution with $(a-1)$ and $(a-1)(b-1)$ degrees of freedom.

To find the minimum size of the experiment which will satisfy given precision requirements, we can in principle proceed as in Section 3.2. But now we must remember that only the degrees of freedom of the corresponding F-statistic influence the power of the test and by this the size of the experiment.

To test the interesting hypothesis that the fixed factor has no influence on the observations, we have $(a-1)$ and $(a-1)(b-1)$ degrees of freedom of numerator and denominator, respectively. Thus the sub-class number n does not influence the degrees of freedom and, therefore, should be chosen as small as possible. If we know that there are no interaction effects, we choose $n = 1$, but if interaction effects may occur, we choose $n = 2$. Because the number a of levels of the factor under test is fixed, we can only choose b as the size of the sample of B-levels to fulfill precision requirements.

By R we solve the problem as follows: We choose $n = 1$ if interaction are of no interest and $n = 2$ otherwise. Then we use the program `size_b.two_way_cross.mixed_model_a_fixed_a()`.

Example 3.3 We want to test the null hypothesis that the six wheat varieties as in Example 3.2 do not differ in their yields. For the experiment we can use a number of plots for each of the varieties at experimental stations (farms, for short) which should be randomly selected from a large universe. The varieties are the levels of a fixed factor A, and the farms are considered now as random levels of a (block) factor B because it is random which ones will be selected for the investigation. Let us take the same precision demands as in Example 3.2: Type I error probability $\alpha = 0.05$, and Type II error probability $\beta = 0.1$ if $a_{\max} - a_{\min} = \delta = \sigma$. What now has to be determined to fulfill the precision requirement is the number b of levels of the random factor B. This is the size of the sample of levels of the factor B from the available set of test stations.

When choosing $n = 1$ (no interaction)

```
> size_b.two_way_cross.mixed_model_a_fixed_a(0.05, 0.1, 1, 6, 1,
+      "maximin")

[1] 35

> size_b.two_way_cross.mixed_model_a_fixed_a(0.05, 0.1, 1, 6, 1,
+      "minimin")

[1] 13
```

When choosing $n = 2$

```
> size_b.two_way_cross.mixed_model_a_fixed_a(0.05, 0.1, 1, 6, 2,
+      "maximin")

[1] 18

> size_b.two_way_cross.mixed_model_a_fixed_a(0.05, 0.1, 1, 6, 2,
+      "minimin")
```

```
[1] 7
```

Thus when there is no interaction, we choose $n = 1$. Our R-program gives the mini-min size $b_l = 13$ and the maxi-min size $b_u = 35$.
When there is interaction, we choose $n = 2$. Our R-program gives the mini-min size $b_l = 7$ and the maxi-min size $b_u = 18$. In principle we can split the product bn into two parts from an economical point of view.

3.3.2 Nested-classification $A \succ B$

Let $\boldsymbol{y}_{ijk}$ be the k-th observation in the factor combination (i, j), i.e., in B_{ij}. Then we have.

$$\boldsymbol{y}_{ijk} = \mu + a_i + b_{j(i)} + \boldsymbol{e}_{ijk} \quad (i = 1, \ldots, a, \ j = 1, \ldots, b_i, \quad k = 1, \ldots, n_{ij}) \tag{3.9}$$

In this equation the symbols are μ is the general mean, a_i the main effects of the a levels A_i of A with $\Sigma a_i = 0$, $b_{j(i)}$ the main effects of the b_i levels B_{ij} of B within A_i with $\Sigma b_{j(i)} = 0$ for each i; interaction effects do not occur in this classification. In A_i, j runs from 1 to b_i; in class $(i, j), k$ runs from 1 to n_{ij}. If $n_{ij} = n$ for all (i, j), we have the case of equal class numbers (replications). Optimal experimental designs in the two-way hierarchical analysis of variance often need equal replications and moreover equal b_i (this is the balanced case).

Levels of A	A_1				A_2				...	A_a			
Levels of B	B_{11}	B_{12}	...	B_{1b}	B_{21}	B_{22}	...	B_{2b}		B_{a1}	B_{a2}	...	B_{ab}
Observations	y_{111} y_{112} ... y_{11n}	y_{121} y_{122} ... y_{12n}		y_{1b1} y_{1b2} ... y_{1bn}	y_{211} y_{212} ... y_{21n}	y_{221} y_{222} ... y_{22n}		y_{2b1} y_{2b2} ... y_{2bn}		y_{a11} y_{a12} ... y_{a1n}	y_{a21} y_{a22} ... y_{a2n}		y_{ab1} y_{ab2} ... y_{abn}

3.3.2.1 Two-way analysis of variance—nested classification—Model I

The representation of a Model I two-way nested analysis of variance is given by the model equation (3.9). Here all the effects are real numbers, i.e., not random. The model is completed by the following conditions: the sums of the a_i and of the $b_{j(i)}$ (for each i) equal zero.
Table 3.5 contains a column containing the expected mean squares $E(\boldsymbol{MS})$ for Model I for the balanced case ($n_{ij} = n, b_i = b$). With the aid of this table we can use the methods described in Section 3.1 to derive the F-tests for the testing of various hypotheses—we will illustrate these in an example below. We write $N = abn$.

We test the null hypothesis that factor A has no effect with the test statistic

TABLE 3.5
Analysis of Variance Table for the Two-Way Nested Classification and the $E(\boldsymbol{MS})$ for Model I—Balanced Case: $n_{ij} = n, b_i = b$.

Source of Variation	$\boldsymbol{SS}$	*df*
Main Effect A	$\boldsymbol{SS}_A = \frac{1}{bn}\sum_i \boldsymbol{Y}_{i..}^2 - \frac{\boldsymbol{Y}_{...}^2}{N}$	$a-1$
Main Effect B within factor A	$\boldsymbol{SS}_{B \text{ in } A} = \frac{1}{n}\sum_{i,j} \boldsymbol{Y}_{ij.}^2 - \frac{1}{bn}\sum_i \boldsymbol{Y}_{i.}^2$	$a(b-1)$
Residual	$\boldsymbol{SS}_R = \sum_{i,j,k} \boldsymbol{y}_{ijk}^2 - \frac{1}{n}\sum_{i,j} \boldsymbol{Y}_{ij,}^2$	$N-ab$
Total	$\boldsymbol{SS}_T = \sum_{i,j,k} \boldsymbol{y}_{ijk}^2 - \frac{\boldsymbol{Y}_{...}^2}{N}$	$N-1$

Source of Variation	$\boldsymbol{MS}$	$E(\boldsymbol{MS})$ Model I
Main Effect A	$\boldsymbol{MS}_A = \frac{\boldsymbol{SS}_A}{a-1}$	$\sigma^2 + \frac{bn}{a-1}\sum_i a_i^2$
Main Effect B within factor A	$\boldsymbol{MS}_{B \text{ in } A} = \frac{\boldsymbol{SS}_{B \text{ in } A}}{ab-a}$	$\sigma^2 + \frac{n}{a(b-1)}\sum_{i,j} b_{j(i)}^2$
Residual	$\boldsymbol{MS}_R = \frac{\boldsymbol{SS}_R}{N-ab}$	σ^2
Total		

Note: We use the convention for writing sums by capitalization of the corresponding letter and replacing suffixes by a dot if summation over it has taken place as has been explained in Table 3.2.

$\boldsymbol{F}_A = \frac{\boldsymbol{MS}_A}{\boldsymbol{MS}_R}$, which under the null hypothesis has an F-distribution with $(a-1)$ and $N-ab = ab(n-1)$ degrees of freedom.

The null hypothesis that factor B has no effect is tested with the test statistic $\boldsymbol{F}_B = \frac{\boldsymbol{MS}_B}{\boldsymbol{MS}_R}$, which under H_0 has an F-distribution with $a(b-1)$ and $N-ab = ab(n-1)$ degrees of freedom.
To find the minimum size of the experiment which will satisfy given precision requirements, we can in principle proceed as in Section 3.2.

By R we solve the problem as follows:

Testing the null hypothesis on factor A:

`size_n.two_way_nested.model_1_test_factor_a()` Testing the null hypothesis on factor B:

`size_n.two_way_nested.model_1_test_factor_b()`.

Example 3.4 We consider an example of the balanced case with $a = 6$ levels of factor A, $b = 4$ levels of factor B within each level of factor A.

a) We wish to determine the number n of observations in the 24 classes required to test the null hypothesis on the factor A so that $\alpha = 0.05;\ \beta = 0.1;\ \delta = a_{\max} - a_{\min} = \sigma$.

b) Further we wish to determine the number n of observations in the 24 classes required to test the null hypothesis on the factor B so that $\alpha = 0.05;\ \beta = 0.1;\ \delta = b_{\max} - b_{\min} = \sigma$.

Testing null hypothesis on factor A:

```
> size_n.two_way_nested.model_1_test_factor_a(0.05, 0.1, 1, 6,
+      4, "maximin")

[1] 9

> size_n.two_way_nested.model_1_test_factor_a(0.05, 0.1, 1, 6,
+      4, "minimin")

[1] 4
```

Testing null hypothesis on factor B:

```
> size_n.two_way_nested.model_1_test_factor_b(0.05, 0.1, 1, 6,
+      4, "maximin")

[1] 51

> size_n.two_way_nested.model_1_test_factor_b(0.05, 0.1, 1, 6,
+      4, "minimin")

[1] 5
```

Our R-program gives in case a the mini-min size $n_{min} = 4$ and the maxi-min size $n_{max} = 9$ and in case b $n_{min} = 5$ and the maxi-min size $n_{max} = 51$.

3.3.2.2 Two-way analysis of variance—nested classification—mixed model, *A* fixed and *B* random

The model equation for this mixed model is

$$\boldsymbol{y}_{ijk} = \mu + a_i + \boldsymbol{b}_{j(i)} + \boldsymbol{e}_{ijk}(i = 1, \ldots, a;\ j = 1, \ldots, b;\ k = 1, \ldots, n) \quad (3.10)$$

The effects μ and a_i are fixed; the remaining ones random. The model is completed by the following side conditions: $var(\boldsymbol{b}_{ij}) = \sigma^2_{b \text{ in } a}$; $var(\boldsymbol{e}_{ijk}) = \sigma^2$;

TABLE 3.6
$E(\boldsymbol{MS})$ for Balanced Nested Mixed Models.

	Mixed Models	
Source of Variation	A fixed B random	A random B fixed
Between A-Levels	$\sigma^2 + n\,\sigma^2_{b \text{ in } a} + \frac{bn}{a-1}\sum_{i=1}^{a} a_i^2$	$\sigma^2 + bn\sigma_a^2$
Between B-Levels within A	$\sigma^2 + n\,\sigma^2_{b \text{ in } a}$	$\sigma^2 + \frac{n}{a(b-1)}\sum_{i,j} b^2_{j(i)}$
Residual	σ^2	σ^2

all the random variables are mutually independent with expectation zero and the a_i sum to zero. The ANOVA table for this model is given in Table 3.5; the expected mean squares are in Table 3.6.

The null hypothesis that the effects of all the levels of factor A are equal is tested using the test statistic:

$$\boldsymbol{F}_A = \frac{\boldsymbol{MS}_A}{\boldsymbol{MS}_{B \text{ in } A}}$$

which under H_0 has an F-distribution with $(a-1)$ and $a(b-1)$ degrees of freedom.
To find the minimum size of the experiment which will satisfy given precision requirements, we can in principle proceed as in Section 3.2. Because in nested models we have no interaction effects, we can fix $n = 1$.

Example 3.5 We consider an example of the balanced case with $a = 6$ levels of the factor A. We wish to determine the number b of levels of the factor B to test the null hypothesis on the factor A so that $\alpha = 0.05$; $\beta = 0.1$; $\delta = a_{\max} - a_{\min} = \sigma_y$.

```
> size_b.two_way_nested.b_random_a_fixed_a(0.05, 0.1, 1, 6,
+         "maximin")

[1] 34

> size_b.two_way_nested.b_random_a_fixed_a(0.05, 0.1, 1, 6,
+         "minimin")

[1] 12
```

Our R-program gives for $n = 1$ the mini-min size $b_{min} = 12$ and the maxi-min size $b_{max} = 34$.

3.3.2.3 Two-way analysis of variance—nested classification—mixed model, B fixed and A random

The model equation for this mixed model is

$$\boldsymbol{y}_{ijk} = \mu + \boldsymbol{a}_i + b_{j(i)} + \boldsymbol{e}_{ijk}(i = 1, \ldots, a;\ j = 1, \ldots, b;\ k = 1, \ldots, n) \quad (3.11)$$

The effects μ and $b_{j(i)}$ are fixed; the remaining effects are random. The model is completed by the following conditions: $var(\boldsymbol{a}_i) = \sigma_a^2$; all the random variables are mutually independent with expectation zero and the $b_{j(i)}$ sum to zero over j for each i.

The ANOVA table for this model is again Table 3.5; the expected mean squares are found in Table 3.6.
The null hypothesis that the effects of all the levels of factor B are equal is tested using the test statistic:

$$\boldsymbol{F}_B = \frac{\boldsymbol{MS}_B}{\boldsymbol{MS}_R}$$

which under H_0 has an F-distribution with $a(b-1)$ and $N - ab$ degrees of freedom.
To find the minimum size of the experiment which will satisfy given precision requirements, we can in principle proceed as in Section 3.2. Here, again the subclass number has some influence on the power of the test. Therefore the R-program asks for proposals for the number a of levels of the random factor A and calculates for these the subclass number. But because the denominator degrees of freedom (Table 3.5) decrease with a, we conjecture that we should choose a as small as possible, $a = 2$, say.

The corresponding R-program is

`size_n.two_way_nested.a_random_b_fixed_b()`.

Example 3.6 We consider an example of the balanced case with $b = 10$ levels of the factor B. We wish to determine the number n in the ab classes for a preselected number of levels of the factor A to test the null hypothesis on the factor B so that $\alpha = 0.05$; $\beta = 0.1$; $\delta = \sigma_y$.

```
> size_n.two_way_nested.a_random_b_fixed_b(0.05, 0.1, 1, 2, 10,
+       "maximin")

[1] 52

> size_n.two_way_nested.a_random_b_fixed_b(0.05, 0.1, 1, 2, 10,
+       "minimin")

[1] 6
```

```
> size_n.two_way_nested.a_random_b_fixed_b(0.05, 0.1, 1, 3, 10,
+       "maximin")

[1] 60

> size_n.two_way_nested.a_random_b_fixed_b(0.05, 0.1, 1, 3, 10,
+       "minimin")

[1] 5

> size_n.two_way_nested.a_random_b_fixed_b(0.05, 0.1, 1, 10, 10,
+       "maximin")

[1] 95

> size_n.two_way_nested.a_random_b_fixed_b(0.05, 0.1, 1, 10, 10,
+       "minimin")

[1] 3
```

Thus we found for $a = 2, n_{max} = 52, n_{min} = 6$; $a = 3, n_{max} = 60, n_{min} = 5$; $a = 10, n_{max} = 95, n_{min} = 3$.

3.4 Three-way layout

In the three-way layout we assume that three factors, say, A, B, C, influence some character y modeled by a random variable $\boldsymbol{y}$.

We are interested here in models with at least one fixed factor and not in models with random factors only, and for determining the minimum size of the experiment, we are interested only in hypotheses about fixed factors.

For three-way analysis of variance, we have the following types of classifications:

Cross-classification: Observations are possible for all combinations of the levels of the three factors A, B and C (Symbol: $A \times B \times C$).
Nested classification: The levels of C are nested within B, and those of B are nested within A (Symbol $A \succ B \succ C$).
Mixed classification: We have two cases: Case 1: A and B are cross-classified, and C is nested within the classes (i, j) (Symbol $(A \times B) \succ C$). Case 2: B is nested in A, and C is cross-classified with all $A \succ B$ combinations (Symbol $(A \succ B) \times C$).

Because there are many cases with at least one fixed factor, we use for abbreviation the following system.

Null hypothesis		Alternative hypothesis	
F-statistic	Degrees of freedom of the numerator	Degrees of freedom of the denominator	Non-centrality parameter

This always stands for the following: we like to test the null hypothesis (in the scheme) against the alternative hypothesis (in the scheme). Following the eight steps in Section 3.1, we find that the F-statistic (in the scheme) follows a non-central F-distribution with numerator degrees of freedom (in the scheme) and denominator degrees of freedom (in the scheme) and non-centrality parameter (in the scheme).

3.4.1 Three-way analysis of variance—cross-classification $A \times B \times C$

The observations y_{ijkl} in the combinations of factor levels are as follows.

Factor Levels	B_1			B_2			$\cdots$	B_b		
	C_1	$\cdots$	C_c	C_1	$\cdots$	C_c	$\cdots$	C_1	$\cdots$	C_c
A_1	y_{1111}		y_{11c1}	y_{1211}		y_{12c1}	$\cdots$	y_{1b11}		y_{1bc1}
	y_{1112}	$\cdots$	y_{11c2}	y_{1212}		y_{12c2}		y_{1b12}		y_{1bc2}
	$\cdots$	$\cdots$	$\cdots$	$\cdots$	$\cdots$	$\cdots$		$\cdots$		$\cdots$
	y_{111n}		y_{11cn}	y_{121n}		y_{12cn}		y_{1b1n}		y_{1bcn}
A_2	y_{2111}		y_{21c1}	y_{2211}		y_{22cl}	$\cdots$	y_{2b11}		y_{2bc1}
	y_{2112}		y_{21c2}	y_{2212}		y_{22c2}		y_{2b12}		y_{2bc2}
	$\cdots$	$\cdots$	$\cdots$	$\cdots$	$\cdots$	$\cdots$		$\cdots$		$\cdots$
	y_{211n}		y_{21cn}	y_{221n}		y_{22cn}		y_{2b1n}		y_{2bcn}
$\vdots$		$\vdots$			$\vdots$		$\vdots$		$\vdots$	
A_a	y_{a111}		y_{a1c1}	y_{a211}		y_{a2c1}	$\cdots$	y_{ab11}		y_{abc1}
	y_{a112}	$\cdots$	y_{a1c2}	y_{a212}		y_{a2c2}		y_{ab12}		y_{abc2}
	$\cdots$	$\cdots$	$\cdots$	$\cdots$	$\cdots$	$\cdots$		$\cdots$		$\cdots$
	y_{a11n}		y_{a1cn}	y_{a21n}		y_{a2cn}		y_{ab1n}		y_{abcn}

We get, without considering random and fixed effects, the model equation

$$\boldsymbol{y}_{ijkl} = \mu + a_i + b_j + c_k + (ab)_{ij} + (ac)_{ik} + (bc)_{jk} + (abc)_{ijk} + \boldsymbol{e}_{ijkl}$$

In this equation μ = the general mean, a_i = the main effects of levels A_i of A, b_j = the main effects of levels B_j of B, c_k = the main effects of the levels C_k of $C, (ab)_{ij}$ = the interaction effects between A_i and $B_j, (ac)_{ik}$ = the interaction effects between A_i and $C_k, (bc)_{jk}$ = the interaction effects between B_j and C_k and $(abc)_{ijk}$ = the interaction effects between A_i, B_j and C_k. Interaction effects are defined only if, in the corresponding combinations of factor levels, observations are available. l runs in class (i, j, k) from 1 to n_{ijk}.

If for all (i, j, k) we have $n_{ijk} = n$, we have the case of equal subclass numbers. Optimal experimental designs in the three-way analysis of variance need to equal subclass numbers.
We consider the following models of the three-way cross-classification (random factors are bold):

Model I:	The levels of all factors are fixed (defined by the investigator) $A \times B \times C$.
Model III:	The levels of A and B are fixed; the levels of C are selected randomly $A \times B \times \boldsymbol{C}$.
Model IV:	The levels of A are fixed; the levels of B and C are selected randomly $A \times \boldsymbol{B} \times \boldsymbol{C}$.

The missing Model II is one with three random factors and will not be discussed here.

The analysis of variance table for all four models is given in Table 3.7

3.4.1.1 Three-way analysis of variance—classification $\boldsymbol{A} \times \boldsymbol{B} \times \boldsymbol{C}$—Model I

The model equation is given by

$$\boldsymbol{y}_{ijkl} = \mu + a_i + b_j + c_k + (ab)_{ij} + (ac)_{ik} + (bc)_{jk} + (abc)_{ijk} + \boldsymbol{e}_{ijkl}$$

All effects are real numbers; i.e., they are not random. The model becomes complete under the following conditions: that the sums of each of $a_i, b_j, c_k, (ab)_{ij}, (ac)_{ik}, (bc)_{jk}$ and of $(abc)_{ijk}$ (separately over each index), equal zero.
The $E(\boldsymbol{MS})$ for all models are given in Table 3.8.

To find the appropriate F-test for testing our hypothesis $H_0 : a_i = 0, \ \forall i$; H_A : at least one $a_i \neq 0$, we demonstrate the algorithm for this model step by step. In the next sections we present only the results.

Step 1: Define the null hypothesis.
All the $\boldsymbol{a_i}$ are zero.
Step 2: Choose the appropriate model (I, II, or mixed).
Model I.
Step 3: Find the $E(\boldsymbol{MS})$ column in the ANOVA table (if there are several such columns, find the one that corresponds to your model).
Table 3.8, second column.
Step 4: In the same table, find the row for the factor that appears in your null hypothesis.
Main effect A.
Step 5: Change the $E(\boldsymbol{MS})$ in this row to what it would be if the null hypothesis were true.

TABLE 3.7
ANOVA Table—Three-Way ANOVA—Cross-Classification.

Source of Variation	SS	d.f.
Main Effect A	$SS_A = \frac{1}{bcn}\sum_i Y_{i...}^2 - \frac{1}{N}Y_{....}^2$	$a-1$
Main Effect B	$SS_B = \frac{1}{acn}\sum_j Y_{.j..}^2 - \frac{1}{N}Y_{....}^2$	$b-1$
Main Effect C	$SS_C = \frac{1}{abn}\sum_k Y_{..k.}^2 - \frac{1}{N}Y_{....}^2$	$c-1$
Interaction $A \times B$	$SS_{AB} = \frac{1}{cn}\sum_{i,j} Y_{ij..}^2 - \frac{1}{bcn}\sum_i Y_{i...}^2 - \frac{1}{acn}\sum_j Y_{.j..}^2 + \frac{1}{N}Y_{....}^2$	$(a-1)(b-1)$
Interaction $A \times C$	$SS_{AC} = \frac{1}{bn}\sum_{i,k} Y_{i.k.}^2 - \frac{1}{bcn}\sum_i Y_{i...}^2 - \frac{1}{abn}\sum_k Y_{..k.}^2 + \frac{1}{N}Y_{....}^2$	$(a-1)(c-1)$
Interaction $B \times C$	$SS_{BC} = \frac{1}{an}\sum_{j,k} Y_{.jk.}^2 - \frac{1}{acn}\sum_j Y_{.j..}^2 - \frac{1}{abn}\sum_k Y_{..k.}^2 + \frac{1}{N}Y_{....}^2$	$(b-1)(c-1)$
Interaction $A \times B \times C$	$SS_{ABC} = SS_T - SS_A - SS_B - SS_C - SS_{AB} - SS_{AC} - SS_{BC} - SS_R$	$(a-1)(b-1)(c-1)$
Residual	$SS_R = \sum_{i,j,k,l} y_{ijkl}^2 - \frac{1}{n}\sum_{i,j,k} Y_{ijk.}^2$	$abc(n-1)$
Total	$SS_T = \sum_{i,j,k,l} y_{ijkl}^2 - \frac{1}{N}Y_{....}^2$	$N-1$

Source of Variation	MS
Main Effect A	$MS_A = \frac{SS_A}{a-1}$
Main Effect B	$MS_B = \frac{SS_B}{b-1}$
Main Effect C	$MS_C = \frac{SS_C}{c-1}$
Interaction $A \times B$	$MS_{AB} = \frac{SS_{AB}}{(a-1)(b-1)}$
Interaction $A \times C$	$MS_{AC} = \frac{SS_{AC}}{(a-1)(c-1)}$
Interaction $B \times C$	$MS_{BC} = \frac{SS_{BC}}{(b-1)(c-1)}$
Interaction $A \times B \times C$	$MS_{ABC} = \frac{SS_{ABC}}{(a-1)(b-1)(c-1)}$
Residual	$MS_R = \frac{SS_R}{abc(n-1)}$
Total	

Note: We use the convention for writing sums by capitalization of the corresponding letter and replacing suffixes by a dot if summation over it has taken place as has been explained in Table 3.2.

$\sigma^2 + \frac{bcn}{a-1}\sum_i a_i^2$ becomes σ^2 if the hypothesis is true.

Step 6: Search in the same table (in the same column) for the row which now has the same $E(\boldsymbol{MS})$ as you found in the fifth step.

Residual

Step 7: The F-value is now the value of the $\boldsymbol{MS}$ of the row you found in the fourth step divided by the value of the $\boldsymbol{MS}$ of the row you found in the sixth step.

TABLE 3.8

Expected Mean Squares for the Three-Way Cross-Classification.

Source of Variation	Model I A, B, C Fixed	Mixed Model C Random (Model III)
Main Effect A	$\sigma^2 + \frac{bcn}{a-1}\sum_i a_i^2$	$\sigma^2 + bn\sigma_{ac}^2 + \frac{bcn}{a-1}\sum_i a_i^2$
Main Effect B	$\sigma^2 + \frac{acn}{b-1}\sum_j b_j^2$	$\sigma^2 + an\sigma_{bc}^2 + \frac{acn}{b-1}\sum_j b_j^2$
Main Effect C	$\sigma^2 + \frac{abn}{c-1}\sum_k c_k^2$	$\sigma^2 + abn\sigma_c^2$
Interaction $A \times B$	$\sigma^2 + \frac{cn}{(a-1)(b-1)}\sum_{ij}(ab)_{ij}^2$	$\sigma^2 + n\sigma_{abc}^2 + \frac{cn}{(a-1)(b-1)}\sum_{ij}(ab)_{ij}^2$
Interaction $A \times C$	$\sigma^2 + \frac{bn}{(a-1)(c-1)}\sum_{i.k}(ac)_{ik}^2$	$\sigma^2 + bn\sigma_{ac}^2$
Interaction $B \times C$	$\sigma^2 + \frac{an}{(b-1)(c-1)}\sum_{jk}(bc)_{jk}^2$	$\sigma^2 + an\sigma_{bc}^2$
Interaction $A \times B \times C$	$\sigma^2 + \frac{n}{(a-1)(b-1)(c-1)}\sum_{ijk}(abc)_{ijk}^2$	$\sigma^2 + n\sigma_{abc}^2$
Residual	σ^2	σ^2

Source of Variation	Mixed Model A Fixed, B, C Random (Model IV)
Main Effect A	$\sigma^2 + n\sigma_{abc}^2 + bn\sigma_{ac}^2 + cn\sigma_{ab}^2 + \frac{bcn}{a-1}\sum_i a_i^2$
Main Effect B	$\sigma^2 + an\sigma_{bc}^2 + acn\sigma_b^2$
Main Effect C	$\sigma^2 + an\sigma_{bc}^2 + abn\sigma_c^2$
Interaction $A \times B$	$\sigma^2 + n\sigma_{abc}^2 + cn\sigma_{ab}^2$
Interaction $A \times C$	$\sigma^2 + n\sigma_{abc}^2 + bn\sigma_{ac}^2$
Interaction $B \times C$	$\sigma^2 + an\sigma_{bc}^2$
Interaction $A \times B \times C$	$\sigma^2 + n\sigma_{abc}^2$
Residual	σ^2

$F_A = \frac{\boldsymbol{MS}_A}{\boldsymbol{MS}_R}$ which under H_0 has an F-distribution with $f_1 = a - 1$ and $f_2 = abc(n-1)$ d.f.

Using now our scheme we write for short

$H_0: a_i = 0$ for all i		$H_A: a_i \neq 0$ for at least one i	
$\boldsymbol{F} = \frac{\boldsymbol{MS_A}}{\boldsymbol{MS_R}}$	$a-1$	$abc(n-1)$	$\lambda = \frac{bcn \sum_{i=1}^{a} a_i^2}{\sigma^2}$

By R we solve the problem as follows: `size_n.three_way_cross.model_1_a()`.

Example 3.7 We consider 6 varieties of potatoes (factor A) treated with 5 types of fertiliser (factor B) at 4 soil types (factor C). We thus have $a = 6, b = 5$ and $c = 4$ and give the precision requirements $\alpha = 0.05; \beta = 0.1$ and $\delta = a_{max} - a_{min} = 0.5\sigma$.

```
> size_n.three_way_cross.model_1_a(0.05, 0.1, 0.5, 6, 5, 4,
+         "maximin")

[1] 7

> size_n.three_way_cross.model_1_a(0.05, 0.1, 0.5, 6, 5, 4,
+         "minimin")

[1] 3
```

The maxi-min size is $n = 7$; thus a total size of the experiment $N = 840$. The tests for the effects of the factor B and C can be obtained by rearranging the letters A, B and C.

We now test interaction effects.

$H_0: (ab)_{ij} = 0$ for all i, j		$H_A: (ab)_{ij} \neq 0$ for at least one i, j	
$\boldsymbol{F} = \frac{\boldsymbol{MS_{AB}}}{\boldsymbol{MS_R}}$	$(a-1)(b-1)$	$abc(n-1)$	$\lambda = \frac{cn \sum_{i=1}^{a} \sum_{j=1}^{b} (ab)_{ij}^2}{\sigma^2}$

The non-centrality parameter for the most and least favourable cases have to be calculated as shown in Section 3.3.1.1

The non-centrality parameter in our example is $\lambda = 4n\frac{6 \cdot 5 \cdot \delta^2}{4\sigma^2} = 30n\frac{\delta^2}{\sigma^2}$.

The R-program for this case is `size_n.three_way_cross.model_1_axb()`.

If in Example 3.7 we would like to test the hypothesis above, we obtain for $\delta = (ab)_{max} - (ab)_{min} = 0.5\sigma$:

```
> size_n.three_way_cross.model_1_axb(0.05, 0.1, 0.5, 6, 5, 4,
+ "maximin")

[1] 53

> size_n.three_way_cross.model_1_axb(0.05, 0.1, 0.5, 6, 5, 4,
+ "minimin")
```

```
[1] 4
```

and we obtain a mini-min size of $n = 4$, a maxi-min size of $n = 53$.

We will not consider the case of second order interactions $A \times B \times C$ because we think that very seldom experimenters calculate the size of an experiment in dependence on precision requirements for hypotheses about the corresponding effects.

3.4.1.2 Three-way analysis of variance—cross classification $A \times B \times C$—Model III

The model equation is given by

$$\boldsymbol{y}_{ijkl} = \mu + a_i + b_j + \boldsymbol{c}_k + (ab)_{ij} + (\boldsymbol{ac})_{ij} + (\boldsymbol{bc})_{jk} + (\boldsymbol{abc})_{ijk} + \boldsymbol{e}_{ijkl}$$

The model becomes complete under the conditions that the random variables on the r.h.s. of the equation with the same suffixes have equal variances, defined in the previous section, and all fixed effects besides μ over j and k and all interaction sum up over each suffix to zero.
The expected mean squares $E(\boldsymbol{MS})$ are given in Table 3.8.

Using now our scheme we write for short

$H_0 : a_i = 0$ for all i		$H_A : a_i \neq 0$ for at least one i	
$\boldsymbol{F} = \frac{\boldsymbol{MS}_A}{\boldsymbol{MS}_{AC}}$	$a-1$	$(a-1)(c-1)$	$\lambda = \frac{bcn \sum_{i=1}^{a} a_i^2}{\sigma_y^2}$

The corresponding R-program is `size_c.three_way_cross.model_3_a()`.

Example 3.8 Let us assume that in Example 3.7 the experiment should be made on randomly selected farms (factor C) all the other things remain unchanged.

R-output for $n = 2$ and $\delta = 0.5\sigma_y$.

```
> size_c.three_way_cross.model_3_a(0.05, 0.1, 0.5, 6, 5, 2,
+ "maximin")

[1] 15

> size_c.three_way_cross.model_3_a(0.05, 0.1, 0.5, 6, 5, 2,
+ "minimin")

[1] 6
```

For $n = 2$ the mini-min size is $c = 6$ and the maxi-min size is $c = 15$.
To test interaction effects we write

$H_0 : (ab)_{ij} = 0$ for all i, j		$H_A : (ab)_{ij} \neq 0$ for at least one i, j	
$F = \frac{MS_{AB}}{MS_{ABC}}$	$(a-1)(b-1)$	$(a-1)(b-1)(c-1)$	$\lambda = \frac{cn\sum_{ij}(ab)_{ij}^2}{\sigma_y^2}$

The corresponding R-program is `size_c.three_way_cross.model_3_axb()`.
When we would like to test these interaction effects in Example 3.8 with $(ab)_{max} - (ab)_{min} = \delta = 0.5\sigma_y$, the R-output is

```
> size_c.three_way_cross.model_3_axb(0.05, 0.1, 0.5, 6, 5, 2,
+   "maximin")

[1] 106

> size_c.three_way_cross.model_3_axb(0.05, 0.1, 0.5, 6, 5, 2,
+ "minimin")

[1] 8
```

and for $n = 2$ we need $c_{min} = 8$ and $c_{max} = 106$, respectively.

3.4.1.3 Three-way analysis of variance—cross classification $A \times \boldsymbol{B} \times \boldsymbol{C}$—Model IV

The model equation is given by

$$\boldsymbol{y}_{ijkl} = \mu + a_i + \boldsymbol{b}_j + \boldsymbol{c}_k + (\boldsymbol{ab})_{ij} + (\boldsymbol{ac})_{ik} + (\boldsymbol{bc})_{jk} + (\boldsymbol{abc})_{ijk} + \boldsymbol{e}_{ijkl}$$

The model becomes complete under the conditions that the random variables on the r.h.s. of the equation with the same suffixes have equal variances and that the a_i adds up to zero. The interaction effects with a in the terms sums up to zero.
The $E(\boldsymbol{MS})$ are given in Table 3.8. Our null hypothesis is
$H_0 : a_i = 0\ \forall i;\ H_A :$ at least one $a_i \neq 0$.
Here we need step 8 of our algorithm too.
$\sigma^2 + n\sigma_{abc}^2 + bn\sigma_{ac}^2 + cn\sigma_{ab}^2 + \frac{bcn}{a-1}\sum_i a_i^2$ becomes $\sigma^2 + n\sigma_{abc}^2 + bn\sigma_{ac}^2 + cn\sigma_{ab}^2$, if all the a_i are zero. There exists no row in Table 3.8 with such an $E(\boldsymbol{MS})$. But the expectation of the linear combination $\boldsymbol{MS}_{A\times B} + \boldsymbol{MS}_{A\times C} - \boldsymbol{MS}_{A\times B\times C}$ is $\sigma^2 + n\sigma_{abc}^2 + bn\sigma_{ac}^2 + cn\sigma_{ab}^2$. An approximate (Satterthwaite) F-test is based on the ratio

$$\boldsymbol{F} = \frac{\boldsymbol{MS}_A}{\boldsymbol{MS}_{A\times B} + \boldsymbol{MS}_{A\times C} - \boldsymbol{MS}_{A\times B\times C}}$$

which is approximately F-distributed with $a-1$ and

$$f = \frac{(\boldsymbol{MS}_{A\times B} + \boldsymbol{MS}_{A\times C} - \boldsymbol{MS}_{A\times B\times C})^2}{\frac{\boldsymbol{MS}^2_{A\times B}}{(a-1)(b-1)} + \frac{\boldsymbol{MS}^2_{A\times C}}{(a-1)(c-1)} + \frac{\boldsymbol{MS}^2_{A\times B\times C}}{(a-1)(b-1)(c-1)}}$$

degrees of freedom.

But the *d.f.* of the denominator depends upon observations. This gives problems for the experimental size determination if these variance components are not known a priori. A practical solution for this by computing bounds for the size needed is given by Wang (2002) and Wang et. al. (2005).
Another approach is that of Rasch, Spangl and Wang (2011).
An exact solution can be given if we assume that either $\sigma^2_{ac} = 0$ (case 1) or $\sigma^2_{ab} = 0$ (case 2). In these cases an exact F-test exists. If $\sigma^2_{ac} = 0$, we have $E(\boldsymbol{MS}_A) = \sigma^2 + n\sigma^2_{abc} + cn\sigma^2_{ab} + \frac{bcn}{a-1}\sum_i a_i^2$, and this is under the null hypothesis of no A-effects equal to $E(\boldsymbol{MS}_{AB}) = \sigma^2 + n\sigma^2_{abc} + cn\sigma^2_{ab}$. Therefore the hypothesis can exactly be tested by $\boldsymbol{F} = \frac{\boldsymbol{MS}_A}{\boldsymbol{MS}_{AB}}$ which is F-distributed with $a-1$ and $(a-1)(b-1)$ d.f. Analogously in the case $\sigma^2_{ab} = 0$, $\boldsymbol{F} = \frac{\boldsymbol{MS}_A}{\boldsymbol{MS}_{AC}}$ is F-distributed with $a-1$ and $(a-1)(c-1)$ d.f.

We now determine the number of levels of the factor B under the assumption that $\sigma^2_{ac} = 0$ or the number of levels of the factor C under the assumption that $\sigma^2_{ac} = 0$ for given precision requirements. Both numbers are identical.

From Rasch, Spangl and Wang (2011) we take the worst and the best simulation results, from the following models for the variance components

Model	σ^2_b	σ^2_c	σ^2_{ab}	σ^2_{ac}	σ^2_{bc}	σ^2_{abc}	σ^2	σ^2_y
1	5		5	5	5		5	25
2	5	5	5	5			5	25
3	5	5	3	3	2	2	5	25
4	3	3	4	4	4	4	3	25

The results are given in Table 3.9.

The R-program is `size_bc.three_way_cross.model_4_a_case1()`.

For $\delta = 0.5\sigma_y$ we get

```
> size_bc.three_way_cross.model_4_a_case1(0.05, 0.1, 0.5, 6, 2,
+      "maximin")

[1] 9

> size_bc.three_way_cross.model_4_a_case1(0.05, 0.1, 0.5, 6, 2,
+      "minimin")

[1] 6
```

For case 2 and $\delta = 0.5\sigma_y$ we get

TABLE 3.9
Values of the Empirical Power in Percentages for $\alpha = 0.05$ and $\alpha = 0.01$, Respectively in Dependence on the Number $b = c$ of Levels of the Two Random Factors and of the Number a of Levels of the Fixed Factor with $n = 2$ and $\delta = \sigma_y = 5$.

$\alpha = 0.05$

$b = c$	Most Favourable Case					Least Favourable Case				
	$a = 3$	$a = 4$	$a = 5$	$a = 6$	$a = 10$	$a = 3$	$a = 4$	$a = 5$	$a = 6$	$a = 10$
5	48.33	75.12	71.06	87.46	96.98	48.33	42.89	40.14	36.97	29.25
6	59.03	84.93	82.36	94.08	99.43	59.03	53.29	49.21	45.45	36.77
7	68.00	91.63	89.28	97.71	99.82	68.00	61.93	57.76	53.83	44.47
8	75.15	95.32	94.02	99.00	99.97	75.15	70.05	65.49	61.97	50.78
10	86.38	98.67	98.27	99.88	100.00	86.38	82.25	78.02	74.38	63.78
12	92.09	99.66	99.50	99.99	100.00	92.09	88.76	86.57	83.76	74.33
14	95.97	99.93	99.91	100.00	100.00	95.97	93.67	92.20	89.85	82.71
16	97.95	99.97	99.97	100.00	100.00	97.95	96.91	95.44	94.12	88.52
18	98.96	99.99	100.00	100.00	100.00	98.96	98.39	97.32	96.63	92.67
20	99.53	100.00	100.00	100.00	100.00	99.53	99.05	98.63	97.83	95.19
25	99.86	100.00	100.00	100.00	100.00	99.86	99.88	99.72	99.65	98.76

$\alpha = 0.01$

	$a = 3$	$a = 4$	$a = 5$	$a = 6$	$a = 10$	$a = 3$	$a = 4$	$a = 5$	$a = 6$	$a = 10$
5	20.43	47.52	44.57	66.70	88.54	20.43	19.21	17.74	15.95	11.20
6	30.49	62.46	59.49	81.35	96.10	30.49	26.94	24.08	21.69	16.07
7	40.38	74.98	71.11	89.96	98.73	40.38	34.78	31.78	28.18	21.60
8	49.22	83.76	80.47	95.10	99.71	49.22	44.08	39.97	36.94	27.26
10	65.66	94.19	92.45	98.93	99.99	65.66	59.88	54.78	49.54	38.85
12	76.80	97.76	97.08	99.84	100.00	76.80	71.44	67.40	63.45	52.07
14	86.45	99.44	99.25	99.95	100.00	86.45	81.18	77.78	74.21	63.07
16	91.45	99.88	99.76	100.00	100.00	91.45	88.31	85.28	82.41	72.04
18	95.21	99.93	99.94	100.00	100.00	95.21	93.05	90.77	88.29	80.09
20	97.08	99.99	99.94	100.00	100.00	97.08	95.46	94.31	92.48	85.58
25	99.34	100.00	100.00	100.00	100.00	99.34	99.07	98.35	97.75	94.86

```
> size_bc.three_way_cross.model_4_a_case2(0.05, 0.1, 0.5, 6, 2,
+       "maximin")

[1] 9

> size_bc.three_way_cross.model_4_a_case2(0.05, 0.1, 0.5, 6, 2,
+       "minimin")

[1] 6
```

Hence for $a_{max} - a_{min} = 0.5\sigma_y$ with $\sigma_y^2 = \sigma_b^2 + \sigma_{ab}^2 + \sigma_{ac}^2 + \sigma_{bc}^2 + \sigma_{abc}^2 + \sigma^2$, we have with $\sigma_{ac}^2 = 0$ that $b_{max} = c_{max} = 9$ and $b_{min} = c_{\min} = 6$, and with $\sigma_{ab}^2 = 0$ that $b_{max} = c_{max} = 9$ and $b_{min} = c_{min} = 6$.

Now we repeat the calculation of the size of the experiment for $a = 6$ and the case $\delta = \sigma_y$

```
> size_bc.three_way_cross.model_4_a_case1(0.05, 0.1, 1, 6, 2,
+ "maximin")

[1] 5

> size_bc.three_way_cross.model_4_a_case1(0.05, 0.1, 1, 6, 2,
+ "minimin")

[1] 4

> size_bc.three_way_cross.model_4_a_case2(0.05, 0.1, 1, 6, 2,
+ "maximin")

[1] 5

> size_bc.three_way_cross.model_4_a_case2(0.05, 0.1, 1, 6, 2,
+ "minimin")

[1] 4
```

Now we can see, how far the size under the condition of missing interaction is from that one for the general case using Table 3.9. The result barely depends on the number a levels of the factor A. The maximin size under the assumption of missing interaction is for $\alpha = 0.01$

a	3	4	5	6	10
Maximin size	6	6	6	6	6

and for $\alpha = 0.05$

a	3	4	5	6	10
Maximin size	5	5	5	5	5

Even the maximin size is too small for the general case. A power of 0.9 (i.e., $\beta = 0.1$) in Table 3.9 in the least favourable case is obtained for the following values of $b = c$ (after interpolation):

for $\alpha = 0.01$

a	3	4	5	6	10
Maximin size	16	17	18	19	22

and for $\alpha = 0.05$

a	3	4	5	6	10
Maximin size	11	13	14	15	17

Therefore we can use the values from our R-program for the special cases for the general case as lower bounds for the minimin and maximin sizes only.

3.4.2 Three-way analysis of variance—nested classification $A \succ B \succ C$

The observations y_{ijkl} in the combinations of factor levels are as follows:

Factor A	A_1						
Factor B	B_{11}			...	B_{1b}		
Factor C	C_{111}	...	C_{11c}	...	C_{1b1}	...	C_{1bc}
Observations	y_{1111}		y_{11c1}		y_{1b11}		y_{1bc1}
	y_{1112}		y_{11c2}		y_{1b12}		y_{1bc2}
	...		...		...		...
	y_{111n}		y_{11cn}		y_{1b1n}		y_{1bcn}

...	A_a						
	B_{a1}			...	B_{ab}		
	C_{a11}	...	C_{a1c}	...	C_{ab1}	...	C_{abc}
	y_{a111}		y_{a1c1}		y_{ab11}		y_{abc1}
	y_{a112}		y_{a1c2}		y_{ab12}		y_{abc2}
	...		...		...		...
	y_{a11n}		y_{a1cn}		y_{ab1n}		y_{abcn}

For the three-way nested classification with at least one fixed factor, we have the following seven models (the missing Model II is traditionally that with all factors random):

Model I	all three factors are fixed
Model III	factor A random, the others fixed
Model IV	factor B random, the others fixed
Model V	factor C random, the others fixed
Model VI	factor A fixed, the others random
Model VII	factor B fixed, the others random
Model VIII	factor C fixed, the others random.

For all models, the Analysis of Variance Table (without a column for expected mean squares) is the same and given as Table 3.10

For all the tests below we choose $\alpha = 0.05, \beta = 0.1$ and $\delta = 0.5\sigma_y$.

We consider the following example with different interpretation of the three factors as fixed or random.

Example 3.9 Let the levels of a factor A be farms in a certain country with herds of milk cattle with the same number of sires and the same number of cows per sire inseminated by this sire. The sires are the levels of a factor B nested within farms and the cows are the levels of a factor C nested within sires and farms. If we consider all farms, sires and cows, we consider all factors as fixed. When we take a random sample of the levels available for a

TABLE 3.10
ANOVA Table of the Three-Way Nested Classification.

Source of Variation	SS	df
Between A-Levels	$SQ_A = \frac{1}{bcn}\sum_i Y^2_{i...} - \frac{Y^2_{....}}{N}$	$a-1$
Between B-Levels within A-Levels	$SQ_{B \text{ in } A} = \frac{1}{cn}\sum_{i,j} Y^2_{ij..} - \frac{1}{bcn}\sum_i Y^2_{i...}$	$a(b-1)$
Between C-Levels within B-Levels	$SQ_{C \text{ in } B} = \frac{1}{n}\sum_{i,j,k} Y^2_{ijk..} - \frac{1}{cn}\sum_i Y^2_{ij..}$	$ab(c-1)$
Residual	$SQ_R = \sum_{i,j,k,l} y^2_{ijkl} - \frac{1}{n}\sum_{i,j,k} Y^2_{ijk.}$	$abc \cdot (n-1)$

Source of Variation	MS
Between A-Levels	$MS_A = \frac{SQ_A}{a-1}$
Between B-Levels within A-Levels	$MS_{B \text{ in } A} = \frac{SQ_{B \text{ in } A}}{a(b-1)}$
Between C-Levels within B-Levels	$MS_{C \text{ in } B} = \frac{SQ_{C \text{ in } B}}{ab(c-1)}$
Residual	$MS_R = \frac{SQ_R}{abc(n-1)}$

Note: We use the convention for writing sums by capitalization of the corresponding letter and replacing suffixes by a dot if summation over it has taken place as has been explained in Table 3.2.

certain factor, this factor will be considered random. The subclass number is the number n of daughters of the cows, whose milk performance in the first lactation is the observed character y. This number n may also always be the same in all classes.

3.4.2.1 Three-way analysis of variance—nested classification—Model I

The model equation is given by $y_{ijkl} = \mu + a_i + b_{j(i)} + c_{k(ij)} + e_{ijkl}$, $i = 1, \dots, a; j = 1, \dots, b; k = 1, \dots, c; l = 1, \dots, n$ with the side conditions $\sum_{i=1}^{a} a_i = 0; \sum_{j=1}^{b} b_{j(i)} = 0; \forall i; \sum_{k=1}^{c} c_{k(i,j)} = 0; \forall i, j.$

The expected mean squares are given in Table 3.11.

We have three fixed factors and can test a hypothesis about each of them.

TABLE 3.11
Expected Mean Squares for Model I.

Source of Variation	$E(\boldsymbol{MS})$
Between A-Levels	$\sigma^2 + \frac{bcn}{a-1}\sum_i a_i^2$
Between B-Levels within A-Levels	$\sigma^2 + \frac{cn}{a(b-1)} \cdot \sum_{i,j} b_{j(i)}^2$
Between C-Levels within B-Levels	$\sigma^2 + \frac{n}{ab(c-1)} \cdot \sum_{i,j,k} c_{k(i,j)}^2$
Residual	σ^2

$H_0 : a_i = 0$ for all i		$H_A : a_i \neq 0$ for at least one i	
$\boldsymbol{F}_A = \frac{\boldsymbol{MS}_A}{\boldsymbol{MS}_R}$	$(a-1)$	$abc(n-1)$	$\lambda = \frac{bcn \sum_{i=1}^{a} a_i^2}{\sigma_y^2}$

Example 3.9 – continued Let us assume that in Example 3.9 all three factors are fixed and the number of levels are $a = 6; b = 5; c = 4$, respectively. We like to test the effects of the factor A against zero with $\alpha = 0.05, \beta = 0.1$ as long as the maximum difference between the effects is not smaller than 0.5σ. How many daughters per cow have to be used in the survey?

By R we solve the problem as follows:
`size_n.three_way_nested.model_1_a()`.

```
> size_n.three_way_nested.model_1_a(0.05, 0.1, 0.5, 6, 5, 4,
+ "maximin")

[1] 7

> size_n.three_way_nested.model_1_a(0.05, 0.1, 0.5, 6, 5, 4,
+ "minimin")

[1] 3
```

Let us consider the same situation as for testing the effects of the factor A but now we would like to test the effects of the factor B against zero.

$H_0 : b_{j(i)} = 0$ for all j		$H_A : b_{j(i)} \neq 0$ for at least one j	
$\boldsymbol{F}_{B \text{ in A}} = \frac{\boldsymbol{MS}_{B \text{ in A}}}{\boldsymbol{MS}_R}$	$a(b-1)$	$abc(n-1)$	$\lambda = \frac{cn \sum b_{j(i)}^2}{\sigma_y^2}$

By R we solve the problem as follows:
`size_n.three_way_nested.model_1_b()`.

Then the number of daughters needed is larger as in the case above.

```
> size_n.three_way_nested.model_1_b(0.05, 0.1, 0.5, 6, 5, 4,
+ "maximin")

[1] 57

> size_n.three_way_nested.model_1_b(0.05, 0.1, 0.5, 6, 5, 4,
+ "minimin")

[1] 4
```

Let us consider the same situation as for testing the effects of the factor A but now we like to test the effects of the factor C against zero. Then the number of daughters needed is larger as in the cases above.

$H_0 : c_{k(ij)} = 0$ for all k		$H_A : c_{k(ij)} \neq 0$ for at least one k	
$\boldsymbol{F}_{C \text{ in B}} = \frac{\boldsymbol{MS}_{C \text{ in B}}}{\boldsymbol{MS}_R}$	$ab(c-1)$	$abc(n-1)$	$\lambda = \frac{n \sum c^2_{k(i,j)}}{\sigma^2_y}$

By R we solve the problem as follows:
`size_n.three_way_nested.model_1_c()`.

```
> size_n.three_way_nested.model_1_c(0.05, 0.1, 0.5, 6, 5, 4,
+ "maximin")

[1] 378

> size_n.three_way_nested.model_1_c(0.05, 0.1, 0.5, 6, 5, 4,
+ "minimin")

[1] 7
```

3.4.2.2 Three-way analysis of variance—nested classification—Model III

The model equation is given by

$\boldsymbol{y}_{ijkl} = \mu + \boldsymbol{a}_i + b_{j(i)} + c_{k(ij)} + \boldsymbol{e}_{ijkl}$, with the side conditions

$\sum_{j=1} b_{j(i)} = 0; \forall i; \sum_{k=1}^{c} c_{k(i,j)} = 0; \forall i, j.$

The expected mean squares are given in Table 3.12.

TABLE 3.12
Expected Mean Squares for Model III.

Source of Variation	$E(\boldsymbol{MS})$
Between A-Levels	$\sigma^2 + bcn\sigma_a^2$
Between B-Levels within A-Levels	$\sigma^2 + \frac{cn}{a(b-1)} \cdot \sum_{i,j} b_{j(i)}^2$
Between C-Levels within B-Levels	$\sigma^2 + \frac{n}{ab(c-1)} \cdot \sum_{i,j,k} c_{k(i,j)}^2$
Residual	σ^2

From this table, we can derive the test statistics

$H_0 : b_{j(i)} = 0$ for all j		$H_A : b_{j(i)} \neq 0$ for at least one j	
$\boldsymbol{F}_{B \text{ in } A} = \frac{\boldsymbol{MS}_{B \text{ in } A}}{\boldsymbol{MS}_R}$	$a(b-1)$	$abc(n-1)$	$\lambda = \frac{cn \sum b_{j(i)}^2}{\sigma_y^2}$

and

$H_0 : c_{k(ij)} = 0$ for all k		$H_A : c_{k(ij)} \neq 0$ for at least one k	
$\boldsymbol{F}_{C \text{ in } B} = \frac{\boldsymbol{MS}_{C \text{ in } B}}{\boldsymbol{MS}_R}$	$ab(c-1)$	$abc(n-1)$	$\lambda = \frac{n \sum_{i,j,k} c_{k(ij)}^2}{\sigma_y^2}$

They are identical with $F_{B \text{ in } A}$ and $F_{C \text{ in } B}$ from Model I. But now we have to determine the pair (a, n) in a proper way.

The size of the experiment can be determined by the program, demonstrated for $a = 6$, in Example 3.10.

Example 3.10 A Model III situation occurs if we draw a random sample from the 1000 farms. We now have to determine the size of this sample and the subclass number n.

We look for the (a, n)-combination so that an is minimum. We have to find this by systematic search; the results for $a = 6$ are already known. The minimum can be found by the program

```
> size_n.three_way_nested.model_3_b(0.05, 0.1, 0.5, 6, 5, 4, for
a = 6 + "maximin")

[1] 57

> size_n.three_way_nested.model_3_b(0.05, 0.1, 0.5, 6, 5, 4,
+ "minimin")

[1] 4
```

2) for testing the null hypothesis about factor C:

Again we draw a random sample from the 1000 farms. We now have again to determine the size of this sample and the subclass number n for given values of $b = 5$ and $c = 4$. The results for $a = 6$ are then

```
> size_n.three_way_nested.model_3_c(0.05, 0.1, 0.5, 6, 5, 4,
+ "maximin")

[1] 378

> size_n.three_way_nested.model_3_c(0.05, 0.1, 0.5, 6, 5, 4,
+ "minimin")

[1] 7
```

3.4.2.3 Three-way analysis of variance—nested classification—Model IV

The model equation is given by

$\boldsymbol{y}_{ijkl} = \mu + a_i + \boldsymbol{b}_{j(i)} + c_{k(ij)} + \boldsymbol{e}_{ijkl}$, with the side conditions $\sum_{k=1}^{c} c_{k(i,j)} = 0$ and $\sum_{i=1}^{a} a_1 = 0$.

The expected mean squares are given in Table 3.13.

TABLE 3.13
Expected Mean Squares for Model IV.

Source of Variation	$E(\boldsymbol{MS})$
Between A-Levels	$\sigma^2 + cn\sigma^2_{b(a)} + \frac{bcn}{a-1}\sum_i^a a_i^2$
Between B-Levels within A-Levels	$\sigma^2 + cn\sigma^2_{b(a)}$
Between C-Levels within B-Levels	$\sigma^2 + \frac{n}{ab(c-1)} \cdot \sum_{i,j,k} c^2_{k(i,j)}$
Residual	σ^2

From this table, we can derive the test statistic for testing

$H_0 : a_i = 0$ for all i		$H_A : a_i \neq 0$ for at least one i	
$\boldsymbol{F}_A = \frac{\boldsymbol{MS}_A}{\boldsymbol{MS}_{B \text{ in } A}}$	$a-1$	$a(b-1)$	$\lambda = \frac{bcn \sum_{i=1}^{a} a_i^2}{\sigma_y^2}$

By R we solve the problem as follows:
`size_n.three_way_nested.model_4_a()`.

Example 3.11 A Model IV situation occurs if we draw in the situation of Example 3.9 random samples from $b = 20$ sires per farm. We now have to determine the size of this sample and the subclass number n. We can do this under the condition that bn is minimum. We proceed as follows. In the program above we first try $b = 2$ and $b = 20$.

For $a_{(max)} - a_{min} = 0.5\sigma_y$ we get

```
> size_n.three_way_nested.model_4_a(0.05, 0.1, 0.5, 6, 2, 4,
+ "maximin")

[1] 40

> size_n.three_way_nested.model_4_a(0.05, 0.1, 0.5, 6, 2, 4,
+ "minimin")

[1] 14
```

further

```
> size_n.three_way_nested.model_4_a(0.05, 0.1, 0.5, 6, 20, 4,
+ "maximin")

[1] NA

> size_n.three_way_nested.model_4_a(0.05, 0.1, 0.5, 6, 20, 4,
+ "minimin")

[1] NA
```

The final (b, n)-combination that minimises bn is found by systematic search through a call to `size_n.three_way_nested.model_4_a` with parameter b set to `NA`.

```
> size_n.three_way_nested.model_4_a(0.05, 0.1, 0.5, 6, NA, 4,
+ "maximin")

minimum value of b*n is  36

$b
[1] 12

$n
[1] 3

> size_n.three_way_nested.model_4_a(0.05, 0.1, 0.5, 6, NA, 4,
+ "minimin")
```

```
minimum value of b*n is  15

$b
[1] 5

$n
[1] 3
```

Finally we can test the hypothesis

$H_0 : c_{k(ij)} = 0$ for all k		$H_A : c_{k(ij)} \neq 0$ for at least one k	
$\boldsymbol{F}_{C \text{ in B}} = \frac{\boldsymbol{MS}_{C \text{ in B}}}{\boldsymbol{MS}_R}$	$ab(c-1)$	$abc(n-1)$	$\lambda = \frac{n \sum c^2_{k(ij)}}{\sigma^2_y}$

The corresponding R-program is `size_n.three_way_nested.model_4_c()`. In the program above we first try $b = 2$ and $b = 20$.

```
> size_n.three_way_nested.model_4_c(0.05, 0.1, 0.5, 6, 2, 4,
+ "maximin")

[1] 262

> size_n.three_way_nested.model_4_c(0.05, 0.1, 0.5, 6, 2, 4,
+ "minimin")

[1] 12
```

further

```
> size_n.three_way_nested.model_4_c(0.05, 0.1, 0.5, 6, 20, 4,
+ "maximin")

[1] 693

> size_n.three_way_nested.model_4_c(0.05, 0.1, 0.5, 6, 20, 4,
+ "minimin")

[1] 4
```

The final (b, n)-combination that minimises bn is found by systematic search through a call to `size_n.three_way_nested.model_4_c` with parameter b set to `NA`.

```
> size_n.three_way_nested.model_4_c(0.05, 0.1, 0.5, 6, NA, 4,
+ "maximin")
```

```
minimum value of b*n is  524

$b
[1] 2

$n
[1] 262

> size_n.three_way_nested.model_4_c(0.05, 0.1, 0.5, 6, NA, 4,
+ "minimin")

minimum value of b*n is  24

$b
[1] 2

$n
[1] 12
```

3.4.2.4 Three-way analysis of variance—nested classification—Model V

The model equation is given by $\boldsymbol{y}_{ijkl} = \mu + a_i + b_{j(i)} + \boldsymbol{c}_{k(ij)} + \boldsymbol{e}_{ijkl}$, with the side conditions $\sum_{i=1}^{a} a_i = 0$; $\sum_{j=1}^{b} b_{j(i)} = 0; \forall i$.

The expected mean squares are given in Table 3.14.

TABLE 3.14
Expected Mean Squares for Model V.

Source of Variation	$E(\boldsymbol{MS})$
Between A-Levels	$\sigma^2 + n\sigma^2_{c(ab)} + \frac{bcn}{a-1}\sum_i^a a_i^2$
Between B-Levels within A-Levels	$\sigma^2 + n\sigma^2_{c(ab)} + \frac{cn}{a-1}\sum_{ij} b_{j(i)}^2$
Between C-Levels within B-Levels	$\sigma^2 + n\sigma^2_{c(ab)}$
Residual	σ^2

From this table, we can derive the test statistic for testing

$H_0 : a_i = 0$ for all k		$H_A : a_i \neq 0$ for at least one i	
$\boldsymbol{F}_A = \frac{\boldsymbol{MS}_A}{\boldsymbol{MS}_{C \text{ in B}}}$	$(a-1)$	$ab(c-1)$	$\lambda = \frac{bcn\sum_{i=1}^{a} a_i^2}{\sigma_y^2}$

By R we solve the problem as follows:
`size_c.three_way_nested.model_5_a()`.

Example 3.12 A Model V situation occurs if we draw random samples from the $c = 30$ cows per sire. We now have to determine the size of this sample and the subclass number n so that cn is minimum. For $n = 2$ we find c

```
> size_c.three_way_nested.model_5_a(0.05, 0.1, 0.5, 6, 5, 2,
+ "maximin")

[1] 14

> size_c.three_way_nested.model_5_a(0.05, 0.1, 0.5, 6, 5, 2,
+ "minimin")

[1] 5
```

We further can test

$H_0 : b_{j(i)} = 0$ for all j		$H_A : b_{j(i)} \neq 0$ for at least one j	
$\boldsymbol{F}_{B \text{ in A}} = \frac{\boldsymbol{MS}_{B \text{ in A}}}{\boldsymbol{MS}_{C \text{ in B}}}$	$a(b-1)$	$ab(c-1)$	$\lambda = \frac{cn \sum b_{j(i)}^2}{\sigma_y^2}$

The corresponding R-program is `size_c.three_way_nested.model_5_b()`.
We now determine the best (c, n) combination for testing the B-effects. For $n = 2$ we find c

```
> size_c.three_way_nested.model_5_b(0.05, 0.1, 0.5, 6, 5, 2,
+ "maximin")

[1] 113

> size_c.three_way_nested.model_5_b(0.05, 0.1, 0.5, 6, 5, 2,
+ "minimin")

[1] 9
```

3.4.2.5 Three-way analysis of variance—nested classification—Model VI

The model equation is given by

$\boldsymbol{y}_{ijkl} = \mu + a_i + \boldsymbol{b}_{j(i)} + \boldsymbol{c}_{k(ij)} + \boldsymbol{e}_{ijkl}$, with the side condition $\sum_{i=1}^{a} a_i = 0$.

The expected mean squares are given in Table 3.15.
From this table, we can derive the test statistic for testing

TABLE 3.15
Expected Mean Squares for Model VI.

Source of Variation	$E(\boldsymbol{MS})$
Between A-Levels	$\sigma^2 + n\sigma^2_{c(ab)} + cn\sigma^2_{b(a)} + \frac{bcn}{a-1}\sum_i a_i^2$
Between B-Levels within A-Levels	$\sigma^2 + n\sigma^2_{c(ab)} + cn\sigma^2_{b(a)}$
Between C-Levels within B-Levels	$\sigma^2 + n\sigma^2_{c(ab)}$
Residual	σ^2

$H_0: a_i = 0$ for all i		$H_A: a_i \neq 0$ for at least one i	
$\boldsymbol{F}_A = \frac{\boldsymbol{MS}_A}{\boldsymbol{MS}_{B \text{ in } A}}$	$a-1$	$a(b-1)$	$\lambda = \frac{bcn\sum_{i=1}^{a} a_i^2}{\sigma_y^2}$

The R-program in this case is `size_b.three_way_nested.model_6_a()`.

Example 3.13 A Model VI situation occurs in our Example 3.9 if we take a random sample of the 20 sires as well as of the 50 cows per sire. We have to determine both the sizes of these two samples and the number n of daughters. We take here $c = 4$ and $n = 2$ as an example.

```
> size_b.three_way_nested.model_6_a(0.05, 0.1, 0.5, 6, 4, 2,
+ "maximin")

[1] 18

> size_b.three_way_nested.model_6_a(0.05, 0.1, 0.5, 6, 4, 2,
+ "minimin")

[1] 7
```

3.4.2.6 Three-way analysis of variance—nested classification—Model VII

The model equation is given by:

$\boldsymbol{y}_{ijkl} = \mu + \boldsymbol{a}_i + b_{j(i)} + \boldsymbol{c}_{k(ij)} + \boldsymbol{e}_{ijkl}$, with the side conditions $\sum_{j=1}^{b} b_{j(i)} = 0; \forall i$.

The expected mean squares are given in Table 3.16.

From this table, we can derive the test statistic for testing

$H_0: b_{j(i)} = 0$ for all j		$H_A: b_{j(i)} \neq 0$ for at least one j	
$\boldsymbol{F}_{B \text{ in } A} = \frac{\boldsymbol{MS}_{B \text{ in } A}}{\boldsymbol{MS}_{C \text{ in } B}}$	$a(b-1)$	$ab(c-1)$	$\lambda = \frac{cn\sum b_{j(i)}^2}{\sigma_y^2}$

The corresponding R-program is `size_c.three_way_nested.model_7_b()`.

TABLE 3.16
Expected Mean Squares for Model VII.

Source of Variation	$E(\boldsymbol{MS})$
Between A-Levels	$\sigma^2 + n\sigma^2_{c(ab)} + bcn\sigma^2_a$
Between B-Levels within A-Levels	$\sigma^2 + n\sigma^2_{c(ab)} + \frac{cn}{a(b-1)} \sum_{i,j} b^2_{j(i)}$
Between C-Levels within B-Levels	$\sigma^2 + n\sigma^2_{c(ab)}$
Residual	σ^2

Example 3.14 A Model VII situation occurs in our Example 3.9 if we take a random sample of the 1000 sires as well as of the 50 cows per sire. We have to determine both the sizes of these two samples and the number n of daughters. Here we take $b = 4$ and $n = 1$.

```
> size_c.three_way_nested.model_7_b(0.05, 0.1, 0.5, 6, 4, 1,
+ "maximin")

[1] 202

> size_c.three_way_nested.model_7_b(0.05, 0.1, 0.5, 6, 4, 1,
+ "minimin")

[1] 18
```

3.4.2.7 Three-way analysis of variance—nested classification—Model VIII

The model equation is given by

$\boldsymbol{y}_{ijkl} = \mu + \boldsymbol{a}_i + \boldsymbol{b}_{j(i)} + c_{k(ij)} + \boldsymbol{e}_{ijkl}$, with the side conditions $\sum_{k=1}^{c} c_{k(ij)} = 0; \forall i, j$.

The expected mean squares are given in Table 3.17.

From this table, we can derive the test statistic for testing

$H_0 : c_{k(ji)} = 0$ for all k		$H_A : c_{k(ij)} \neq 0$ for at least one k	
$\boldsymbol{F}_{C \text{ in B}} = \frac{\boldsymbol{MS}_{C \text{ in B}}}{\boldsymbol{MS}_R}$	$ab(c-1)$	$abc(n-1)$	$\lambda = \frac{n \sum c^2_{k(ji)}}{\sigma^2_y}$

The corresponding R-program is `size_n.three_way_nested.model_8_c()`.

Example 3.15 A Model VIII situation occurs in our Example 3.9 if we take a random sample of the 1000 farms as well as of the 20 sires per selected farm. We have to determine both the sizes of these two samples and the number n of daughters.

TABLE 3.17
Expected Mean Squares for Model VIII.

Source of Variation	$E(\boldsymbol{MS})$
Between A-Levels	$\sigma^2 + cn\sigma^2_{b(a)} + bcn\sigma^2_a$
Between B-Levels within A-Levels	$\sigma^2 + cn\sigma^2_{b(a)}$
Between C-Levels within B-Levels	$\sigma^2 + \frac{n}{ab(c-1)} \cdot \sum_{ijk} c^2_{k(i,j)}$
Residual	σ^2

For the case $a = 6$, $b = 5$ and $c = 4$

```
> size_n.three_way_nested.model_8_c(0.05, 0.1, 0.5, 6, 5, 4,
+ "maximin")

[1] 378

> size_n.three_way_nested.model_8_c(0.05, 0.1, 0.5, 6, 5, 4,
+ "minimin")

[1] 7
```

3.4.3 Three-way analysis of variance—mixed classification $(A \times B) \succ C$

The observations y_{ijkl} in the combinations of factor levels are as follows.

Levels	B_1			B_2			...	B_b		
A_1	C_{111}		C_{11c}	C_{121}		C_{12c}	...	C_{1b1}		C_{1bc}
	y_{1111}		y_{11c1}	y_{1211}		y_{12c1}	...	y_{1b11}		y_{1bc1}
	y_{1112}	...	y_{11c2}	y_{1212}		y_{12c2}		y_{1b12}		y_{1bc2}
	...	...	...	...	...	...		...		...
	y_{111n}		y_{11cn}	y_{121n}		y_{12cn}		y_{1b1n}		y_{1bcn}
A_2	C_{211}		C_{21c}	C_{221}		C_{22c}	...	C_{2b1}		C_{2bc}
	y_{2111}		y_{21c1}	y_{2211}		y_{22cl}	...	y_{2b11}		y_{2bc1}
	y_{2112}		y_{21c2}	y_{2212}		y_{22c2}		y_{2b12}		y_{2bc2}
	...	...	...	...	...	...		...		...
	y_{211n}		y_{21cn}	y_{221n}		y_{22cn}		y_{2b1n}		y_{2bcn}
⋮		⋮			⋮		⋮		⋮	
A_a	C_{a11}		C_{a1c}	C_{a21}		C_{a2c}	...	C_{ab1}		C_{abc}
	y_{a111}	...	y_{a1c1}	y_{a211}		y_{a2c1}		y_{ab11}		y_{abc1}
	y_{a112}	...	y_{a1c2}	y_{a212}		y_{a2c2}		y_{ab12}		y_{abc2}
	...		...	...		...		...		...
	y_{a11n}		y_{a1cn}	y_{a21n}		y_{a2cn}		y_{ab1n}		y_{abcn}

The model equation is given by

$\boldsymbol{y}_{ijkl} = \mu + a_i + b_j + (ab)_{ij} + c_{k(ij)} + \boldsymbol{e}_{ijkl}$, with the side conditions

$\sum_{i=1}^{a} a_i = 0; \sum_{j=1}^{b} b_j = 0; \sum_{k=1}^{c} c_{k(ij)} = 0; \forall i, j, \sum_{i=1}^{a} (ab)_{ij} = 0; \forall j, \sum_{j=1}^{b} (ab)_{ij} = 0; \forall i.$

We always assume $var(\boldsymbol{e}_{ijkl}) = \sigma^2 \ \forall i, j, k, l$.

Besides the above model with fixed effects, we have four mixed models in this mixed clasification. The ANOVA table is independent of the model and given in Table 3.18. For all our models we use in the examples of tests $\alpha = 0.05, \beta = 0.1$ and $\delta = 0.5\sigma_y$.

TABLE 3.18
ANOVA Table for Three-Way Mixed Classification $(A \times B) \succ C$.

Source of Variation	SS	df
Between A-Levels	$SS_A = \frac{1}{bcn} \sum_i \boldsymbol{Y}_{i...}^2 - \frac{1}{N} \boldsymbol{Y}_{....}^2$	$a-1$
Between B-Levels	$SS_B = \frac{1}{acn} \sum_j \boldsymbol{Y}_{.j..}^2 - \frac{1}{N} \boldsymbol{Y}_{....}^2$	$b-1$
Between C-Levels within $A \times B$-Levels	$SS_{C \text{ in } AB} = \frac{1}{n} \sum_{i,j,k} \boldsymbol{Y}_{ijk.}^2 - \frac{1}{cn} \sum_i \boldsymbol{Y}_{ij..}^2$	$ab(c-1)$
Interaction $A \times B$	$SS_{AB} = \frac{1}{cn} \sum_{i,j} \boldsymbol{Y}_{ij..}^2 - \frac{1}{bcn} \sum_i \boldsymbol{Y}_{i...}^2 - \frac{1}{acn} \sum_j \boldsymbol{Y}_{.j..}^2 + \frac{1}{N} \boldsymbol{Y}_{....}^2$	$(a-1) \cdot (b-1)$
Residual	$SS_R = \sum_{i,j,k,l} \boldsymbol{y}_{ijkl}^2 - \frac{1}{n} \sum_{i,j,k} \boldsymbol{Y}_{ijk.}^2$	$abc(n-1)$

Source of Variation	MS
Between A-Levels	$MS_A = \frac{SS_A}{a-1}$
Between B-Levels	$MS_B = \frac{SS_B}{b-1}$
Between C-Levels within $A \times B$-Levels	$MS_{C \text{ in } AB} = \frac{SS_{C \text{ in } AB}}{ab(c-1)}$
Interaction $A \times B$	$MS_{AB} = \frac{SS_{AB}}{(a-1)(b-1)}$
Residual	$MS_R = \frac{SS_R}{abc(n-1)}$

Note: We use the convention for writing sums by capitalization of the corresponding letter and replacing suffixes by a dot if summation over it has taken place as has been explained in Table 3.2.

3.4.3.1 Three-way analysis of variance—mixed classification $(A \times B) \succ C$—Model I

The model equation is given by $\boldsymbol{y}_{ijkl} = \mu + a_i + b_j + (ab)_{ij} + c_{k(ij)} + \boldsymbol{e}_{ijkl}$ and the appropriate side conditions.

All effects are real numbers; i.e., they are not random. The model becomes complete under the following conditions that, as well as the sum of each of $a_i, b_j, c_{k(ij)}$ and of the $(ab)_{ij}$ (separately over each index) equals zero.

The expected mean squares are given in Table 3.19.

TABLE 3.19
Expected Mean Squares for Model I.

Source of Variation	Model I $E(\boldsymbol{MS})$
Between A-Levels	$\sigma^2 + \frac{bcn}{a-1}\sum_i a_i^2$
Between B-Levels	$\sigma^2 + \frac{acn}{b-1}\sum_j b_j^2$
Between C-Levels within $A \times B$	$\sigma^2 + \frac{n}{ab(c-1)} \cdot \sum_{i,j,k} c_{k(i,j)}^2$
Interaction $A \times B$	$\sigma^2 + \frac{cn}{(a-1)(b-1)} \cdot \sum_{ij} (ab)_{ij}^2$
Residual	σ^2

From this table, we can derive the test statistic for testing

$H_0: a_i = 0$ for all i		$H_A: a_i \neq 0$ for at least one i	
$\boldsymbol{F} = \frac{\boldsymbol{MS}_A}{\boldsymbol{MS}_R}$	$a-1$	$abc(n-1)$	$\lambda = \frac{bcn \sum_{i=1}^{a} a_i^2}{\sigma_y^2}$

The corresponding R-program is
`size_n.three_way_mixed_ab_in_c.model_1_a()`.

Example 3.16 Let a varieties of tomatoes be the levels of a factor A and b farms be the levels of a factor B. In each farm each variety is grown in c plots, the levels of a factor C in each $A \times B$-combination. For the fixed factors we use $a = 6, b = 5, c = 4$.

```
> size_n.three_way_mixed_ab_in_c.model_1_a(0.05, 0.1, 0.5, 6, 5,
+     4, "maximin")

[1] 7

> size_n.three_way_mixed_ab_in_c.model_1_a(0.05, 0.1, 0.5, 6, 5,
+     4, "minimin")

[1] 3
```

We now test hypotheses about the fixed factor B.

$H_0: b_j = 0 \forall j$		H_A: at least one $b_j \neq 0$	
$F = \frac{MS_B}{MS_R}$	$b-1$	$abc(n-1)$	$\lambda = \frac{acn \sum_{j=1}^{b} b_j^2}{\sigma_y^2}$

```
> size_n.three_way_mixed_ab_in_c.model_1_b(0.05, 0.1, 0.5, 6, 5,
+     4, "maximin")

[1] 6

> size_n.three_way_mixed_ab_in_c.model_1_b(0.05, 0.1, 0.5, 6, 5,
+     4, "minimin")

[1] 3
```

From Table 3.19, we can also derive the test statistic for testing

$H_0: (ab)_{ij} = 0 \forall i, j$		H_A at least one $(ab)_{ij} \neq 0$	
$F = \frac{MS_{A \times B}}{MS_R}$	$(a-1)(b-1)$	$abc(n-1)$	$\lambda = \frac{cn \sum (ab)_{ij}^2}{\sigma_y^2}$

The corresponding R-program is
`size_n.three_way_mixed_ab_in_c.model_1_axb()`.

```
> size_n.three_way_mixed_ab_in_c.model_1_axb(0.05, 0.1, 0.5, 6,
+     5, 4, "maximin")

[1] 53

> size_n.three_way_mixed_ab_in_c.model_1_axb(0.05, 0.1, 0.5, 6,
+     5, 4, "minimin")

[1] 4
```

From Table 3.19, we can also derive the test statistic for testing

$H_0: c_{k(ij)} = 0 \forall k, i, j$		H_A at least one $c_{k(ij)} \neq 0$	
$F = \frac{MS_{C \text{ in } AB}}{MS_R}$	$ab(c-1)$	$abc(n-1)$	$\lambda = \frac{n \sum_{ijk} c_{k(ij)}^2}{\sigma_y^2}$

The corresponding R-program is
`size_n.three_way_mixed_ab_in_c.model_1_c()`.

In our example is

```
> size_n.three_way_mixed_ab_in_c.model_1_c(0.05, 0.1, 0.5, 6, 5,
+     4, "maximin")
```

```
[1] 378

> size_n.three_way_mixed_ab_in_c.model_1_c(0.05, 0.1, 0.5, 6, 5,
+     4, "minimin")

[1] 7
```

3.4.3.2 Three-way analysis of variance—mixed classification $(A \times B) \succ C$—Model III

The model equation is given by

$$\boldsymbol{y}_{ijkl} = \mu + a_i + \boldsymbol{b}_j + (\boldsymbol{ab})_{ij} + c_{k(ij)} + \boldsymbol{e}_{ijkl}$$

The effects a_i, and $c_{k(ij)}$ are real numbers; i.e., they are not random. The model becomes complete under the following conditions: that the sum of each of a_i, and $c_{k(ij)}$ (separately over each index) equals zero and that the random variables on the r.h.s. of the equation for all suffixes have equal variances and are independent.

The expected mean squares are given in Table 3.20.

TABLE 3.20
Expected Mean Squares for Model III.

Source of Variation	$E(\boldsymbol{MS})$
Between A-Levels	$\sigma^2 + \frac{bcn}{a-1}\sum_i a_i^2 + cn\sigma_{ab}^2$
Between B-Levels	$\sigma^2 + acn\sigma_b^2$
Between C-Levels within $A \times B$	$\sigma^2 + \frac{n}{ab(c-1)} \cdot \sum_{i,j,k} c_{k(i,j)}^2$
Interaction $A \times B$	$\sigma^2 + cn\sigma_{ab}^2$
Residual	σ^2

From this table, we can derive the test statistic for testing

$H_0 : a_i = 0$ for all i		$H_A : a_i \neq 0$ at least one i	
$\boldsymbol{F}_A = \frac{\boldsymbol{MS}_A}{\boldsymbol{MS}_{AB}}$	$a-1$	$(a-1)(b-1)$	$\lambda = \frac{bcn\sum_{i=1}^{a} a_i^2}{\sigma_y^2}$

The corresponding R-program is
`size_b.three_way_mixed_ab_in_c.model_3_a()`.

For the case $a = 6$, $c = 5$ and $n = 1$

```
> size_b.three_way_mixed_ab_in_c.model_3_a(0.05, 0.1, 0.5, 6, 5,
+     1, "maximin")

[1] 28
```

```
> size_b.three_way_mixed_ab_in_c.model_3_a(0.05, 0.1, 0.5, 6, 5,
+     1, "minimin")

[1] 10
```

From Table 3.20, we can also derive the test statistic for testing

$H_0 : c_{k(ji)} = 0 \forall k, j, i$		H_A at least one $c_{k(ij)} \neq 0$	
$F = \frac{MS_{C \text{ in } AB}}{MS_R}$	$ab(c-1)$	$abc(n-1)$	$\lambda = \frac{n \sum_{ijk} c^2_{k(ij)}}{\sigma^2_y}$

The corresponding R-program is
`isize_n.three_way_mixed_ab_in_c.model_3_c()`.

Example 3.17 We consider the situation of Example 3.16 but now we draw a random sample from the 100 farms. By our program we have to determine a proper combination of the number of selected farms and the plot size n.

For the case $a = 6$, $b = 5$ and $c = 4$

```
> size_n.three_way_mixed_ab_in_c.model_3_c(0.05, 0.1, 0.5, 6, 5,
+     4, "maximin")

[1] 378

> size_n.three_way_mixed_ab_in_c.model_3_c(0.05, 0.1, 0.5, 6, 5,
+     4, "minimin")

[1] 7
```

3.4.3.3 Three-way analysis of variance—mixed classification $(A \times B) \succ C$—Model IV

The model equation is given by

$$\boldsymbol{y}_{ijkl} = \mu + \boldsymbol{a}_i + \boldsymbol{b}_j + (\boldsymbol{ab})_{ij} + c_{k(ij)} + \boldsymbol{e}_{ijkl}$$

The effects $c_{k(ij)}$ are real numbers; i.e., they are not random. The model becomes complete under the following conditions: that the sum of $c_{k(ij)}$ equals zero and that the random variables on the r.h.s. of the equation for all suffixes have equal variances and are independent.

The variance of the $\boldsymbol{y}_{ijkl}$ is given by $var(\boldsymbol{y}_{ijkl}) = \sigma^2_y = \sigma^2_a + \sigma^2_b + \sigma^2_{ab} + \sigma^2$.

The expected mean squares are given in Table 3.21.

From this table, we can derive the test statistic for testing

$H_0 : c_{k(ji)} = 0 \forall k, j, i$		H_A at least one $c_{k(ij)} \neq 0$	
$F = \frac{MS_{C \text{ in } AB}}{MS_R}$	$ab(c-1)$	$abc(n-1)$	$\lambda = \frac{n \sum_{ijk} c^2_{k(ij)}}{\sigma^2_y}$

TABLE 3.21
Expected Mean Squares for Model IV.

Source of Variation	$E(\boldsymbol{MS})$
Between A-Levels	$\sigma^2 + bcn\sigma_a^2 + cn\sigma_b^2$
Between B-Levels	$\sigma^2 + acn\sigma_b^2 + cn\sigma_{ab}^2$
Between C-Levels within $A \times B$	$\sigma^2 + \frac{n}{ab(c-1)} \cdot \sum_{ijk} c_{k(ij)}^2$
Interaction $A \times B$	$\sigma^2 + cn\sigma_{ab}^2$
Residual	σ^2

Example 3.18 We consider the situation of Example 3.16 but now we draw a random sample from the 100 farms and also from the 50 varieties. By our program we have to determine a proper combination of the number of selected farms, varieties and the plot size n.

For the case $a = 6$, $b = 5$ and $c = 4$

```
> size_n.three_way_mixed_ab_in_c.model_4_c(0.05, 0.1, 0.5, 6, 5,
+     4, "maximin")

[1] 378

> size_n.three_way_mixed_ab_in_c.model_4_c(0.05, 0.1, 0.5, 6, 5,
+     4, "minimin")

[1] 7
```

3.4.3.4 Three-way analysis of variance—mixed classification $(A \times B) \succ C$—Model V

The model equation is given by

$$\boldsymbol{y}_{ijkl} = \mu + a_i + b_j + (ab)_{ij} + \boldsymbol{c}_{k(ij)} + \boldsymbol{e}_{ijkl}$$

The effects a_i, b_j and $(ab)_{ij}$ are real numbers; i.e., they are not random. The model becomes complete under the following conditions: that the sum of each of a_i, b_j and of the $(ab)_{ij}$ (separately over each index) equals zero and that the random variables on the r.h.s. of the equation for all suffixes have expectation zero, have equal variances and are independent.
The variance of the $\boldsymbol{y}_{ijkl}$ is given by

$$var(\boldsymbol{y}_{ijkl}) = \sigma_y^2 = \sigma_{c(ab)}^2 + \sigma^2.$$

The expected mean squares are given in Table 3.22.

TABLE 3.22
Expected Mean Squares for Model V.

Source of Variation	$E(\boldsymbol{MS})$
Between A-Levels	$\sigma^2 + \frac{bcn}{a-1}\sum_i a_i^2 + n\sigma^2_{c(ab)}$
Between B-Levels	$\sigma^2 + \frac{acn}{b-1}\sum_j b_j^2 + n\sigma^2_{c(ab)}$
Between C-Levels within $A \times B$	$\sigma^2 + n\sigma^2_{c(ab)}$
Interaction $A \times B$	$\sigma^2 + n\sigma^2_{c(ab)} + \frac{cn}{(a-1)(b-1)}\sum_{i,j}(ab)^2_{ij}$
Residual	σ^2

From this table, we can derive the test statistic for testing

$H_0 : a_i = 0$ for all i		$H_A : a_i \neq 0$ for at least one i	
$\boldsymbol{F}_A = \frac{\boldsymbol{MS}_A}{\boldsymbol{MS}_{C \text{ in } AB}}$	$a-1$	$ab(c-1)$	$\lambda = \frac{bcn\sum_{i=1}^{a} a_i^2}{\sigma_y^2}$

The corresponding R-program is
`size_c.three_way_mixed_ab_in_c.model_5_a()`.

Example 3.19 We consider the situation of Example 3.16 but now we draw a random sample from the plots in each farm x variety-combination. By our program we have to determine a proper combination of the number of selected plots and the plot size n.

```
> size_c.three_way_mixed_ab_in_c.model_5_a(0.05, 0.1, 0.5, 6, 5,
+      1, "maximin")

[1] 27

> size_c.three_way_mixed_ab_in_c.model_5_a(0.05, 0.1, 0.5, 6, 5,
+      1, "minimin")

[1] 9
```

$H_0 : bj = 0 \forall j$		$H_A : b_j \neq 0$ for at least one j	
$\boldsymbol{F}_B = \frac{\boldsymbol{MS}_B}{\boldsymbol{MS}_{C \text{ in } AB}}$	$b-1$	$ab(c-1)$	$\lambda = \frac{acn\sum_j b_j^2}{\sigma_y^2}$

For the case $a = 6$, $b = 5$ and $n = 1$

```
> size_c.three_way_mixed_ab_in_c.model_5_b(0.05, 0.1, 0.5, 6, 5,
+      1, "maximin")

[1] 21
```

```
> size_c.three_way_mixed_ab_in_c.model_5_b(0.05, 0.1, 0.5, 6, 5,
+      1, "minimin")
```

```
[1] 9
```

$H_0 : (ab)_{ij} = 0 \forall i,j$		H_A at least one $(ab)_{ij} \neq 0$	
$F = \frac{MS_{A \times B}}{MS_{C \text{ in } AB}}$	$(a-1)(b-1)$	$ab(c-1)$	$\lambda = \frac{cn \sum_{ij} (ab)_y^2}{\sigma_y^2}$

```
> size_c.three_way_mixed_ab_in_c.model_5_axb(0.05, 0.1, 0.5, 6,
+      5, 1, "maximin")
```

```
[1] 210
```

```
> size_c.three_way_mixed_ab_in_c.model_5_axb(0.05, 0.1, 0.5, 6,
+      5, 1, "minimin")
```

```
[1] 15
```

3.4.3.5 Three-way analysis of variance—mixed classification $(A \times B) \succ C$—Model VI

The model equation is given by

$$\boldsymbol{y}_{ijkl} = \mu + \boldsymbol{a}_i + b_j + (\boldsymbol{ab})_{ij} + \boldsymbol{c}_{k(ij)} + \boldsymbol{e}_{ijkl}$$

The effects b_j are real numbers; i.e., they are not random. The model becomes complete under the following conditions: that the sum of b_j equals zero and that the random variables on the r.h.s. of the equation for all suffixes have expectation zero, have equal variances and are independent.

The variance of the $\boldsymbol{y}_{ijkl}$ is given by $var(\boldsymbol{y}_{ijkl}) = \sigma_y^2 = \sigma_a^2 + \sigma_{ab}^2 + \sigma_{c(ab)}^2 + \sigma^2$. The expected mean squares are given in Table 3.23.

From this table, we can derive the test statistic for testing

$H_0 : b_j = 0$ for all j		$H_A : b_j \neq 0$ for at least one j	
$F = \frac{MS_B}{MS_{AB}}$	$b-1$	$(a-1)(b-1)$	$\lambda = \frac{acn \sum_{j=1}^{b} b_j^2}{\sigma_y^2}$

Example 3.20 We consider the situation of Example 3.16, but now we draw a random sample from the farms and the plots in each farm × variety-combination. By our program we have to determine a proper combination of the number of selected farms, plots and the plot size n.

TABLE 3.23
Expected Mean Squares for Model VI.

Source of Variation	$E(\boldsymbol{MS})$
Between A-Levels	$\sigma^2 + bcn\sigma_a^2 + n\sigma_{c(ab)}^2$
Between B-Levels	$\sigma^2 + \frac{acn}{b-1}\sum_j b_j^2 + n\sigma_{c(ab)}^2 + cn\sigma_{ab}^2$
Between C-Levels within $A \times B$	$\sigma^2 + n\sigma_{c(ab)}^2$
Interaction $A \times B$	$\sigma^2 + n\sigma_{c(ab)}^2 + cn\sigma_{ab}^2$
Residual	σ^2

```
> size_c.three_way_mixed_ab_in_c.model_6_b(0.05, 0.1, 0.5, 6, 5,
+     1, "maximin")

[1] 27

> size_c.three_way_mixed_ab_in_c.model_6_b(0.05, 0.1, 0.5, 6, 5,
+     1, "minimin")

[1] 11
```

3.4.4 Three-way analysis of variance—mixed classification $(A \succ B) \times C$

The observations y_{ijkl} in the combinations of factor levels are as follows

Factor A	A_1			...	A_a		
Factor B	B_{11}	...	B_{1b}		B_{a1}	...	B_{ab}
C_1	y_{1111} y_{1112} $\vdots$ y_{111n}	...	y_{1b11} y_{1b12} $\vdots$ y_{1b1n}	...	y_{a111} y_{a112} $\vdots$ y_{a11n}	...	y_{ab11} y_{ab12} $\vdots$ y_{ab1n}
...	...	...	...	...	...	...	...
C_c	y_{11c1} y_{11c2} $\vdots$ y_{11cn}	...	y_{1bc1} y_{1bc2} $\vdots$ y_{1bcn}	...	y_{a1c1} y_{a1c2} $\vdots$ y_{a1cn}	...	y_{abc1} y_{abc2} $\vdots$ y_{abcn}

The model equation is given by

$$\boldsymbol{y}_{ijkl} = \mu + a_i + \boldsymbol{b}_{j(i)} + c_k + (ac)_{ik} + (\boldsymbol{bc})_{jk(i)} + \boldsymbol{e}_{ijkl}$$

We assume $var(e_{ijkl}) = \sigma^2 \; \forall i, j, k, l$. Besides the model with fixed effects only, we have six mixed models.

Model I	all three factors are fixed
Model III	factor A random, the other fixed
Model IV	factor B random, the other fixed
Model V	factor C random, the other fixed
Model VI	factor A fixed, the other random
Model VII	factor B fixed, the other random
Model VIII	factor C fixed, the other random.

The ANOVA table is the same for all models and is given by Table 3.24. For all the tests below we choose $\alpha = 0.05, \beta = 0.1$ and $\delta = 0.5\sigma_y$.

3.4.4.1 Three-way analysis of variance—mixed classification $(A \succ B) \times C$—Model I

The model equation is given by

$$\boldsymbol{y}_{ijkl} = \mu + a_i + b_{j(i)} + c_k + (ac)_{ik} + (bc)_{jk(i)} + \boldsymbol{e}_{ijkl}$$

All effects are real numbers; i.e., they are not random. The model becomes complete under the conditions that the sum of all these effects (separately over each index) equals zero. The variance of the $\boldsymbol{y}_{ijkl}$ is given by $var(\boldsymbol{y}_{ijkl}) = \sigma_y^2 = \sigma^2$. The analysis of variance table is given in Table 3.24.

The expected mean squares are given in Table 3.25.

Example 3.21 Case A: Let us assume that we have $a = 200$ cocks (factor A) each paired with $b = 100$ hens (factor B). We would like to observe the birth weight y of the chickens out of each pairing. If we investigate male and female chickens separately, we have a cross classified factor C with $c = 2$ levels. If we consider all levels, we have the Model I situation and have to determine the numbers of chickens which have to be weighed. We use case A if factor C is fixed.

Case B: A psychological test may consist of $c = 200$ items (levels of a factor C) each applied to $b = 300$ testees (levels of a factor B) which are split into a groups (levels of a factor A). The observations y are binary: 1 if the item was solved by the testee and 0 otherwise. We use case B if factor A is fixed. In the R-commands we do not use the parameters of this example but only demonstrate how to use the programs with smaller numbers of levels $a = 6$, $b = 5$, $c = 4$.

From Table 3.25, we can derive the test statistic for testing

$H_0 : a_i = 0$ for all i		$H_A : a_i \neq 0$ for at least one i	
$\boldsymbol{F} = \frac{\boldsymbol{MS}_A}{\boldsymbol{MS}_R}$	$a-1$	$abc(n-1)$	$\lambda = \frac{bcn \sum_{i=1}^{a} a_i^2}{\sigma_y^2}$

TABLE 3.24
ANOVA Table for the Three-Way Analysis of Variance—Mixed Classification $(A \succ B) \times C$.

Source of Variation	SS	df
Between A-Levels	$SS_A = \frac{1}{bcn}\sum_i \boldsymbol{Y}^2_{i...} - \frac{1}{N}\boldsymbol{Y}^2_{....}$	$a-1$
Between B-Levels within A-Levels	$SS_{B \text{ in } A} = \frac{1}{cn}\sum_{i,j} \boldsymbol{Y}^2_{ij..} - \frac{1}{bcn}\sum_i \boldsymbol{Y}^2_{i...}$	$a(b-1)$
Between C-Levels	$SS_C = \frac{1}{abn}\sum_k \boldsymbol{Y}^2_{..k.} - \frac{1}{N}\boldsymbol{Y}^2_{....}$	$c-1$
Interaction $A \times C$	$SS_{AC} = \frac{1}{bn}\sum_{i,k} \boldsymbol{Y}^2_{i.k.} - \frac{1}{bcn}\sum_i \boldsymbol{Y}^2_{i..} - \frac{1}{abn}\sum_k \boldsymbol{Y}^2_{.k.} + \frac{1}{N}\boldsymbol{Y}^2_{i...}$	$(a-1)(c-1)$
Interaction $B \times C$ within A	$SS_{BC \text{ in } A} = \frac{1}{n}\sum_{i,j,k} \boldsymbol{Y}^2_{ijk.} - \frac{1}{cn}\sum_{i,j} \boldsymbol{Y}^2_{ij..} - \frac{1}{bn}\sum_{j,k} \boldsymbol{Y}^2_{.j.k} + \frac{1}{bcn}\sum_i \boldsymbol{Y}^2_{i...}$	$a(b-1)(c-1)$
Residual	$SS_R = \sum_{i,j,k,l} \boldsymbol{y}^2_{ijkl} - \frac{1}{n}\sum_{i,j,k} \boldsymbol{Y}^2_{ijk.}$	$N-abc$

Source of Variation	$(\boldsymbol{MS})$
Between A-Levels	$\frac{\boldsymbol{SS}_A}{a-1}$
Between B-Levels within A-Levels	$\frac{\boldsymbol{SS}_{B \text{ in } A}}{a(b-1)}$
Between C-Levels	$\frac{\boldsymbol{SS}_C}{(c-1)}$
Interaction $A \times C$	$\frac{\boldsymbol{SS}_{AC}}{(a-1)(c-1)}$
Interaction $B \times C$ within A	$\frac{\boldsymbol{SS}_{BC \text{ in } A}}{a(b-1)(c-1)}$
Residual	$\frac{\boldsymbol{SS}_R}{N-abc}$

Note: We use the convention for writing sums by capitalization of the corresponding letter and replacing suffixes by a dot if summation over it has taken place as has been explained in Table 3.2.

TABLE 3.25
Expected Mean Squares for Model I.

Source of Variation	$E(\boldsymbol{MS})$
Between A-Levels	$\sigma^2 + \frac{bcn}{a-1}\sum_i a_i^2$
Between B-Levels within A-Levels	$\sigma^2 + \frac{cn}{a(b-1)}\sum_{i,j} b_{j(i)}^2$
Between C-Levels	$\sigma^2 + \frac{abn}{c-1}\sum_k c_k^2$
Interaction $A \times C$	$\sigma^2 + \frac{bn}{(a-1)(c-1)}\sum_{i,k}(ac)_{ik}^2$
Interaction $B \times C$ within A	$\sigma^2 + \frac{n}{a(b-1)(c-1)} \cdot \sum_{i,j,k}(bc)_{jk(i)}^2$
Residual	σ^2

The corresponding R-program is
`size_n.three_way_mixed_cxbina.model_1_a()`.

```
> size_n.three_way_mixed_cxbina.model_1_a(0.05, 0.1, 0.5, 6, 5,
+     4, "maximin")

[1] 7

> size_n.three_way_mixed_cxbina.model_1_a(0.05, 0.1, 0.5, 6, 5,
+     4, "minimin")

[1] 3
```

From Table 3.25, we can also derive the test statistic for testing

$H_0: b_{j(i)} = 0$ for all j		$H_A: b_{j(i)} \neq 0$ for at least one j	
$\boldsymbol{F} = \frac{\boldsymbol{MS}_{B \text{ in } A}}{\boldsymbol{MS}_R}$	$a(b-1)$	$abc(n-1)$	$\lambda = \frac{cn\sum b_{j(i)}^2}{\sigma_y^2}$

The corresponding R-program is
`size_n.three_way_mixed_cxbina.model_1_b()`.

```
> size_n.three_way_mixed_cxbina.model_1_b(0.05, 0.1, 0.5, 6, 5,
+     4, "maximin")

[1] 57

> size_n.three_way_mixed_cxbina.model_1_b(0.05, 0.1, 0.5, 6, 5,
+     4, "minimin")
```

```
[1] 4
```

From Table 3.25, we can also derive the test statistic for testing

$H_0: c_k = 0 \forall k$		H_A at least one $c_k \neq 0$	
$\boldsymbol{F} = \frac{\boldsymbol{MS_C}}{\boldsymbol{MS_R}}$	$c-1$	$abc(n-1)$	$\lambda = \frac{abn \sum_{k=1}^{c} c_k^2}{\sigma_y^2}$

The corresponding R-program is
`size_n.three_way_mixed_cxbina.model_1_c()`.

```
> size_n.three_way_mixed_cxbina.model_1_c(0.05, 0.1, 0.5, 6, 5,
+     4, "maximin")

[1] 4

> size_n.three_way_mixed_cxbina.model_1_c(0.05, 0.1, 0.5, 6, 5,
+     4, "minimin")

[1] NA
```

We can also test interaction effects

$H_0: (ac)_{ij} = 0 \forall i, k$		H_A at least one $(ac)_{ik} \neq 0$	
$\boldsymbol{F} = \frac{\boldsymbol{MS_{A \times C}}}{\boldsymbol{MS_R}}$	$(a-1)(c-1)$	$abc(n-1)$	$\lambda = \frac{bn \sum_{ik}(ac)_{ik}^2}{\sigma_y^2}$

The corresponding R-program is
`size_n.three_way_mixed_cxbina.model_1_axc()`.

```
> size_n.three_way_mixed_cxbina.model_1_axc(0.05, 0.1, 0.5, 6,
+     5, 4, "maximin")

[1] 38

> size_n.three_way_mixed_cxbina.model_1_axc(0.05, 0.1, 0.5, 6,
+     5, 4, "minimin")

[1] 4
```

We can also test interaction effects

$H_0: (bc)_{jk(i)} = 0 \forall i, j, k$		H_A at least one $(bc)_{j(i)k} \neq 0$	
$\boldsymbol{F} = \frac{\boldsymbol{MS_{BxC \text{ in } A}}}{\boldsymbol{MS_R}}$	$a(b-1)(c-1)$	$abc(n-1)$	$\lambda = \frac{n \sum (bc)_{jk(i)}^2}{\sigma_y^2}$

```
> size_n.three_way_mixed_cxbina.model_1_bxc(0.05, 0.1, 0.5, 6,
+     5, 4, "maximin")
```

```
[1] 345

> size_n.three_way_mixed_cxbina.model_1_bxc(0.05, 0.1, 0.5, 6,
+     5, 4, "minimin")

[1] 7
```

3.4.4.2 Three-way analysis of variance—mixed classification $(A \succ B) \times C$—Model III

The model equation is given by

$$y_{ijkl} = \mu + a_i + b_{j(i)} + c_k + (ac)_{ik} + (bc)_{jk(i)} + e_{ijkl}$$

The effects $b_{j(i)}$ and c_k and $(bc)_{jk(i)}$ are real numbers; i.e., they are not random. The model becomes complete under the following conditions: the sum of each of $b_{j(i)}$ and of the $(bc)_{jk(i)}$ (separately over each index) equals zero and the random variables on the r.h.s. of the equation for all suffixes have expectation zero, have equal variances and are independent.

The expected mean squares are given in Table 3.26.

The variance of the y_{ijkl} is given by $var(y_{ijkl}) = \sigma_y^2 = \sigma_a^2 + \sigma_{ac}^2 + \sigma^2$.

TABLE 3.26
Expected Mean Squares for Model III.

Source of Variation	$E(MS)$
Between A-Levels	$bcn\sigma_a^2 + \sigma^2$
Between B-Levels within A-Levels	$\frac{cn}{a(b-1)} \sum_{i,j} b_{j(i)}^2 + \sigma^2$
Between C-Levels	$\frac{abn}{c-1} \sum_k c_k^2 + bn\sigma_{ac}^2 + \sigma^2$
Interaction $A \times C$	$bn\sigma_{ac}^2 + \sigma^2$
Interaction $B \times C$ within A	$\frac{n}{a(b-1)(c-1)} \cdot \sum_{i,j,k} (bc)_{jk(i)}^2 + \sigma^2$
Residual	σ^2

From Table 3.26, we can derive the test statistic for testing

$H_0 : b_{j(i)} = 0 \forall i, j$		$H_A : b_{j(i)} \neq 0$ for at least one j	
$F = \frac{MS_{B \text{ in } A}}{MS_R}$	$a(b-1)$	$abc(n-1)$	$\lambda = \frac{cn \sum b_{j(i)}^2}{\sigma_y^2}$

The corresponding R-program is
`size_n.three_way_mixed_cxbina.model_3_b()`.

For the case $a = 6$, $b = 5$, $c = 4$.

```
> size_n.three_way_mixed_cxbina.model_3_b(0.05, 0.1, 0.5, 6, 5,
+     4, "maximin")

[1] 57

> size_n.three_way_mixed_cxbina.model_3_b(0.05, 0.1, 0.5, 6, 5,
+     4, "minimin")

[1] 4
```

From Table 3.26, we can also derive the test statistic for testing

$H_0 : c_k = 0 \forall k$		H_A for at least one $c_k \neq 0$	
$F = \frac{MS_C}{MS_{AC}}$	$c-1$	$(a-1)(c-1)$	$\lambda = \frac{abn \sum_{k=1}^{c} c_k^2}{\sigma_y^2}$

The corresponding R-program is
`size_a.three_way_mixed_cxbina.model_3_c()`.
For the case $b = 5$, $c = 4$ and $n = 1$

```
> size_a.three_way_mixed_cxbina.model_3_c(0.05, 0.1, 0.5, 5, 4,
+     1, "maximin")

[1] 25

> size_a.three_way_mixed_cxbina.model_3_c(0.05, 0.1, 0.5, 5, 4,
+     1, "minimin")

[1] 13
```

$H_0 : (bc)_{jk(i)} = 0$ for all j,k		H_A : at least one $(bc)_{jk(i)} \neq 0$	
$F = \frac{MS_{BC \text{ in } A}}{MS_R}$	$a(b-1)(c-1)$	$abc(n-1)$	$\lambda = \frac{n \sum_{ijh} (bc)^2_{jk(i)}}{\sigma_y^2}$

```
> size_n.three_way_mixed_cxbina.model_3_bxc(0.05, 0.1, 0.5, 6,
+     5, 4, "maximin")

[1] 345

> size_n.three_way_mixed_cxbina.model_3_bxc(0.05, 0.1, 0.5, 6,
+     5, 4, "minimin")

[1] 7
```

3.4.4.3 Three-way analysis of variance—mixed classification $(A \succ B) \times C$—Model IV

The model equation is given by

$$\boldsymbol{y}_{ijkl} = \mu + a_i + \boldsymbol{b}_{j(i)} + c_k + (ac)_{ik} + (\boldsymbol{bc})_{jk(i)} + \boldsymbol{e}_{ijkl}$$

The effects a_i and c_k are real numbers; i.e., they are not random. The model becomes complete under the following conditions: that the sum of each of a_i and of the c_k and $(ac)_{ik}$ equals zero and that the random variables on the r.h.s. of the equation for all suffixes have expectation zero, have equal variances and are independent.

The expected mean squares are given in Table 3.27.

TABLE 3.27
Expected Mean Squares for Model IV.

Source of Variation	$\boldsymbol{E(MS)}$
Between A-Levels	$\frac{bcn}{a-1}\sum\limits_i a_i^2 + cn\sigma_{b(a)}^2 + \sigma^2$
Between B-Levels within A-Levels	$cn\sigma_{b(a)}^2 + \sigma^2$
Between C-Levels	$\frac{abn}{c-1}\sum\limits_k c_k^2 + n\sigma_{bc(a)}^2 + \sigma^2$
Interaction $A \times C$	$\frac{bn}{(a-1)(c-1)}\sum\limits_{i,k}(ac)_{ik}^2 + n\sigma_{bc(a)}^2 + \sigma^2$
Interaction $B \times C$ within A	$n\sigma_{bc(a)}^2 + \sigma^2$
Residual	σ^2

From Table 3.27, we can derive the test statistic for testing

$H_0 : a_i = 0$ for all i		$H_A : a_i \neq 0$ for at least one i	
$\boldsymbol{F} = \frac{\boldsymbol{MS}_A}{\boldsymbol{MS}_{B \text{ in } A}}$	$a-1$	$a(b-1)$	$\lambda = \frac{bcn\sum\limits_{k=1}^{a} a_i^2}{\sigma_y^2}$

The corresponding R-program is
`size_b.three_way_mixed_cxbina.model_4_a()`.
For the case $a = 6$, $c = 4$, $n = 1$

```
> size_b.three_way_mixed_cxbina.model_4_a(0.05, 0.1, 0.5, 6, 4,
+     1, "maximin")

[1] 34

> size_b.three_way_mixed_cxbina.model_4_a(0.05, 0.1, 0.5, 6, 4,
+     1, "minimin")
```

```
[1] 12
```

From Table 3.27, we can also derive the test statistic for testing

$H_0 : c_k = 0$ for all k		H_A:$c_k \neq 0$ for at least one k	
$F = \frac{MS_C}{MS_{BC \text{ in } A}}$	$c-1$	$a(b-1)(c-1)$	$\lambda = \frac{abn \sum_{k=1}^{c} c_k^2}{\sigma_y^2}$

The coresponding R-program is
size_b.three_way_mixed_cxbina.model_4_c().

```
> size_b.three_way_mixed_cxbina.model_4_c(0.05, 0.1, 0.5, 6, 4,
+      1, "maximin")

[1] 20

> size_b.three_way_mixed_cxbina.model_4_c(0.05, 0.1, 0.5, 6, 4,
+      1, "minimin")

[1] 10
```

We can also test interaction effects

$H_0 : (ac)_{ik} = 0$ for all i, k		H_A : for at least one $(ac)_{ik} \neq 0$	
$F = \frac{MS_{A\times C}}{MS_{B\times C \text{ in } A}}$	$(a-1)(c-1)$	$a(b-1)(c-1)$	$\lambda = \frac{bn \sum_{i,k} (ac)_{ik}^2}{\sigma_y^2}$

The corresponding R-program is
size_b.three_way_mixed_cxbina.model_4_axc().

```
> size_b.three_way_mixed_cxbina.model_4_axc(0.05, 0.1, 0.5, 6,
+      4, 1, "maximin")

[1] 190

> size_b.three_way_mixed_cxbina.model_4_axc(0.05, 0.1, 0.5, 6,
+      4, 1, "minimin")

[1] 17
```

3.4.4.4 Three-way analysis of variance—mixed classification $(A \succ B) \times C$—Model V

The model equation is given by

$$\boldsymbol{y}_{ijkl} = \mu + a_i + \boldsymbol{b}_{j(i)} + \boldsymbol{c}_k + (\boldsymbol{ac})_{ik} + (\boldsymbol{bc})_{jk(i)} + \boldsymbol{e}_{ijkl}$$

The effects a_i and $b_{j(i)}$ are real numbers; i.e., they are not random. The model becomes complete under the following conditions: that the sum of each of a_i and of the $b_{j(i)}$ equals zero and that the random variables on the r.h.s. of the equation for all suffixes have expectation zero, have equal variances and are independent.

The expected mean squares are given in Table 3.28.

TABLE 3.28
Expected Mean Squares for Model V.

Source of Variation	$\boldsymbol{E(MS)}$
Between A-Levels	$\frac{bcn}{a-1}\sum_i a_i^2 + bn\sigma_{ac}^2 + \sigma^2$
Between B-Levels within A-Levels	$\frac{cn}{a-1}\sum_{ij} b_{j(i)}^2 + n\sigma_{bc(a)}^2 + \sigma^2$
Between C-Levels	$abn\sigma_c^2 + \sigma^2$
Interaction $A \times C$	$bn\sigma_{ac}^2 + \sigma^2$
Interaction $B \times C$ within A	$n\sigma_{bc(a)}^2 + \sigma^2$
Residual	σ^2

The variance of the $\boldsymbol{y}_{ijkl}$ is given by

$$var(\boldsymbol{y}_{ijkl}) = \sigma_y^2 = \sigma_c^2 + \sigma_{ac}^2 + \sigma_{bc(a)}^2 + \sigma^2$$

From Table 3.28, we can also derive the test statistic for testing

$H_0: a_i = 0\ \forall i$		$H_A: a_i \neq 0$ for at least one $i \neq 0$	
$\boldsymbol{F}_A = \frac{\boldsymbol{MS}_A}{\boldsymbol{MS}_{A\times C}}$	$a-1$	$(a-1)(c-1)$	$\lambda = \frac{bcn\sum a_i^2}{\sigma_y^2}$

By R we solve the problem as follows for $a = 6$, $b = 5$ and $n = 2$:

```
> size_c.three_way_mixed_cxbina.model_5_a(0.05, 0.1, 0.5, 6, 5,
+      2, "maximin")

[1] 15

> size_c.three_way_mixed_cxbina.model_5_a(0.05, 0.1, 0.5, 6, 5,
+      2, "minimin")

[1] 6
```

From Table 3.27, we can also derive the test statistic for testing

$H_0: b_{j(i)} = 0$ for all i, j		$H_A: b_{j(i)} \neq 0$ for at least one j	
$F = \frac{MS_{B \text{ in } A}}{MS_{B \times C \text{ in } A}}$	$a(b-1)$	$a(b-1)(c-1)$	$\lambda = \frac{cn \sum b_{j(i)}^2}{\sigma_y^2}$

The corresponding R-program is
`size_c.three_way_mixed_cxbina.model_5_b()`.

```
> size_c.three_way_mixed_cxbina.model_5_b(0.05, 0.1, 0.5, 6, 5,
+     2, "maximin")

[1] 113

> size_c.three_way_mixed_cxbina.model_5_b(0.05, 0.1, 0.5, 6, 5,
+     2, "minimin")

[1] 9
```

3.4.4.5 Three-way analysis of variance—mixed classification $(A \succ B) \times C$—Model VI

The model equation is given by

$$\boldsymbol{y}_{ijkl} = \mu + a_i + \boldsymbol{b}_{j(i)} + \boldsymbol{c}_k + (\boldsymbol{ac})_{ik} + (\boldsymbol{bc})_{jk(i)} + \boldsymbol{e}_{ijkl}$$

The effects a_i are real numbers; i.e., they are not random. The model becomes complete under the following conditions: that the sum of a_i equals zero and that the random variables on the r.h.s. of the equation for all suffixes have expectation zero, have equal variances and are independent.

The variance of the $\boldsymbol{y}_{ijkl}$ is given by

$$var(\boldsymbol{y}_{ijkl}) = \sigma_y^2 = \sigma_{b(a)}^2 + \sigma_c^2 + \sigma_{ac}^2 + \sigma_{bc(a)}^2 + \sigma^2$$

The expected mean squares are given in Table 3.29.

From Table 3.29, we can also derive the test statistic for testing

$H_0: a_i = 0$ for all i		$H_A: a_i \neq 0$ for at least one i	
$\boldsymbol{F}_{appr}$	$a-1$	f	$\lambda = \frac{bcn \sum_{i=1}^{a} a_i^2}{\sigma_y^2}$

Here

$\boldsymbol{F}_{appr} = \frac{\boldsymbol{MS}_A}{\boldsymbol{MS}_{B \text{ in } A} + \boldsymbol{MS}_{A \times C} - \boldsymbol{MS}_{B \times C \text{ in } A}}$

is approximately F-distributed with $a-1$ and

$f = \frac{(MS_{B \text{ in } A} + MS_{A \times C} - MS_{B \times C \text{ in } A})^2}{\frac{MS_{B \text{ in } A}^2}{a(b-1)} + \frac{MS_{A \times C}^2}{(a-1)(c-1)} + \frac{MS_{B \times C \text{ in } A}^2}{a(b-1)(c-1)}}$

degrees of freedom.

TABLE 3.29
Expected Mean Squares for Model VI.

Source of Variation	$\boldsymbol{E(MS)}$
Between A-Levels	$\frac{bcn}{a-1}\sum_i a_i^2 + cn\sigma_{b(a)}^2 + bn\sigma_{ac}^2 + n\sigma_{bc(a)}^2 + \sigma^2$
Between B-Levels within A-Levels	$cn\sigma_{b(a)}^2 + n\sigma_{bc(a)}^2 + \sigma^2$
Between C-Levels	$abn\sigma_c^2 + n\sigma_{bc(a)}^2 + \sigma^2$
Interaction $A \times C$	$bn\sigma_{ac}^2 + n\sigma_{bc(a)}^2 + \sigma^2$
Interaction $B \times C$ within A	$n\sigma_{bc(a)}^2 + \sigma^2$
Residual	σ^2

An exact solution can be given if we assume that either $\sigma_{ac}^2 = 0$ or $\sigma_{b(a)}^2 = 0$. In these cases an exact F-test exists. For $\sigma_{ac}^2 = 0$, $\boldsymbol{F}_{AB} = \frac{\boldsymbol{MS}_A}{\boldsymbol{MS}_{B \text{ in } A}}$ is F-distributed with $a-1$ and $a(b-1)$ d.f.. Analogously in the case $\sigma_{b(a)}^2 = 0$, $\boldsymbol{F}_{AC} = \frac{\boldsymbol{MS}_A}{\boldsymbol{MS}_{AC}}$ is F-distributed with $a-1$ and $(a-1)(c-1)$ d.f. We now determine the number of levels of the factor B within A-levels under the assumption that $\sigma_{ac}^2 = 0$ or the number of levels of the factor C under the assumption that $\sigma_{b(a)}^2 = 0$ for given precision requirements.

Case 1: Assumption is that $\sigma_{ac}^2 = 0$ and $b = c$. We take here $a = 6$ and $n = 2$.

```
> size_bc.three_way_mixed_cxbina.model_6_a_case1(0.05, 0.1, 0.5,
+       6, 2, "maximin")

[1] 9

> size_bc.three_way_mixed_cxbina.model_6_a_case1(0.05, 0.1, 0.5,
+       6, 2, "minimin")

[1] 6
```

Case 2: Assumption is that $\sigma_{b(a)}^2 = 0$ and $b = c$. We take here $a = 6$ and $n = 2$.

```
> size_bc.three_way_mixed_cxbina.model_6_a_case2(0.05, 0.1, 0.5,
+       6, 2, "maximin")

[1] 9

> size_bc.three_way_mixed_cxbina.model_6_a_case2(0.05, 0.1, 0.5,
+       6, 2, "minimin")

[1] 6
```

3.4.4.6 Three-way analysis of variance—mixed classification $(A \succ B) \times C$—Model VII

The model equation is given by

$$\boldsymbol{y}_{ijkl} = \mu + \boldsymbol{a}_i + b_{j(i)} + \boldsymbol{c}_k + (\boldsymbol{ac})_{ik} + (\boldsymbol{bc})_{jk(i)} + \boldsymbol{e}_{ijkl}$$

The effects $b_{j(i)}$ are real numbers; i.e., they are not random. The model becomes complete under the following conditions: that the sum of each of $b_{j(i)}$, equals zero and that the random variables on the r.h.s. of the equation for all suffixes have expectation zero, have equal variances and are independent.

The variance of the $\boldsymbol{y}_{ijkl}$ is given by

$$var(\boldsymbol{y}_{ijkl}) = \sigma_y^2 = \sigma_a^2 + \sigma_c^2 + \sigma_{ac}^2 + \sigma_{bc(a)}^2 + \sigma^2$$

The expected mean squares are given in Table 3.30.

TABLE 3.30
Expected Mean Squares for Model VII.

Source of Variation	$E(\boldsymbol{MS})$
Between A-Levels	$bcn\sigma_a^2 + bn\sigma_{ac}^2 + \sigma^2$
Between B-Levels within A-Levels	$\frac{cn}{a(b-1)} \sum_{i,j} b_{j(i)}^2 + n\sigma_{bc(a)}^2 + \sigma^2$
Between C-Levels	$abn\sigma_c^2 + bn\sigma_{ac}^2 + \sigma^2$
Interaction $A \times C$	$bn\sigma_{ac}^2 + \sigma^2$
Interaction $B \times C$ within A	$n\sigma_{bc(a)}^2 + \sigma^2$
Residual	σ^2

From Table 3.30, we can derive the test statistic for testing

$H_0 : b_{j(i)} = 0$ for all i, j		$H_A : b_{j(i)} \neq 0$ for at least one j	
$\boldsymbol{F} = \frac{\boldsymbol{MS}_{B \text{ in } A}}{\boldsymbol{MS}_{B\times C \text{ in } A}}$	$a(b-1)$	$a(b-1)(c-1)$	$\lambda = \frac{cn \sum b_{j(i)}^2}{\sigma_y^2}$

The corresponding R-program is
`size_c.three_way_mixed_cxbina.model_7_b()`.

```
> size_c.three_way_mixed_cxbina.model_7_b(0.05, 0.1, 0.5, 6, 5,
+      2, "maximin")

[1] 113

> size_c.three_way_mixed_cxbina.model_7_b(0.05, 0.1, 0.5, 6, 5,
+      2, "minimin")

[1] 9
```

3.4.4.7 Three-way analysis of variance—mixed classification $(A \succ B) \times C$—Model VIII

The model equation is given by

$$\boldsymbol{y}_{ijkl} = \mu + \boldsymbol{a}_i + \boldsymbol{b}_{j(i)} + c_k + (\boldsymbol{ac})_{ik} + (\boldsymbol{bc})_{jk(i)} + \boldsymbol{e}_{ijkl}$$

The effects c_k are real numbers; i.e., they are not random. The model becomes complete under the following conditions: that the sum of each of c_k (separately over each index) equals zero and that the random variables on the r.h.s. of the equation for all suffixes have expectation zero, have equal variances and are independent.

The variance of the $\boldsymbol{y}_{ijkl}$ is given by

$$var(\boldsymbol{y}_{ijkl}) = \sigma_y^2 = \sigma_a^2 + \sigma_{b(a)}^2 + \sigma_{ac}^2 + \sigma_{bc(a)}^2 + \sigma^2.$$

The expected mean squares are given in Table 3.31.

TABLE 3.31
Expected Mean Squares for Model VIII.

Source of Variation	$E(\boldsymbol{MS})$
Between A-Levels	$bcn\sigma_a^2 + cn\sigma_{b(a)}^2 + \sigma^2$
Between B-Levels within A-Levels	$cn\sigma_{b(a)}^2 + \sigma^2$
Between C-Levels	$\frac{abn}{c-1}\sum_k c_k^2 + bn\sigma_{ac}^2 + n\sigma_{bc(a)}^2 + \sigma^2$
Interaction $A \times C$	$bn\sigma_{ac}^2 + n\sigma_{bc(a)}^2 + \sigma^2$
Interaction $B \times C$ within A	$n\sigma_{bc(a)}^2 + \sigma^2$
Residual	σ^2

From Table 3.31, we can derive the test statistic for testing

$H_0 : c_k = 0 \forall k$		$H_A : c_k \neq 0$ for at least one k	
$\boldsymbol{F} = \frac{\boldsymbol{MS}_C}{\boldsymbol{MS}_{A\times C}}$	$c-1$	$(a-1)(c-1)$	$\lambda = \frac{abn \sum_{k=1}^{c} c_k^2}{\sigma_y^2}$

The corresponding R-programs are
`size_a.three_way_mixed_cxbina.model_7_c()`
`size_b.three_way_mixed_cxbina.model_7_c()` and
`size_ab.three_way_mixed_cxbina.model_7_c()`.

We demonstrate the last program; the end result gives the numbers **a**, **b**, **c**, **n**. We take here $c = 5$ and $n = 2$.

```
> size_ab.three_way_mixed_cxbina.model_7_c(0.05, 0.1, 0.5, 5, 2,
+      "maximin")

[1] 102.3107
[1]  640.22636 2046.21429   32.01132    2.00000    5.00000    2.00000
[1] 652.46056 640.22636  21.74869   3.00000   5.00000   2.00000
[1] 664.84148 640.22636  16.62104   4.00000   5.00000   2.00000
[1] 677.29580 640.22636  13.54592   5.00000   5.00000   2.00000
[1] 689.89262 640.22636  11.49821   6.00000   5.00000   2.00000
[1] 702.58970 640.22636  10.03700   7.00000   5.00000   2.00000
[1] 715.37116 640.22636   8.94214   8.00000   5.00000   2.00000
[1] 728.426883 640.226361   8.093632   9.000000   5.000000   2.000000
[1] 741.485806 640.226361   7.414858  10.000000   5.000000   2.000000
[1] 754.660086 640.226361   6.860546  11.000000   5.000000   2.000000
[1] 767.959563 640.226361   6.399663  12.000000   5.000000   2.000000
[1] 780.895064 640.226361   6.006885  13.000000   5.000000   2.000000
[1] 794.157344 640.226361   5.672552  14.000000   5.000000   2.000000
[1] 807.432627 640.226361   5.382884  15.000000   5.000000   2.000000
[1] 821.469612 640.226361   5.134185  16.000000   5.000000   2.000000
[1] 835.166283 640.226361   4.912743  17.000000   5.000000   2.000000
[1] 848.17081 640.22636   4.71206  18.00000   5.00000   2.00000
[1] 861.650236 640.226361   4.535001  19.000000   5.000000   2.000000
[1] 876.102106 640.226361   4.380511  20.000000   5.000000   2.000000
[1] 890.093901 640.226361   4.238542  21.000000   5.000000   2.000000
[1] 902.849493 640.226361   4.103861  22.000000   5.000000   2.000000
[1] 917.613114 640.226361   3.989622  23.000000   5.000000   2.000000
[1] 931.833455 640.226361   3.882639  24.000000   5.000000   2.000000
[1] 944.165848 640.226361   3.776663  25.000000   5.000000   2.000000
[1] 959.760129 640.226361   3.691385  26.000000   5.000000   2.000000
[1] 971.908634 640.226361   3.599662  27.000000   5.000000   2.000000
[1] 986.111743 640.226361   3.521828  28.000000   5.000000   2.000000
[1] 1001.952441  640.226361   3.455008  29.000000   5.000000   2.000000
[1] 1013.552549  640.226361   3.378508  30.000000   5.000000   2.000000
[1] 1026.250231  640.226361   3.310485  31.000000   5.000000   2.000000
[1] 1037.773529  640.226361   3.243042  32.000000   5.000000   2.000000
[1] 1047.223108  640.226361   3.173403  33.000000   5.000000   2.000000
[1] 1053.159764  640.226361   3.097529  34.000000   5.000000   2.000000
[1] 1053.422736  640.226361   3.009779  35.000000   5.000000   2.000000
[1] 1045.069453  640.226361   2.902971  36.000000   5.000000   2.000000
[1] 1024.696061  640.226361   2.769449  37.000000   5.000000   2.000000
[1] 989.501814 640.226361   2.603952  38.000000   5.000000   2.000000
[1] 939.189206 640.226361   2.408177  39.000000   5.000000   2.000000
[1] 877.814017 640.226361   2.194535  40.000000   5.000000   2.000000
[1] 813.612964 640.226361   1.984422  41.000000   5.000000   2.000000
[1] 33  2  5  2

> size_ab.three_way_mixed_cxbina.model_7_c(0.05, 0.1, 0.5, 5, 2,
+      "minimin")

[1] 40.92429
```

```
[1] 270.91832 818.48572  13.54592    2.00000    5.00000    2.00000
[1] 283.588406 270.918321   9.452947   3.000000   5.000000   2.000000
[1] 296.594322 270.918321   7.414858   4.000000   5.000000   2.000000
[1] 309.710686 270.918321   6.194214   5.000000   5.000000   2.000000
[1] 322.973051 270.918321   5.382884   6.000000   5.000000   2.000000
[1] 336.827213 270.918321   4.811817   7.000000   5.000000   2.000000
[1] 350.440843 270.918321   4.380511   8.000000   5.000000   2.000000
[1] 363.844066 270.918321   4.042712   9.000000   5.000000   2.000000
[1] 377.666339 270.918321   3.776663  10.000000   5.000000   2.000000
[1] 392.295789 270.918321   3.566325  11.000000   5.000000   2.000000
[1] 405.421019 270.918321   3.378508  12.000000   5.000000   2.000000
[1] 417.133009 270.918321   3.208715  13.000000   5.000000   2.000000
[1] 421.369094 270.918321   3.009779  14.000000   5.000000   2.000000
[1] 403.621939 270.918321   2.690813  15.000000   5.000000   2.000000
[1] 351.125607 270.918321   2.194535  16.000000   5.000000   2.000000
[1] 292.481914 270.918321   1.720482  17.000000   5.000000   2.000000
[1] 14  2  5  2
```

4

Sample Size Determination in Model II of Regression Analysis

Regression coefficients are like the expectation of a random variable location parameter. Therefore, we can use the methodology from Chapter two to determine the sample sizes needed for a given precision. But there are two totally different situations concerning the underlying model. This fact is seldom recognised in analysing data because the standard software packages, contrary to the situation in ANOVA, make no difference between the models when determining regression analysis. The reason is that the least squares approach leads to the same formula for both models for estimating the regression coefficients, but the standard errors are different for these estimators in the two models.

4.1 Introduction

In regression analysis with one response variable (regressand) and one cause variable (regressor) we have two basic situations.

Model I:

The model equation is given by

$$\boldsymbol{y}_i = f(x_i, \theta) + \boldsymbol{e}_i \tag{4.1}$$

where θ is a parameter vector with p components. We assume that the error terms $\boldsymbol{e}_i$ are i.i.d. with expectation zero. Thus the vector of the $\boldsymbol{y}_i$ is a random sample. Because in this model the values x_i are real numbers and not random, we can choose them in an optimal way, concerned only that at least p of them are different. In the linear case $p = 2$, and the equation becomes $\boldsymbol{y}_i = \beta_0 + \beta_1 x_i + \boldsymbol{e}_i$.

The determination of an experiment which is based on Model I of regression analysis depends on the selection of the x_i-values. The choice of these x_i-values is part of the design of the experiment and will be treated together with the determination of the minimum size for a given choice of these x_i-values in Chapter eight.

Model II:

Let $(\boldsymbol{x}, \boldsymbol{y})$ be two-dimensional normally distributed with expectation vector $\begin{pmatrix} \mu_x \\ \mu_y \end{pmatrix}$ and covariance matrix $\begin{pmatrix} \sigma_x^2 & \sigma_{xy} \\ \sigma_{xy} & \sigma_y^2 \end{pmatrix}$ and take a random sample $(\boldsymbol{x}_1, \boldsymbol{y}_1)$; $(\boldsymbol{x}_2, \boldsymbol{y}_2)$; $\ldots$; $(\boldsymbol{x}_n, \boldsymbol{y}_n)$ of size n from this distribution. Then
$E(\boldsymbol{y}_i|\boldsymbol{x}_i) = \beta_0 + \beta_1 \boldsymbol{x}_i$ and
$E(\boldsymbol{x}_i|\boldsymbol{y}_i) = \beta_0^* + \beta_1^* \boldsymbol{y}_i$.

First we have to decide between one of these two models. W.l.o.g. lets us use the first equation. Contrary to Model I in Chapter eight, we now have a new parameter, namely the correlation coefficient $\rho = \frac{\sigma_{xy}}{\sigma_x \sigma_y}$ with the covariance σ_{xy} between $\boldsymbol{x}$ and $\boldsymbol{y}$ and the two standard deviations. We can construct confidence intervals or test hypotheses about these parameters or use selection procedures. Whereas the tests and confidence intervals for the regression coefficients for Model II look like those for Model I, the expected width of the intervals and the power of the tests differ from those for Model I. However, this is just important for the determination of minimum sizes. In this chapter we concentrate mainly on the correlation coefficient but give also some results for the regression equation. If we test $H_0 : \rho = \frac{\sigma_{xy}}{\sigma_x \sigma_y} = 0$, this means $H_0 : \sigma_{xy} = 0$ and therefore also $H_0 : \beta_1 = \frac{\sigma_{xy}}{\sigma_x^2} = 0$. Most of the sample size formulae are based on approximations.

We consider a random sample $\begin{pmatrix} \boldsymbol{x}_1 & \boldsymbol{x}_2 & \ldots & \boldsymbol{x}_n \\ \boldsymbol{y}_1 & \boldsymbol{y}_2 & \ldots & \boldsymbol{y}_n \end{pmatrix}$ from a bivariate normal distribution with covariance matrix $\begin{pmatrix} \sigma_x^2 & \sigma_{xy} \\ \sigma_{xy} & \sigma_y^2 \end{pmatrix}$ by which we would like to estimate the parameters of the model:

$$\boldsymbol{y}_i = \beta_0 + \beta_1 \boldsymbol{x}_i + \boldsymbol{e}_i \tag{4.2}$$

under the assumption that the $\boldsymbol{x}_i$ and the $\boldsymbol{e}_i$ are a system of independent random variables; $\sigma^2 = var(\boldsymbol{e}_i)$ is the residual variance in the model equation.

4.2 Confidence intervals

In this section we give confidence intervals for the parameters in Model II of regression analysis.

The point estimator of $\beta_1 = \frac{\sigma_{xy}}{\sigma_x^2}$ is given by

$$b_1 = \frac{\sum_{i=1}^{n} x_i y_i - \frac{\left(\sum_{i=1}^{n} x_i\right)\cdot\left(\sum_{i=1}^{n} y_i\right)}{n}}{\sum_{i=1}^{n} x_i^2 - \frac{\left(\sum_{i=1}^{n} x_i\right)^2}{n}} = \frac{s_{xy}(n-1)}{s_x^2(n-1)} = \frac{s_{xy}}{s_x^2}$$

4.2.1 Confidence intervals for the slope

With $s_y^2 = \frac{\sum_{i=1}^{n} y_i^2 - \frac{\left(\sum_{i=1}^{n} y_i\right)^2}{n}}{n-1}$ an exact $(1-\alpha)$-confidence interval for β_1 is given (Bartlett 1933) by

$$\left[b_1 - t\left(n-2;\ 1-\frac{\alpha}{2}\right)\sqrt{\frac{1}{n-2}\left(\frac{s_y^2}{s_x^2} - b_1^2\right)};\right.$$

$$\left. b_1 + t\left(n-2;\ 1-\frac{\alpha}{2}\right)\sqrt{\frac{1}{n-2}\left(\frac{s_y^2}{s_x^2} - b_1^2\right)}\right] \tag{4.3}$$

The minimal sample size for a $(1-\alpha)$-confidence interval for β_1 whose half expected width is not larger than δ can be obtained from

$$n = 2 + \left\lceil \frac{t^2\left(n-2;\ 1-\frac{\alpha}{2}\right)\cdot\left(\frac{\sigma_y^2}{\sigma_x^2} - \beta_1^2\right)}{\delta^2} \right\rceil \tag{4.4}$$

But we have to be careful that the inequality

$$0 \le \frac{\sigma_y^2}{\sigma_x^2} - \beta_1^2 \tag{4.5}$$

always holds.

Example 4.1 To determine the sample size for a confidence estimation of the slope in (4.2) using (4.4) and checking (4.5) we need prior information. Let us determine the number of students needed to estimate the slope in the regression of body height (y) on shoe size (x) both in *cm*. This can easily be done by asking students in a classroom. Prior information is taken from Example 5.1 in Rasch et al. (2007b) where values of 25 students in Wageningen (The Netherlands) in 1996 are given. From these data, we find the following estimates:

$s_x^2 \approx 3.7, s_y^2 \approx 72.5, s_{xy} \approx 11.4$, and this means $b_1 = 11.4/3.7 = 3.081$, because $0 < \frac{72.5}{3.7} - \left(\frac{11.4}{3.7}\right)^2$.

Starting the iteration with $n_0 = \infty$, we get $n_1 = 63$, $n_2 = 66$, $n_3 = 65$ and $n_4 = 66$ which remains unchanged in further iterations; thus we have to ask 66 students for their measurements.

The calculation is done by the R-function size_n.regII.ci_b1().

Introducing the parameters we now compute n:

```
> cov <- matrix(c(3.7, 11.4, 11.4, 72.5), 2, 2)
> n <- size_n.regII.ci_b1(cov, alpha = 0.05, delta = 0.8)

[1] 63
[1] 66
[1] 65
[1] 66

> n

[1] 66
```

4.2.2 A confidence interval for the correlation coefficient

An approximate $(1-\alpha)$-confidence interval for the correlation coefficient $\rho = \frac{\sigma_{xy}}{\sigma_x \sigma_y}$ is given by $[\boldsymbol{r}_l; \boldsymbol{r}_u]$ with

$\boldsymbol{r}_v = \frac{e^{2\boldsymbol{z}_v}-1}{e^{2\boldsymbol{z}_v}+1}, v = l, u$ and $\boldsymbol{z}_l = \boldsymbol{z} - \frac{u(1-\frac{\alpha}{2})}{\sqrt{n-3}}$; $\boldsymbol{z}_u = \boldsymbol{z} + \frac{u(1-\frac{\alpha}{2})}{\sqrt{n-3}}$; $\boldsymbol{z} = \frac{1}{2}\ln\frac{1+\boldsymbol{r}}{1-\boldsymbol{r}}$ and $\boldsymbol{r}$ is the sample correlation coefficient. The sample size n is needed so that $P(\boldsymbol{r} \leq \rho + \delta_1) \geq 1 - \frac{\alpha}{2}$ as well as $P(\boldsymbol{r} \geq \rho - \delta_2) \geq 1 - \frac{\alpha}{2}$ is $n = \max(n_1, n_2)$ with

$$n_i = \left\lceil \frac{u^2\left(1-\frac{\alpha}{2}\right)}{\Delta_i^2} \right\rceil; \; i = 1, 2 \tag{4.6}$$

and $\Delta_1 = \frac{1}{2}\ln\frac{1+\rho+\delta_1}{1-\rho-\delta_1} - \frac{1}{2}\ln\frac{1+\rho}{1-\rho}$; $\Delta_2 = \frac{1}{2}\ln\frac{1+\rho}{1-\rho} - \frac{1}{2}\ln\frac{1+\rho-\delta_2}{1-\rho+\delta_2}$. The sample size is maximal if $\rho = 0$. When no prior information for ρ is known, we can use the sample size for this worst case. If an upper bound for $|\rho|$ is known, we use this bound and determine the corresponding size (Rasch et al., 2008; 4/32/1101).

Example 4.2 Let $\delta_1 = 0.07$; $\alpha = 0.05$. Then we need the following sample sizes in dependence on ρ and δ_2. In the case that $\rho < 0$, in the table below interchange δ_1 and δ_2. The values of δ_2 in the table are chosen in such a way that $n = n_1 = n_2$.

ρ	0.9	0.8	0.7	0.6	0.5	0.4	0.3	0.2	0.1	0
δ_2	0.208	0.102	0.087	0.081	0.077	0.075	0.073	0.072	0.071	0.07
n	13	73	168	281	402	893	1073	1212	1309	1358

Example 4.3 For the problem in Example 4.1, we would like to construct a 95% confidence interval for the correlation coefficient so that the half expected width is not larger than $\delta_1 = \delta_2 = 0.1$. From the shoe data from Wageningen, we use as prior information $r \approx 0.7$. By the R-program below, we get $n = 127$. When we construct a 95% confidence interval for the correlation coefficient for $\rho = 0$ so that $\delta_1 = 0.1$; $\delta_2 = 0.2$, we obtain $n = 382$.

```
> size_n.regII.ci_rho(cov, delta1 = 0.1, delta2 = 0.1, alpha =
+ 0.05, rho = 0.7)

[1] 127

> size_n.regII.ci_rho(cov, delta1 = 0.1, delta2 = 0.2, alpha =
+ 0.05)

[1] 382
```

4.2.3 Confidence intervals for partial correlation coefficients

If $\boldsymbol{y}, \boldsymbol{x}_1, \boldsymbol{x}_2, \ldots, \boldsymbol{x}_k;\ k \geq 2$ is $(k+1)$-dimensional normally distributed, we can construct approximate confidence intervals and determine the sample size for the partial correlation coefficients by replacing ρ with $\rho_{yx_i|x_1,\ldots,x_{i-1},x_{i+1},x_k}$ and analogously replacing r and further n by $n-k+1$.

Example 4.4 With $k = 2$, a confidence interval for a $\rho_{yx_1|x_2}$ is obtained by replacing $\boldsymbol{z} = \ln \frac{1+\boldsymbol{r}}{1-\boldsymbol{r}}$ in Section 4.2.2 by $\boldsymbol{z} = \ln \frac{1+\boldsymbol{r}_{yx_1|x_2}}{1-\boldsymbol{r}_{yx_1|x_2}}$, where $\boldsymbol{r}_{yx_1|x_2}$ is the partial sample correlation coefficient and continuing as described there. In the precision requirements, we write $\Delta_1 = \frac{1}{2}\ln\frac{1+\rho_{yx_1|x_2}+\delta_1}{1-\rho_{yx_1|x_2}-\delta_1} - \frac{1}{2}\ln\frac{1+\rho_{yx_1|x_2}}{1-\rho_{yx_1|x_2}}$ and $\Delta_2 = \frac{1}{2}\ln\frac{1+\rho_{yx_1|x_2}}{1-\rho_{yx_1|x_2}} - \frac{1}{2}\ln\frac{1+\rho_{yx_1|x_2}-\delta_2}{1-\rho_{yx_1|x_2}+\delta_2}$ and obtain the sample size as $n_i = \left\lceil \frac{u^2\left(1-\frac{\alpha}{2}\right)}{\Delta_i^2} \right\rceil;\ i = 1,2$

We can use the R-program from Section 4.2.2.

4.2.4 A confidence interval for $E(\boldsymbol{y}|x) = \beta_0 + \beta_1 x$

A $(1-\alpha)$-confidence interval for $E(\boldsymbol{y}|x) = \beta_0 + \beta_1 x$ for any fixed x is given by

$$\left[\boldsymbol{b}_0 + \boldsymbol{b}_1 x - t(n-2;\ 1-\frac{\alpha}{2})\boldsymbol{s}\sqrt{\frac{1}{n} + \frac{(x-\overline{x})^2}{SS_x}}\ ;\right.$$

$$\left. \boldsymbol{b}_0 + \boldsymbol{b}_1 x + t(n-2;\ 1-\frac{\alpha}{2})\boldsymbol{s}\sqrt{\frac{1}{n} + \frac{(x-\overline{x})^2}{SS_x}} \right] \tag{4.7}$$

(Graybill, 1961; p. 122).
To construct a $(1-\alpha)$-confidence interval for $E(\boldsymbol{y}|x) = \beta_0 + \beta_1 x$ in such a way that the half expected width of the confidence interval (4.7) does not exceed δ, we need a minimum sample size (sampled from the bivariate normal) of

$$n = \left\lceil \left[\frac{\Gamma\left(\frac{n-1}{2}\right)\sqrt{2}}{\Gamma\left(\frac{n-2}{2}\right)\sqrt{n-2}} \right]^2 \left(\frac{\sigma}{\delta}\right)^2 t^2\left(n-2;\ 1-\frac{\alpha}{2}\right)\left[1 + \frac{n(x-\overline{x})^2}{SS_x}\right] \right\rceil$$

$$\approx \left\lceil \left(\frac{\sigma}{\delta}\right)^2 t^2\left(n-2;\ 1-\frac{\alpha}{2}\right)\left[1 + \frac{(x-\overline{x})^2}{\sigma_x^2}\right] \right\rceil \tag{4.8}$$

for either $x = x_{\max}$ or $x = x_{\min}$. Of course we again have to use Procedure II in Section 2.2.1.2, to solve the equation.
When no prior information about σ_x is known, we estimate it by $\frac{x_{\max}-x_{\min}}{6}$.

Example 4.5 For the problem in Example 4.1, we wish to construct a 95% confidence interval for the regression line so that the expectation of the square of the maximum half-width of the confidence interval is not larger than $\delta = \sigma$. From the data from Wageningen, we use as prior information $\overline{x} \approx 40$; $\overline{y} \approx 170$; $s_y^2 \approx 72.5$; $s_x^2 \approx 3.7$. When we assume for the regressor shoe sizes between 34 and 46 as realistic, then we obtain from (4.8) $n \approx \left\lceil t^2(n-2;\ 0.975)\left[1 + \frac{(x-\overline{x})^2}{3.7}\right]\right\rceil$.

This is maximal for either $x = x_{\max} = 46$ or $x = x_{\min} = 34$ and becomes $n \approx \left\lceil t^2(n-2;\ 0.975)\left[1 + \frac{6^2}{3.7}\right]\right\rceil$.
The solution is $n = 44$, and this is the size we should choose. When no prior information about σ_x is known, we estimate it by $\sigma_x = \frac{x_{\max}-x_{\min}}{6} = \frac{46-34}{6} = 2$. This leads to $n \approx \left\lceil t^2(n-2;\ 0.975)\left[1 + \frac{(6)^2}{4}\right]\right\rceil = \lceil 10t^2(n-2,\ 0.975)\rceil$. Here the solution is $n = 41$.

```
> size_n.regII.ci_exp(mx = 40, my = 170, sx2 = 3.7, sy2 = 72.5,
+       xmin = 34, xmax = 46, alpha = 0.05)
```

```
[1] 44

> size_n.regII.ci_exp(mx = 40, my = 170, sy2 = 72.5, xmin = 34,
+      xmax = 46, alpha = 0.05)

[1] 41
```

4.3 Hypothesis testing

We give the sample sizes for testing hypotheses for correlation coefficients and the slope of the regression line.

4.3.1 Comparing the correlation coefficient with a constant

We wish to test $H_0 : \rho = \rho_0$: against one of the following alternative hypotheses:

a) $H_A : \rho > \rho_0$ (one-sided alternative)

b) $H_A : \rho < \rho_0$ (one-sided alternative)

c) $H_A : \rho \neq \rho_0$ (two-sided alternative)

While an exact test exists for $\rho_0 = 0$ which is based on a test-statistic with a known distribution; this is not the case for $\rho \neq 0$.
The respective test-statistic $\boldsymbol{u}$ is for the general case only asymptotically normally distributed. The test needs to transform ρ_0 as well as r as follows: $z_0 = \frac{1}{2}\ln\frac{1+\rho_0}{1-\rho_0} - \frac{\rho_0}{2(n-1)}$ and $\boldsymbol{z} = \frac{1}{2}\ln\frac{1+\boldsymbol{r}}{1-\boldsymbol{r}}$. Then $\boldsymbol{u} = (\boldsymbol{z} - z_0)\sqrt{n-3}$ is the test-statistic which is asymptotically standard normally distributed. Often the term $-\frac{\rho_0}{2(n-1)}$ in z_0 is neglected.
We reject H_0 if

a) $u > u(1-\alpha)$

b) $u < -u(1-\alpha)$

c) $|u| > u\left(1 - \frac{\alpha}{2}\right)$

The minimum sample size for a risk of the second kind β if

a) $\frac{1}{2}\ln\frac{1+\rho}{1-\rho} - \frac{1}{2}\ln\frac{1+\rho_0}{1-\rho_0} \geq \delta$

b) $\frac{1}{2}\ln\frac{1+\rho}{1-\rho} - \frac{1}{2}\ln\frac{1+\rho_0}{1-\rho_0} \le \delta$

c) $|\frac{1}{2}\ln\frac{1+\rho}{1-\rho} - \frac{1}{2}\ln\frac{1+\rho_0}{1-\rho_0}| \ge \delta$

is approximately given by

$$n = \left\lceil \frac{(u_P + u_{1-\beta})^2}{\delta^2} \right\rceil + 3 \tag{4.9}$$

with $P = 1 - \alpha$ in the cases a and b and $P = 1 - \alpha/2$ in case c.

Example 4.6 For the problem in Example 4.1, we wish to test the hypothesis that $\rho = 0.6$ against $\rho > 0.6$ so that the risk of the first kind is $\alpha = 0.05$ and of the second kind is not larger than 0.2 as long as $\rho - 0.6 < 0.1$. From (4.9) we obtain $\left\lceil \frac{\{1.6449+0.8416\}^2}{0.1^2} \right\rceil + 3 = 622$ and have to ask 622 students for their data.

```
> size_n.regII.test_rho(side = "one", alpha = 0.05, beta = 0.2,
+     delta = 0.1)

[1] 622
```

4.3.2 Comparing two correlation coefficients

Let $(\boldsymbol{x}_i, \boldsymbol{y}_i)(i = 1, 2)$ be a bivariate normally distributed random variable with expectation vector $\begin{pmatrix} \mu_x \\ \mu_y \end{pmatrix}$ and covariance matrix $\begin{pmatrix} \sigma_x^2 & \sigma_{xy} \\ \sigma_{xy} & \sigma_y^2 \end{pmatrix}$ and take a random sample $\{(\boldsymbol{x}_{i1}, \boldsymbol{y}_{i1}); (\boldsymbol{x}_{i2}, \boldsymbol{y}_{i2}); \ldots; (\boldsymbol{x}_{in_i} \boldsymbol{y}_{in_i})\}$ of sizes n_i from the distribution.
We want to test the hypothesis:
$H_0 : \rho_1 = \rho_2$ against the alternative hypothesis:
$H_A : \rho_1 \neq \rho_2$.
An approximate test (for $n_i > 50$) is based on the two sample correlation coefficients r_i $(i = 1, 2,)$, their transforms $z_i = \frac{1}{2}\ln\frac{1+\boldsymbol{r}_i}{1-\boldsymbol{r}_i}$ and the test-statistic

$$\boldsymbol{u} = (\boldsymbol{z}_1 - \boldsymbol{z}_2)\sqrt{\frac{(n_1 - 3)\cdot(n_2 - 3)}{(n_1 - 3) + (n_2 - 3)}}$$

H_0 is rejected if $|u| > u\left(1 - \frac{\alpha}{2}\right)$.
The precision requirement for the sample size determination is based on the transformed population correlation coefficients $z_i^* = \frac{1}{2}\ln\frac{1+\rho_i}{1-\rho_i}$; $i = 1, 2$ by fixing the second kind of risk to be below β as long as $|z_1^* - z_2^*| \ge \delta$. The minimum sizes are obtained for $n = n_1 = n_2$ as (Rasch et al., 2008)

$$n = \left\lceil \frac{2\left\{u\left(1-\frac{\alpha}{2}\right)+u(1-\beta)\right\}^2}{\delta^2} + 3 \right\rceil \qquad (4.10)$$

Example 4.7 For the problem in Example 4.1, we want to test the hypothesis that the correlation coefficient between shoe size and body height for female and male students is the same against a two-sided alternative so that the risk of the first kind is $\alpha = 0.05$ and that of the second kind is not larger than $\beta = 0.2$ as long as the two correlation coefficients differ at least by $\delta = 0.1$. From (4.10) we obtain

$$n = \left\lceil \frac{2\{1.96+0.8461\}^2}{0.01} + 3 \right\rceil = 1573$$

This means that measurements from 1573 male and 1573 female students are needed.

```
> size_n.regII.test_rho2(alpha = 0.05, beta = 0.2, delta = 0.1)
```

```
[1] 1573
```

4.3.3 Comparing the slope with a constant

Because $|\rho\frac{\sigma_y}{\sigma_x} - \rho_0\frac{\sigma_y}{\sigma_x}| = |\beta_1 - \beta_{10}| \geq \delta$, we can test the null hypothesis H_0 : $\beta_1 = \beta_{10}$ with the methods of Section 4.3.1 and determine the minimum sample size as described there.

4.3.4 Test of parallelism of two regression lines

We consider the situation of Section 4.3.1 with the two slopes $\beta_1 = \frac{\sigma_{1xy}}{\sigma_{1x}^2}$; $\beta_2 = \frac{\sigma_{2xy}}{\sigma_{2x}^2}$. We wish to test
$H_0 : \beta_1 = \beta_2$ against the alternative hypothesis:
$H_A : \beta_1 \neq \beta_2$.
Under the null hypothesis, the two regression lines are parallel. This is equivalent with the problem described in Section 4.3.2; in both cases the equality of the two covariances is tested. Therefore we can use the procedures described there.

4.4 Selection procedures

Let us assume that we have $a > 1$ bivariate or multivariate normal distributions with unknown mean vectors and unknown covariance matrices. Let θ_i; $i = 1, \ldots, a$ be the squared multiple correlation coefficient of the first variate on the set of the $k-1$ remaining variates in the i-th distribution; if $k = 2$, it is the squared product moment correlation coefficient. We wish to select the $0 < t < a$ distributions with the largest (squared) multiple correlation coefficients so that the probability of an incorrect selection is below a given bound $\beta(\theta)$ as long as the t largest coefficients are not in the indifference zone but in what we call the preference zone. Here $\theta^T = (\theta_1, \ldots, \theta_a) \in \Omega$ is the vector of the a coefficients θ_i; $i = 1, \ldots, a$. For more details see Rizvi and Solomon (1973). An exact solution for the minimum sample size (equal for all distributions) was given by Khursheed et al. (1976) for the k-variate normal distribution with any $k > 1$. The case of bivariate populations has been discussed by Kenneth (1977). For $k > 2$ the problem for selecting the largest multiple correlation coefficient is solved, and for $k = 2$ we obtain the special case of the squared product moment correlation coefficient. The analysis is again simple by selecting the distribution with the largest (squared) sample correlation coefficient.

We obtain the minimal sample size for selecting the t largest of $a > t$ correlation coefficients by drawing from each of the a distributions a sample $\begin{pmatrix} \boldsymbol{y}_{i11} & \boldsymbol{y}_{i21} & \cdots & \boldsymbol{y}_{ik1} \\ \boldsymbol{y}_{i12} & \boldsymbol{y}_{i22} & \cdots & \boldsymbol{y}_{ik2} \\ \cdots & & \cdots & \\ \boldsymbol{y}_{i1n} & \boldsymbol{y}_{i2n} & \cdots & \boldsymbol{y}_{ikn} \end{pmatrix}$ $(i = 1, \ldots, a)$ of equal size $n > k + 2$. The minimal sample size n is a function of a, t, k, β and γ, δ in the definition of the indifference zone and can be determined as follows. Let $\theta_{[a]} \geq \theta_{[a-1]} \geq \cdots \geq \theta_{[1]}$ be the ordered set of parameters of $\theta^T = (\theta_1, \ldots, \theta_a) \in \Omega$, in our case the squares of the multiple correlation coefficients.

Definition 4.1 We define the indifference zone as $\Omega^* = \Omega_1 \cap \Omega_2 \subset \Omega$ with $\Omega_1 = \{\theta \in \Omega : 1 - \theta_{[a-t]} \geq \delta(1 - \theta_{[a-t+1]})\}$; $\Omega_2 = \{\theta \in \Omega;\ \theta_{[a-t+1]}) \geq \gamma\theta_{[a-1]}\}$ and $\gamma > 1$ and $\delta > 1$.

To explain this definition we give an example.

Example 4.8 Let $k = 2$ and $t = 1$ and $\theta^T = (\theta_1, \ldots, \theta_a) = (\rho_1^2, \ldots, \rho_a^2) \in \Omega$; $\rho_{[a]}^2 = 0.7 = \theta_{[a]}; \rho_{[a-1]}^2 = 0.3 = \theta_{[a-1]}$ be the largest and second largest squared correlation coefficients, respectively. Then we obtain with $\gamma = 1.5$ and $\delta = 1.2$, $\Omega_1 = \{\theta \in \Omega : 0.7 \geq \theta \geq 1.2 \cdot 0.3 = 0.36\}$; $\Omega_2 = \{\theta \in \Omega : 0.7 \geq \theta \geq 1.5 \cdot 0.3 = 0.45\}$ as our indifference region the interval $\Omega^* = \Omega_1 \cap \Omega_2 = [0.45;\ 0.7]$. The preference zone is then the set of all $\theta_{[a]} > 0.7$

and/or $\theta_{[a-1]} < 0.45$. The region Ω^* is the upper left part in the $(\theta_{[a-1]}, \theta_{[a]})$ plane (see Figure 4.1).

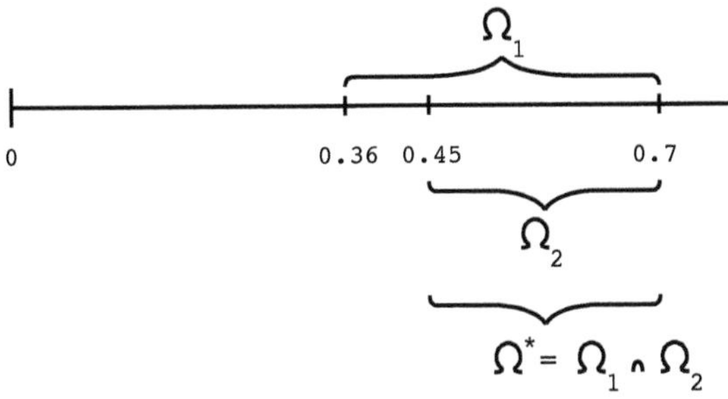

FIGURE 4.1
The indifference zone for Example 4.8 in the parameter space [0; 1]

Khursheed et al. (1976) have shown that the probability of an incorrect selection is maximised at the boundary of Ω^*. The worst case is thus on the boundary where we have to find the minimum sample size to fulfill the precision requirements.

For a given $\beta(\theta)$ we can calculate the minimal number of n fulfilling the conditions above.

Example 4.8 continued: With the values in Example 4.8 above, we calculate the minimum sample size for a probability of an incorrect selection not larger than 0.05; 0.01 and 0.005, respectively, for $k = 2$. We use our R-program for the calculations.

```
> size.prob.selection(a=4, k=2, alpha=0.05, beta
+=c(0.05,0.01,0.005),gamma=1.5, delta=1.2)
```

5

Sequential Designs

5.1 Introduction

Sequential design or sequential analysis means using the following statistical approach.

A random experiment consists of a sequence of not necessarily independent random variables $\boldsymbol{y}_1, \boldsymbol{y}_2, \ldots$ of an unknown size at the beginning of the experimentation. The statistical method includes two elements:

- a stopping rule, with two possibilities after the observation of n of the elements $\boldsymbol{y}_1, \boldsymbol{y}_2, \ldots, \boldsymbol{y}_n$ in the sequence (for each n): either continue experimentation by observing $\boldsymbol{y}_{n+1}$ or stop experimentation with $\boldsymbol{y}_1, \boldsymbol{y}_2, \ldots, \boldsymbol{y}_n$. The decision of the stopping rule usually depends on all experimental units observed so far.

- a decision rule with fixing an action after stopping experimentation.

The experimenter observes realisations of the sequence. Many problems can be solved as well sequentially as with a size fixed in advanced, as discussed in Chapters two through four. But there are cases where only a sequential solution exists.

Example 5.1 In a Bernoulli sequence with probability $p = P(\text{success}) = 1 - P$ (no success); $0 < p < 1$, we have to estimate $\frac{1}{p}$ unbiasedly, for a fixed sample size, such an estimator does not exist. In a sequential approach an unbiased estimator is the smallest $\boldsymbol{n}$ for which a success occurs.

Example 5.2 Let $\boldsymbol{y}$ be normally distributed with expectation μ and standard deviation σ, both unknown. We wish to construct a confidence interval with a predetermined length L and given confidence coefficient $1 - \alpha$. This is impossible with a fixed sample size. We have seen in Section 2.2.1.2 that such an interval has a random length $2\boldsymbol{H}$.

But Stein (1945) derived a two-stage sequential method which can solve the problem. Even if there are only two stages, the size of the experiment is a random variable. The procedure is as follows:

Stage 1: Take a pre-sample $\boldsymbol{y}_1, \boldsymbol{y}_2, \ldots, \boldsymbol{y}_{n_0}$ of size $n_0 > 1$ and estimate expectation and variance by

$$\overline{\boldsymbol{y}}_0 = \frac{1}{n_0}\sum_{i=1}^{n_0} \boldsymbol{y}_i \text{ and } \boldsymbol{s}_0^2 = \frac{1}{n_0-1}\sum_{i=1}^{n_0}(\boldsymbol{y}_i - \overline{\boldsymbol{y}}_0)^2$$

Determine the smallest integer $\boldsymbol{n}$ so that (with z defined later) $\boldsymbol{n} > n_0; \boldsymbol{n} > \frac{\boldsymbol{s}_0^2}{z}$ Now select $\boldsymbol{n} + 1 - n_0$ real numbers fulfilling the conditions:

$$a_0 + \sum_{k=n_0+1}^{\boldsymbol{n}} a_k = 1; \; \frac{1}{\boldsymbol{n}_0} a_0^2 + \sum_{k=n_0+1}^{\boldsymbol{n}} a_k^2 = z, \text{ respectively.}$$

Remark: Simulations done by the authors show that it is reasonable to choose the size n_0 of the sample in the first stage between 10 and 30. The real numbers $a_0, a_{n_0+1}, \ldots, a_n$ are the weights of the linear combination of the mean of the pre-sample and the observations in the second stage used in the confidence interval. Stein showed that just these coefficients guarantee the value fixed for the confidence coefficient.

Stage 2: Take a further sample $\boldsymbol{y}_{n_0+1}, \boldsymbol{y}_{n_0+2}, \ldots, \boldsymbol{y}_n$ of size $n - n_0$ and calculate $\boldsymbol{l}_n = a_0\overline{\boldsymbol{y}}_0 + \sum_{k=n_0+1}^{n} a_k \boldsymbol{y}_k$. If we wish to construct an interval with length L, choose z in Stage 1 equal to $\frac{L}{2} = \sqrt{z} \cdot t\left(n_0 - 1; 1 - \frac{\alpha}{2}\right)$. The $(1-\alpha)$-confidence interval with fixed length L is now given by $\left[\boldsymbol{l}_n - \frac{L}{2}; \boldsymbol{l}_n + \frac{L}{2}\right]$. The proof is given by Stein (1945). An approximation of the method in Stein's paper was used in Section 2.1.1 in the third step of Procedure I.

The two R-functions `add_size.mean.two_stage` and `confint.mean.two_stage` serve the computation of the minimum required additional sample size and the resulting confidence interval of fixed length L, respectively. We demonstrate their use with a small numerical example involving random samples from normal distributions.

Suppose we have a pre-sample named `pres`, of size $n_0 = 15$, and we want to find a confidence interval of length $L = 0.9$.

```
> pres = rnorm(15, 7.6, 0.6)
> L = 0.9
> alpha = 0.05
> pres

 [1] 7.556632 6.886941 8.391619 8.103063 7.909580 7.617286
 7.468438 9.259466
 [9] 7.104280 7.286309 8.782789 7.487819 7.768615 8.344544
 7.067557

> add_size.mean.two_stage(L, alpha, pres)

$`Significance level`
```

```
[1] 0.05

$`Length of confidence interval`
[1] 0.9

$`Length of presample`
[1] 15

$`Standard deviation of presample`
[1] 0.6689689

$`Number of additional observations`
[1] 1
```

Taking the requested $n - n_0 = 1$ additional observations, we arrive at the following confidence interval

```
> v = add_size.mean.two_stage(L, alpha, pres)$
+ "Number of additional observations"
> confint.mean.two_stage(L, alpha, pres, rnorm(v, 7.6, 0.6))

$`Significance level`
[1] 0.05

$`Length of confidence interval`
[1] 0.9

$`Length of presample`
[1] 15

$`Number of additional observations`
[1] 1

$`Total number of observations`
[1] 16

$`confidence interval`
[1] 7.266733 8.166733
```

Example 5.3 The secretary problem

Gardner (1960) formulated the problem: What is the optimal stopping rule in selecting the best secretary out of n candidates if a rejected candidate cannot be brought back? It is assumed that the candidates appear in a random sequence one after the other. We can determine for the ith candidate its rank relative to the $i - 1$ predecessors (see also Flood, 1958).

The solution was found by Dynkin (1963a,b) who tells us to reject the first $k = \lceil \frac{n}{e} \rceil$ candidates (e is the base of the natural logarithm) and to select the next relative best one compared with the candidates already seen.

We have seen that sequential designs can be used for both point and interval estimation as well as ranking and selection problems. For further selection problems we refer to Bechhofer, Kiefer and Sobel (1968) and Ghosh and Sen (1991). Hypothesis testing has not been discussed until now but will be dealt with in the following sections.

5.2 Wald's sequential likelihood ratio test (SLRT) for one-parametric exponential families

The first paper on sequential analysis (or design) was written during World War II by Abraham Wald (1943). The first easily accessible publication was Wald (1944) and the first book was Wald (1947). Wald developed the so-called probability ratio test, now called the likelihood ratio test, a term we will use in future.

All results are given—as usual in this book—without proof. The proofs in this section are given in Wald (1947).

Parts of the text below are written using an unpublished manuscript by Schneider (1994), whom we thank for permission to use it. Let us assume that we have a sequence $\boldsymbol{y}_1, \boldsymbol{y}_2, \ldots$ of i.i.d. random variables distributed like $\boldsymbol{y}$ with likelihood function $f(y;\theta)$. We wish to test the null hypothesis

$H_0 : \theta = \theta_0[f(y;\theta) = f(y;\theta_0)]$ against the alternative

$H_A : \theta = \theta_1[f(y;\theta) = f(y;\theta_1)]; \theta_0 \neq \theta_1; \theta \in \Omega \subset \mathbb{R}$.

The likelihood ratio after n observations is then given by

$$\lambda_n = \prod_{i=1}^{n} \frac{f(y_i;\theta)}{f(y_i;\theta_0)}; n > 1 \tag{5.1}$$

Wald proposed a test by defining two real constants $B < A$ and to accept H_0 if $\lambda_n \leq B$, to accept H_A if $\lambda \geq A$ and to continue experimentation if $B < \lambda_n < A$.

The values A and B are to be fixed in such a way that pre-given risks are guaranteed.

Two questions arise:

— How to choose A, B?

— What is the average size $E(\boldsymbol{n}|\theta)$ of the sequence $\boldsymbol{y}_1, \boldsymbol{y}_2 \dots$ of random size?

Wald used an approximation for the values A,B. If the nominal risks of the first and second kind are given by α_{nom} and β_{nom}, respectively, then the actual risks fulfill (following Wald) the relations

$$\alpha_{act} \leq \frac{1}{A} = \alpha_{nom}; \beta_{act} \leq B = \beta_{nom} \tag{5.2}$$

which means that a test with A and B given by the equalities in (5.2) is conservative. This leads us to the Wald bounds A,B defined by

$$A = \frac{1-\beta}{\alpha}; B = \frac{\beta}{1-\alpha} \tag{5.3}$$

and thus defines his **sequential likelihood ratio test (SLRT)**. The SLRT corresponds with a random walk between two parallel lines.

In all sequential procedures data collecting must be stopped as soon as the continuation region is left. Otherwise the risks fixed in advance cannot be guaranteed. This means that after each observation the data collected so far must be analysed. In group sequential procedures—not discussed in this book—special techniques have been developed to solve such problems. If the continuation region is left and sampling is continued, we have a so-called overshooting what leads to the problems mentioned above.

Example 5.4 Let us assume that the nominal risks of the first and second kind are 0.05 and 0.1, respectively. Then we obtain from (5.3) $A = 18$ and $B = 0.1052632$. Thus we have to continue as long as $0.1052632 < \lambda_n < 18$. In a coordinate system with n on the abscissa and λ_n on the ordinate, the continuing region lies between **two** parallel straight lines, and the process may not end for finite n.

But Wald could show that this sequential likelihood ratio test ends with probability 1, but no upper bound for n is known.

The approximate power function of the SLRT is given by

$$\pi(\theta) \approx \frac{1 - \left(\frac{\beta}{1-\alpha}\right)^h}{\left(\frac{1-\beta}{\alpha}\right)^h - \left(\frac{\beta}{1-\alpha}\right)^h} \text{ for } h \neq 0 \tag{5.4}$$

In (5.4) h is a function (whose existence was shown by Wald (1947)) defined uniquely by

$$\int \left(\frac{f(y,\theta_1)}{f(y,\theta_0)}\right)^{h(\theta)} \cdot f(y,\theta)dy = 1$$

especially in the discrete case by

$$\sum_y \left(\frac{f(y,\theta_1)}{f(y,\theta_0)}\right)^{h(\theta)} \cdot f(y,\theta) = 1$$

Wald could further show that by the sequential likelihood ratio test amongst all sequential tests with no higher risks than α_{nom} and β_{nom}, the average sample size (also called average sample number ASN) is minimised as long as one of the two hypotheses above is true.

If the expectations of all random variables occurring in the SLRT (especially $E(\boldsymbol{n})$) exist, then Wald obtained a formula for $E(\boldsymbol{n})$. To show it we need some new symbols.

$$\boldsymbol{z} = \ln \frac{f(\boldsymbol{y},\theta_1)}{f(\boldsymbol{y},\theta_0)};\ \boldsymbol{z}_i = \ln \frac{f(\boldsymbol{y}_i,\theta_1)}{f(\boldsymbol{y}_i,\theta_0)} \tag{5.5}$$

so that $\ln \boldsymbol{\lambda}_n = \Sigma \boldsymbol{z}_i$.

If $E(|\boldsymbol{z}|) < \infty$, Wald showed that

$$E(\boldsymbol{n}|\theta) \approx \frac{\pi(\theta)\ln A + [1-\pi(\theta)]\ln B}{E(\boldsymbol{z}|\theta)},\ \text{if } E(\boldsymbol{z}|\theta) \neq 0 \tag{5.6}$$

Formulae (5.4) and (5.6) are exact: the experiment ends if one of the equality signs in (5.2) occurs in the last step. This is seldom the case for discrete distributions; often we have to stop if the inequality sign is relevant. In these cases we have gone too far, and this is called an "overshoot".

An approximate formula for $E(\boldsymbol{n})$ is given by (Wijsman, 1991)

$$E(\boldsymbol{n}|\theta_0) \approx \frac{1}{E(\boldsymbol{z}|\theta_0)}\left[\frac{A-1}{A-B}\ln B + \frac{1-B}{A-B}\ln A\right] \tag{5.7}$$

and in general

$$E(\boldsymbol{n}|\theta) \approx \frac{1}{E(\boldsymbol{z}|\theta_0)}\left[\frac{(A-1)\cdot B}{A-B}\ln B + \frac{A\cdot(1-B)}{A-B}\ln A\right] \tag{5.8}$$

If there is no overshoot, the equations are exact. In the case of discrete distributions from an exponential family we often have overshoot. For this case, Young (1994) derived exact values for the power and the ASN.

We repeat here the definition of an exponential family.

Definition 5.1 The distribution of a random variable $\boldsymbol{y}$ with parameter vector $\theta^T = (\theta_1, \theta_2, \ldots, \theta_p)$ belongs to a k-parametric exponential family if its likelihood function can be written in the form

$$f(y, \theta) = h(y)\ exp\left(\sum_{i=1}^{k} \eta_i(\theta) \cdot T_i(y) - B(\theta)\right)$$

where η_i and B are real functions and B does not depend on y. The function $h(y)$ is non-negative and does not depend on θ. The exponential family is in canonical form with the natural parameters η_i if it is written as

$$f(y, \eta) = h(y) exp(\sum_{i=1}^{k} \eta_i \cdot T_i(y) - A(\eta)) \text{ with } \eta^T = (\eta_1, \eta_2, \ldots, \eta_k)$$

We know that in exponential families the first derivative of $A(\theta)$ at $\theta = 0$ gives us the expectation, and the second derivative at $\theta = 0$ gives us the variance of $\boldsymbol{y}$.

Examples of exponential family members are the normal distribution and the Bernoulli, binomial, negative binomial, geometric and Poisson distributions. If the distribution is continuous, Wald (1947) showed that for a θ^* for which $E(Z|\theta^*) = 0$, it follows that in (5.4) $h(\theta^*) = 0$, and in this case we get

$$E(\boldsymbol{n}|\theta^*) \approx \frac{|\ln A| \cdot |\ln B|}{E(\boldsymbol{z}^2|\theta^*)} \text{ if } h(\theta^*) = 0 \tag{5.9}$$

Example 5.5 If θ in Definition 5.1 is a real-valued parameter, then we have $p = k = 1$ and $f(y, \theta) = h(y)e^{y\eta - A(\eta)}$. Remember that we have to test the null hypothesis $H_0 : \theta = \theta_0 (\eta = \eta_0)$ against the alternative $H_A : \theta = \theta_1; (\eta = \eta_1)$ $\theta_0 < \theta_1; \theta \in \Omega \subset \mathbb{R}$ with $\eta_i = \eta(\theta_i); i = 0, 1$. Rename the hypotheses if $\theta_0 > \theta_1$ or reverse the inequalities. The z_i defined above are (realisation-wise) $z_i = (\eta_1 - \eta_0)y_i - [A(\eta_1) - A(\eta_0)]$ and sampling will continue as long as $\ln B < (\eta_1 - \eta_0)\Sigma y_i - n[A(\eta_1) - A(\eta_0)] < \ln A$ or as long as (for $\eta_1 - \eta_0 > 0$)

$$\frac{\ln B + n[A(\eta_1) - A(\eta_0)]}{(\eta_1 - \eta_0)} < \Sigma y_i < \frac{\ln A + n[A(\eta_1) - A(\eta_0)]}{(\eta_1 - \eta_0)}$$

Example 5.6 We consider the Bernoulli distribution (when $\theta = p$ is the probability that a certain event E occurs) with likelihood function $f(\boldsymbol{y}, p) = p^y(1-p)-y; 0 < p < 1; y = 0, 1$. These distributions are an exponential family

$f(\boldsymbol{y},\theta) = e^{y[\ln\frac{\theta}{1-\theta}]+\ln(1-\theta)}$ $0<\theta<1; y=0,1$ with $B(\theta) = -\ln(1-\theta); \eta = \ln\frac{\theta}{1-\theta}; \theta = \frac{e^\eta}{1+e^\eta}; A[\eta] = \ln(1+e^\eta)$. We wish to test the hypothesis: $H_0: p = p_0$ against $H_A: p = p_1$. We obtain

$$\begin{aligned}\boldsymbol{z} &= \ln\frac{f(\boldsymbol{y},p_1)}{f(\boldsymbol{y},p_0)} = \boldsymbol{y}\left[\ln\frac{p_1}{1-p_1} - \ln\frac{p_0}{1-p_0}\right] + [\ln(1-p_1) - \ln(1-p_0)] \\ &= \boldsymbol{y}\left[\ln\frac{p_1}{p_0} - \ln\frac{1-p_1}{1-p_0}\right] + \left[\ln\frac{1-p_1}{1-p_0}\right]\end{aligned}$$

and $\ln\boldsymbol{\lambda}_n = \sum_{i=1}^{n}\boldsymbol{y}_i\left[\ln\frac{p_1}{p_0} - \ln\frac{1-p_1}{1-p_0}\right] + n\cdot\left[\ln\frac{1-p_1}{1-p_0}\right]$

Therefore we obtain for any $0<p<1$: $E(\boldsymbol{z}|p) = p\ln\frac{p_1}{p_0} + (1-p)\ln\frac{1-p_1}{1-p_0}$ observing that $E(\boldsymbol{y}) = p$.

The decision rule now means that sampling continues as long as

$$\frac{\ln B + n\ln\frac{1-p_0}{1-p_1}}{\ln\frac{p_1(1-p_0)}{p_0(1-p_1)}} < \sum_{i=1}^{n} y_i < \frac{\ln A + n\ln\frac{1-p_0}{1-p_1}}{\ln\frac{p_1(1-p_0)}{p_0(1-p_1)}}$$

We can write the continuation region as $a_0 + bn < r < a_1 + bn$ with $r = \sum_{i=1}^{n} y_i$ as the number of the occurrences of the event E,

$$b = \frac{\ln\frac{1-p_0}{1-p_1}}{\ln\frac{p_1(1-p_0)}{p_0(1-p_1)}}, a_0 = \frac{\ln B}{\ln\frac{p_1(1-p_0)}{p_0(1-p_1)}} = \frac{\ln\frac{\beta}{1-\alpha}}{\ln\frac{p_1(1-p_0)}{p_0(1-p_1)}}; a_1 = \frac{\ln A}{\ln\frac{p_1(1-p_0)}{p_0(1-p_1)}} = \frac{\ln\frac{1-\beta}{\alpha}}{\ln\frac{p_0(1-p_1)}{p_0(1-p_1)}}$$

In this form it is easy to see that two straight lines are the borders of the continuation region.

For the power function in (5.4) h is given as a function of p by

$$p = \frac{1-\left(\frac{1-p_1}{1-p_0}\right)^h}{\left(\frac{p_1}{p_0}\right)^h - \left(\frac{1-p_1}{1-p_0}\right)^h}, \text{ if } h \neq 0$$

If $h = 0$, we have $p = b$ and $\pi(p) = \frac{|a_0|}{a_1+|a_0|}$. For the expected sample size we obtain from (5.8)

$$E(\boldsymbol{n}|p) \approx \frac{\pi(p)\ln A + (1-\pi(p))\ln B}{p\ln\frac{p_1}{p_0} + (1-p)\ln\left(\frac{1-p_1}{1-p_0}\right)} \text{ and } E(\boldsymbol{n}|p) \approx \frac{|\ln A|\cdot|\ln B|}{\ln\frac{p_1}{p_0}\cdot\ln\left(\frac{1-p_1}{1-p_0}\right)}$$

If we put $p_1 = 0.6, p_0 = 0.5$, it follows $\ln\boldsymbol{\lambda}_n = 0.40547\sum_{i=1}^{n}\boldsymbol{y}_i - 0.2231\cdot n$.

The decision rule now means that sampling continues as long as

$$\frac{\ln B + 0.2231n}{0.40547} < \sum_{i=1}^{n} y_i < \frac{\ln A + 0.2231n}{0.40547}$$

For $\alpha = 0.05$ and $\beta = 0.1$ ($A = 18$ and $B = 0.10526$) sampling continues as long as $-2.2513 + 0.2231n < 0.40547 \sum_{i=1}^{n} y_i < 2.8904 + 0.2231n$.

Further we have $b = \frac{0.22314}{0.40547} = 0.55; a_0 = \frac{-2.2513}{0.40547} = -5.55232; a_1 = \frac{2.8904}{0.40547} = 7.1284$.

The power function is given by (5.4) where h and p are linked through $p = \frac{1-0.8^h}{1.2^h - 0.8^h}$. The expected size is then

$$E(\boldsymbol{n}|p) \approx \frac{2.8904\pi(p) - 2.2513(1-\pi(p))}{0.1823p + 0.2231(1-p)} \quad \text{for } h \neq 0$$

We can show the procedure as a graph between two parallel lines in the $\left(n; \sum_{i=1}^{n} y_i\right)$-plane with slope 0.5503 and intercepts -5.55 and 7.13, respectively.

In one-parametric exponential families we have seen that sampling will continue as long as $\ln B < (\eta_1 - \eta_0)\Sigma y_i - nA_{10} < \ln A$ or, if $\eta_1 - \eta_0 > 0$, as long as

$$b_L^n = \frac{\ln B + nA_{10}}{\eta_1 - \eta_0} < \sum y_i < \frac{\ln A + nA_{10}}{\eta_1 - \eta_0} = b_u^n \tag{5.10}$$

with lower and upper bounds b_L^n and b_u^n, respectively, and $A_{10} = A(\eta_1) - A(\eta_0)$. We assume w.l.o.g. the case $\eta_1 - \eta_0 > 0$.

In the discrete case, we have a step function as the random walk between the two parallel lines, and we cannot guarantee that in the last step we will meet the Wald bounds exactly—some overshoot may occur. In such cases an algorithm given by Young (1994) may apply; it is described below.

Assume that the test has not terminated prior to the n-th observation. The probability of observing a particular value t_n of the random variable $\boldsymbol{t}_n = \sum \boldsymbol{y}_i$ after observing n sample units is the sum of the probability sequence for which $b_l^i \leq \boldsymbol{t}_i \leq b_u^i; i = 1, 2, \ldots, n-1$ and $\boldsymbol{t}_n = t_n$. We can write this probability as

$$\begin{aligned} P(\boldsymbol{t}_n = t) &= \sum_{j=b_l^{n-1}}^{b_u^{n-1}} P(\boldsymbol{t}_n = t_n | \boldsymbol{t}_{n-1} = j) \cdot P(\boldsymbol{t}_{n-1} = j) \\ &= \sum_{j=b_l^{n-1}}^{b_u^{n-1}} f(t_n - j; \theta) \cdot P(\boldsymbol{t}_{n-1} = j). \end{aligned}$$

Starting with $P(\boldsymbol{t}_0 = 0) = 1$ we can calculate all probabilities by recursion. For a fixed n the probability of accepting the alternative hypothesis at the n-th observation is

$$P(\boldsymbol{t}_n > b_u^n) = \sum_{j=b_l^{n-1}}^{b_u^{n-1}} \sum_{k=b_l^n-j+1}^{\infty} f(k;\theta) \cdot P(\boldsymbol{t}_{n-1} = j) =$$

$$\sum_{j=b_l^{n-1}}^{b_u^{n-1}} [F(b_u^n - j - 1;\theta)] \cdot P(\boldsymbol{t}_{n-1} = j) \tag{5.11}$$

where F is the distribution function.

For a fixed n the probability of accepting the null hypothesis at the n-th observation is

$$\begin{aligned} P(\boldsymbol{t}_n > b_l^n) &= \sum_{j=b_l^{n-1}}^{b_u^{n-1}} \sum^{k=b_l^n-j-1} f(k;\theta) \cdot P(\boldsymbol{t}_{n-1} = j) \\ &= \sum_{j=b_l^{n-1}}^{b_u^{n-1}} [F(b_u^n - j - 1;\theta)] \cdot P(\boldsymbol{t}_{n-1} = j) \end{aligned} \tag{5.12}$$

The power function is given by $\sum_{i=1}^{n} P(\boldsymbol{t}_i < b_l^i)$ if the procedure terminates at step n, and the probability for that event equals $P(\boldsymbol{t}_n < b_l^n) + P(\boldsymbol{t}_n > b_u^n)$.

A computer program is described in Young (1994; p. 631).

Table 5.1 shows parameters of some discrete distributions.

TABLE 5.1
Parameters and Link Functions of Some Discrete Distributions.

Distribution	Likelihood Function	$\eta(\theta)$	$A(\theta)$
Bernoulli	$\theta^y(1-\theta)^{1-y}$	$\eta = \ln\frac{\theta}{1-\theta}$	$\ln(1-e^{\eta})$
Binomial	$\theta^y(1-\theta)^{n-y}$	$\eta = \ln\frac{\theta}{1-\theta}$	$n \cdot \ln(1-e^{\eta})$
Multinomial	$\frac{n!}{y_1! \cdot \ldots \cdot y_k!} \prod_{i=1}^{k} \theta_i^y$	$\eta_j = \ln\frac{\theta_j}{\theta_k}$; $j = 1, \ldots, k-1$	$n \cdot \ln\left(1 + \sum_{j=1}^{k-1} e^{\eta_j}\right)$
Poisson	$\frac{\theta^y \cdot e^{-\theta}}{y!}$	$\eta = \ln\theta$	e^{η}

In the R-program the approach is general. We program for $\eta(\theta)$ and $A(\theta)$, so if the user has a distribution from an exponential other than our examples (which will be demonstrated below), he can insert the corresponding functions.

In the examples below we use one-sided hypotheses and always $\alpha = 0.05; \beta = 0.1$ and $\delta = \theta_1 - \theta_0 = 0.1$.

Example 5.7 Normal distribution with known variance

If $\boldsymbol{y}$ is $N(\mu; \sigma^2)$-distributed with known σ^2, we have $\boldsymbol{z} = \ln \frac{1}{2\sigma^2}[2\boldsymbol{y}(\mu_1 - \mu_0) + \mu_0^2 - \mu_1^2]$ and $E(\boldsymbol{z}|\mu) = \ln \frac{1}{2\sigma^2}[2\mu(\mu_1 - \mu_0) + \mu_0^2 - \mu_1^2]$ when we wish to test

$H_0 : \mu = \mu_0$ against the alternative

$H_A : \mu = \mu_1; \mu_0 \neq \mu_1; \mu \in \mathbb{R}^1$

Especially if we put $\mu_0 - \mu_1 = \sigma$, we obtain for $h(\theta)$ in (5.4)

$\int \left(e^{-\frac{2y\sigma - \sigma^2}{2\sigma^2}}\right)^{h(\mu)} \cdot \frac{1}{\sigma\sqrt{2\pi}} e^{-\frac{1}{2\sigma^2}(y-\mu)^2} dy = 1$ as well as $E(\boldsymbol{n}|\mu)$ and $\pi(\theta)$ by our R-program as a function of μ.

5.3 Test about means for unknown variances

The case of normal distributions with unknown variance is of much more practical importance. But in this case we have a two-parametric exponential family and must adapt the procedure described in Section 5.2 to this case.

If a distribution has $k > 1$ parameters like the normal distribution with unknown variance, then we have $k - 1$ nuisance parameters, and the results of Section 5.2 cannot be applied straightforwardly.

We now have an unknown parameter vector $\theta^T = (\theta_1, \ldots, \theta_k); k \geq 2$ of an exponential family and wish to test

$H_0 : \phi(\theta) \leq \phi_0; \phi \in \mathbb{R}^1$ against

$H_A : \phi(\theta) \geq \phi_1; \phi \in \mathbb{R}^1$

or some two-sided pair of hypotheses like

$H_0 : \phi(\theta) = \phi_0; \phi \in \mathbb{R}^1$ against

$H_A : \phi(\theta) \neq \phi_1; \phi \in \mathbb{R}^1$.

In this book we concentrate on the univariate normal distribution; the corresponding SLRT is called the sequential t-test.

5.3.1 The sequential t-test

For the univariate normal distributions of a random variable $\boldsymbol{y}$, we have a two-parameter exponential family with parameter vector $\theta^T = (\mu; \sigma^2)$ and log-likelihood function $l(\mu;\sigma^2) = -\ln\sqrt{2\pi} - \ln\sigma - \frac{1}{2}(y-\mu)^2/\sigma^2$. If we put $\phi(\theta) = \frac{\mu}{\sigma}$, we may test

$H_0 : \frac{\mu}{\sigma} \le \phi_0$ against
$H_A : \frac{\mu}{\sigma} \ge \phi_1$ or
$H_0 : \frac{\mu}{\sigma} = \phi_0$ against
$H_A : \frac{\mu}{\sigma} \ne \phi_1$.

When we would replace the nuisance parameter by its estimator from the last observed stage (as we did in Chapter two for fixed sample size), λ_n in (5.1) no longer can be treated as a likelihood ratio.

We solve the problem as follows. We look for a sequence $\boldsymbol{u}_1 = u_1(\boldsymbol{y}_1)$; $\boldsymbol{u}_2 = u_2(\boldsymbol{y}_1;\ \boldsymbol{y}_2);\ldots$ such that for each $n > 1$ the joint likelihood function $f_u^{(n)}(u_1, u_2, \ldots, u_n; \phi)$ of $(\boldsymbol{u}_1, \boldsymbol{u}_2, \ldots, \boldsymbol{u}_n)$ depends on θ only through $\phi(\theta)$. Then we can apply the theory of Section 5.2 with

$$\lambda_n^* = \prod_{i=1}^{n} \frac{f_u^{(n)}(u_i;\phi_1)}{f_u^{(n)}(u_i;\phi_0)}; n > 1 \tag{5.13}$$

in place of $\lambda_n^* = \prod_{i=1}^{n} \frac{f(y_i;\theta_1)}{f(y_i;\theta_0)}; n > 1$ in (5.1).

How to choose the sequence $\boldsymbol{u}_1 = u_1(\boldsymbol{y}_1); \boldsymbol{u}_2(\boldsymbol{y}_1;\boldsymbol{y}_2)$ will be explained in the next sections.

Lehmann (1959) formulated the principle of invariant tests. If we multiply μ and σ by a positive real number c, the two hypotheses $H_0 : \frac{\mu}{\sigma} \le \phi_0$ and $H_A : \frac{\mu}{\sigma} \ge \phi_1$ and also the two-sided version remain unchanged; they are invariant against affine transformations. Moreover the transformed random variables $\boldsymbol{y}_i^* = c\boldsymbol{y}_i$ are normally distributed with expectation $c\mu$ and standard deviation $c\sigma$. Thus the family of possible distributions of $\boldsymbol{y}_1, \boldsymbol{y}_2, \ldots, \boldsymbol{y}_n$ is for each $n \ge 1$ the same as the family of possible distributions of $c\boldsymbol{y}_1, c\boldsymbol{y}_2, \ldots, c\boldsymbol{y}_n$. This means that the hypotheses as well as the family of distributions is invariant against affine transformations.

The sequential t-test can now, following Eisenberg and Ghosh (1991), be performed in the following way.

Specialise in (5.13) for our normal case to

$$\lambda_n^* = e^{\frac{1}{2}(n-v_n^2)(\phi_n^2-\phi_0^2)} \frac{\int_0^\infty t^{n-1} e^{\frac{1}{2}(t-v_n\phi_1)^2} dt}{\int_0^\infty t^{n-1} e^{\frac{1}{2}(t-v_n\phi_0)^2} dt} \tag{5.14}$$

and solve the equations (uniquely) $\lambda_n^* = \frac{\beta}{1-\alpha}$ and $\lambda_n^* = \frac{1-\beta}{\alpha}$ and call the solutions v_l^n and v_u^n, respectively. Calculate the value of

$$v_n = \sum_{i=1}^{n} y_i / \sqrt{\sum_{i=1}^{n} y_i^2} \tag{5.15}$$

and continue sampling as long as $v_l^n < v_n < v_u^n$ and accept H_0 if $v_n \leq v_l^n$ and reject H_0 if $v_n \geq v_u^n$.

The name "sequential t-test" stems from the fact that there is a connection between v_n and the test-statistic t in (2.31) for $\mu_0 = 0$.

5.3.2 Approximation of the likelihood function for the construction of an approximate t-test

Using simple functions $\boldsymbol{z}$ and $\boldsymbol{v}$ of the likelihood function allows us to construct relatively simple sequential tests and especially closed tests with a maximum number of observations.

Let the sequence $\boldsymbol{y}_1, \boldsymbol{y}_2, \ldots$ of i.i.d. random variables be distributed like $\boldsymbol{y}$ with likelihood function $f(y;\theta)$. We expand $l(y;\theta) = \ln f(y;\theta)$ in a Taylor series at $\theta = 0$:

$$l(y;\theta) = l(y;0) + \theta \cdot l_\theta(y;0) + \frac{1}{2}\theta^2 l_{\theta\theta}(y;0) + O(\theta^3) \tag{5.16}$$

In (5.16) we used the abbreviations

$$z = l_\theta(y;0) = \frac{\partial \ln(y;\theta)}{\partial \theta}|_{\theta=0} \tag{5.17}$$

and

$$-v = l_{\theta\theta}(y;0) = \frac{\partial^2 \ln(y;\theta)}{\partial \theta^2}|_{\theta=0} \tag{5.18}$$

By neglecting the term $O(\theta^3)$ we get the approximation

$$l(y;\theta) = const + \theta \cdot z - \frac{1}{2}\theta^2 v \tag{5.19}$$

If the likelihood function also depends, besides on θ, on a vector of nuisance parameters $\tau^T = (\tau_1, \ldots, \tau_k)$, Whitehead (1997) proposed to replace this vector by its maximum likelihood estimate.

Now the likelihood function is given by $f(y; \theta, \tau)$, and we have $l(y; \theta, \tau) = \ln f(y; \theta, \tau)$.

For the maximum likelihood estimate of τ we write $\tilde{\tau}(\theta)$ and for $\theta = 0$ we write $\tilde{\tau} = \tilde{\tau}(0)$. The maximum likelihood estimate of τ is the solution of the simultaneous equations $\frac{\partial}{\partial \tau} l(y; \theta, \tau) = 0$. Expanding $\tilde{\tau}(\theta)$ in a Taylor series at $\theta = 0$ gives us

$$\tilde{\tau}(\theta) = \tilde{\tau} + \theta \cdot \frac{\partial}{\partial \theta} \tilde{\tau}(\theta)|_{\theta=0} + O(\theta^2) \tag{5.20}$$

The vector $\tilde{\tau}_\theta = \tilde{\tau}(\theta)$ of the first derivative of $\tilde{\tau}(\theta)$ with respect to θ can be written in matrix form with the matrix of second derivatives $M_{\tau\tau}(y, \theta, \tilde{\tau}(\theta))$ of $\ln f(y; \theta, \tau)$ with respect to τ_i and τ_j at $\tau = \tilde{\tau}$.

After some mathematical derivations (see Whitehead, 1997), with

$$\begin{aligned} l_{\theta\theta}(y; 0, \tilde{\tau}) &= \frac{\partial^2 l(y, \theta, \tilde{\tau})}{\partial \theta^2}|_{\theta=0} \\ l_{\theta_\tau}(y; 0, \tilde{\tau}) &= \frac{\partial^2 l(y, \theta, \tau)}{\partial \theta \cdot \partial \tau}|_{\theta=0; \tau=\tilde{\tau}} \\ l_\theta(y; 0, \tilde{\tau}) &= \frac{\partial l(y, \theta, \tilde{\tau})}{\partial \theta}|_{\theta=0} \end{aligned}$$

we can write

$$z = l_\theta(y; 0, \tilde{\tau}) \tag{5.21}$$

and

$$v = -l_{\theta\theta}(y; 0, \tilde{\tau}) - l_{\theta\tau}(y; 0, \tilde{\tau})^T \cdot M_{\tau\tau}(y, \theta, \tilde{\tau}(\theta)) \cdot l_{\theta\tau}(y; 0, \tilde{\tau}) \tag{5.22}$$

With the z– and v–values in (5.17) and (5.18) if there is no nuisance parameter, or (5.21) and (5.22) if nuisance parameter(s) is (are) present, we can uniquely write approximate SLRTs.

After the observation of n elements $\boldsymbol{y}_1, \boldsymbol{y}_2, \ldots, \boldsymbol{y}_n$ distributed i.i.d. like $\boldsymbol{y}$ with log likelihood function, we write the z-function in (5.17) and (5.21), respectively, as $z_n = \sum_{i=1}^{n} z_i = \sum_{i=1}^{n} l_\theta(y_i, \theta, \tilde{\tau})$, whereby we drop the estimate of the nuisance parameter if there is none. Here, z_n is called the efficient score and characterises the deviation from the null hypothesis.

The v-function can be written analogously. It is connected with the Fisher information matrix $I(\theta) = -\sum_{i=1}^{n} E_\theta \left\{ \frac{\partial^2 l(\boldsymbol{y}_i, \theta, \tilde{\tau})}{\partial \theta^2} \right\} = n i(\theta)$ where $\boldsymbol{i}(\theta) = -\frac{\partial^2 l(\boldsymbol{y}_i, \theta, \tilde{\tau})}{\partial \theta^2}$ is the information from a single observation and $i(\theta) = E_\theta(\boldsymbol{i}(\theta))$.

Now we can write $v = nE[\boldsymbol{i}(\theta)]_{\theta=0}$.

From the fact that likelihood estimators are asymptotically normally distributed, we approximately can state that $\boldsymbol{z} = \sum_{i=1}^{n} \boldsymbol{z}_i = \sum_{i=1}^{n} l_\theta(\boldsymbol{y}_i; 0, \tilde{\tau})$ is asymptotically normally distributed with expectation θv and variance v.

After the observation of n elements $\boldsymbol{y}_1, \boldsymbol{y}_2, \ldots, \boldsymbol{y}_n$ distributed i.i.d. like y with log likelihood function, we write the v-functions (5.18) and (5.22) as $v_n = \sum_{i=1}^{n} \{-l_{\theta\theta}((y; 0, \tilde{\tau})) - l_{\theta\tau}(y; 0, \tilde{\tau})^T \cdot M_{\tau\tau}(y, \theta, \tilde{\tau}(\theta)) \cdot \ln_{\theta\tau}(y; 0, \tilde{\tau})\}$. We wish to test the null hypothesis $H_0 : \theta = 0$ against the alternative $H_A : \theta = \theta_1 \neq 0; \theta_1 \in \Omega \subset \mathbb{R}$. Then the approximate SLRT is defined as follows: continue observation as long as

$a_l = \frac{1}{\theta_1} \ln \frac{\beta}{1-\alpha} < z_n - bv_n < a_u = \frac{1}{\theta_1} \ln \frac{1-\beta}{\alpha}$

with $b = \frac{1}{2}\theta_1$ and accept $H_A : \theta = \theta_1 > 0$ if $z_n - bv_n > a_u = \frac{1}{\theta_1} \ln \frac{1-\beta}{\alpha}$ and accept $H_A : \theta = \theta_1 < 0$ if $z_n - bv_n < a_l = \frac{1}{\theta_1} \ln \frac{\beta}{1-\alpha}$; otherwise accept H_0.

The power function is given by

$$\tau(\theta) \approx \frac{1 - \left(\frac{\beta}{1-\alpha}\right)^{1-2\frac{\theta}{\theta_1}}}{\left(\frac{1-\beta}{\alpha}\right)^{1-2\frac{\theta}{\theta_1}} - \left(\frac{\beta}{1-\alpha}\right)^{1-2\frac{\theta}{\theta_1}}} \text{ for } \theta \neq 0.5\theta_1 \text{ and by}$$

$$\tau(\theta) \approx \frac{\ln\left(\frac{1-\alpha}{\beta}\right)}{\ln\left(\frac{1-\beta}{\alpha}\right) + \ln\left(\frac{1-\alpha}{\beta}\right)} \text{ for } \theta = 0.5\theta_1.$$

The expected sample size is given by

$$E(\boldsymbol{n}|\theta) = \frac{\ln \frac{1-\beta}{\alpha}\left\{1 - \left(\frac{\beta}{1-\alpha}\right)^{1-2\frac{\theta}{\theta_1}}\right\} - \ln \frac{\beta}{1-\alpha}\left\{1 - \left(\frac{1-\beta}{\alpha}\right)^{1-2\frac{\theta}{\theta_1}}\right\}}{\left[\theta_1(\theta - 0.5\theta_1)\left\{\left(\frac{1-\beta}{\alpha}\right)^{1-2\frac{\theta}{\theta_1}} - \left(\frac{\beta}{1-\alpha}\right)^{1-2\frac{\theta}{\theta_1}}\right\}\right]} \text{ if } \theta \neq 0.5\theta_1 \text{ and by}$$

$$E(\boldsymbol{n}|\theta) = \frac{1}{\theta_1^2} \ln \frac{1-\beta}{\alpha} \cdot \ln \frac{\beta}{1-\alpha} \text{ if } \theta = 0.5\theta_1.$$

All we have to do is to program the bounds and derive the z_n and v_n for the corresponding distribution whose parameter is under test.

Remark: Restricting the null hypothesis $H_0 : \mu = 0$ to the value 0 is no loss of a generality. If we have to test $H_0 : \mu = \mu_0 \neq 0$, we subtract the value μ_0 from all our observations and are back to the special form of the null hypothesis.

Let us consider again the situation of the univariate normal distribution with unknown variance but now using the approximation discussed in this section. At first we consider the pair of hypotheses

$H_0 : \mu = 0; \sigma^2$ arbitrary against the alternative

$H_A : \mu = \mu_1, \mu_1 \neq 0; \mu \in \Omega \subset \mathbb{R}$.

We further have $\tau = \sigma^2$ and obtain with $\theta = \frac{\mu}{\sigma}$ from the general formulae above

$$l_\theta(y_1; y_2; \dots, y_n; \theta) = -\frac{1}{2}n \cdot \ln(2\pi\sigma^2) - \frac{1}{2\sigma^2}\sum_{i=1}^{n}(y_i - \mu)^2$$
$$= -\frac{1}{2}n \cdot \ln(2\pi\sigma^2) - \frac{1}{2}\sum_{i=1}^{n}\left(\frac{y_i}{\sigma} - \theta\right)^2 \quad (5.23)$$

We write the efficient score as

$$z_n = \frac{\sum_{i=1}^{n} y_i}{\sqrt{\frac{1}{n}\sum_{i=1}^{n} y_i^2}} \quad (5.24)$$

The v-function can analogously be written as

$$v_n = n - \frac{z_n^2}{2n} \quad (5.25)$$

Example 5.8 We choose $\alpha = 0.05$ and $\beta = 0.1$. We wish to test the null hypothesis

$H_0 : \mu = 0$ against the alternative

$H_A : \mu = \mu_1 = 10$.

Then we obtain $a_l = \frac{1}{10}\ln\frac{0.1}{0.95} = -0.2251$; $a_u = \frac{1}{10}\ln\frac{0.9}{0.05} = 0.2890$; $b = \frac{1}{2}\mu_1 = 5$. The values of z_n and v_n can be derived as

$$z_n = \frac{\sum_{i=1}^{n} y_i}{\sqrt{\frac{1}{n}\sum_{i=1}^{n} y_i^2}}; v_n = n - \frac{z_n^2}{2n}$$

In Table 5.2 for some distributions, z_n and v_n are given.

5.3.3 Approximate tests for binary data

Let the observations $y_1; y_2; \dots, y_n$ of an experiment be realisations of independently Bernoulli distributed random variables $\boldsymbol{y}_i; i = 1, \dots, n$, all having the same probability $0 < p < 1$ for a certain event (called success).

We want to test the hypothesis

$H_0 : p = p_0$ against (w.l.o.g.)

TABLE 5.2
The Formulae for z_n and v_n for Some Distributions.

Distribution	Log Likelihood	Hypotheses	z_n	v_n
Normal, σ known	$-\frac{n}{2}\ln(2\pi)\sigma^2 - \frac{1}{2}\sum_{i=1}^{n}\left[\frac{y_i}{\sigma} - \mu/\sigma\right]^2$	$H_0 : \mu = 0$ $H_A : \mu = \mu_1$	$z_n = \frac{\sum_{i=1}^{n} y_i}{\sigma}$	$v_n = n$
Normal, σ unknown	$-\frac{n}{2}\ln(2\pi)\sigma^2 - \frac{1}{2}\sum_{i=1}^{n}\left[\frac{y_i}{\sigma} - \mu/\sigma\right]^2$	$H_0 : \mu = 0$ $H_A : \mu = \mu_1$	$z_n = \frac{\sum_{i=1}^{n} y_i}{\sqrt{\frac{1}{n}\sum_{i=1}^{n} y_i^2}}$	$v_n = n - \frac{z_n^2}{(2n)}$
Bernoulli	$const + r\ln\frac{p}{(1-p)} + n\ln(1-p)$	$H_0 : p = p_0$	$z_n = r - np_0$	$v_n = np_0(1-p_0)$

$H_A : p = p_1 > p_0$.

If r of the n observations show a "1" = success and therefore $n - r$ are failures = "0", then the likelihood function is $l(y_1, \ldots, y_n; p) = const + r \cdot \ln\left(\frac{p}{1-p}\right) + n \cdot \ln(1-p)$. We reparametrise the parameter p by $\theta = \ln \frac{p(1-p_0)}{p_0(1-p)}$ and then the likelihood function becomes

$$l(y_1, \ldots, y_n; \theta) = const + r \cdot (\theta + \lambda) + n \cdot \ln(1 + e^{\theta+\lambda}); \lambda = \ln \frac{p_0}{1-p_0}$$

From this we get

$$l(y_1, \ldots, y_n; \theta) = r - n\frac{e^{\theta+\lambda}}{1 + e^{\theta+\lambda}}; l_{\theta\theta}(y_1, \ldots, y_n; \theta) = -n\frac{e^{\theta+\lambda}}{(1 + e^{\theta+\lambda})^2}$$

Therefore we obtain $z_n = r - np_0; v_n = np_0(1 - p_0)$.

Analogously, we can derive the z- and v-values for other distributions with a twice differentiable likelihood function.

5.3.4 Approximate tests for the two-sample problem

We have two populations with location parameters ϑ_x, ϑ_y and a common nuisance parameter ψ. Our purpose is to take sequentially two independent random samples $\boldsymbol{x}_{11}, \ldots, \boldsymbol{x}_{1n_x}$ and $\boldsymbol{y}_{21}, \ldots, \boldsymbol{x}_{2n_y}$ of sizes n_x and n_y from the two populations in order to test the null hypothesis

$$H_0 : \vartheta_x = \vartheta_y$$

against one of the following one- or two-sided alternative hypotheses

a) $H_A : \vartheta_x > \vartheta_y$

b) $H_A : \vartheta_x < \vartheta_y$

c) $H_A : \vartheta_x \neq \vartheta_y$

The following reparametrisation of the problem is recommended:

Put $\theta = \frac{1}{2}(\vartheta_1 - \vartheta_2)$; $\varphi = \frac{1}{2}(\vartheta_1 + \vartheta_2)$. The overall likelihood function for both samples is now

$$l(\boldsymbol{x}, \boldsymbol{y}; \theta; \phi, \psi) = l^{(x)}(\boldsymbol{x}; \vartheta_1, \psi) + l^{(y)}(\boldsymbol{y}; \vartheta_2, \psi)$$

Here ψ is the nuisance parameter depending on the variances of the two populations. We consider here the case of equal variances.

Neglecting for simplicity the arguments we obtain for the derivatives are

$$l_\theta = l_{\vartheta_1}^{(x)} - l_{\vartheta_2}^{(y)}, l_\phi = l_{\vartheta_1}^{(x)} + l_{\vartheta_2}^{(y)}, l_\psi = l_\psi^{(x)} + l_\psi^{(y)}$$

$$l_{\theta\theta} = l_{\phi\phi} = l_{\vartheta_1\vartheta_1}^{(x)} + l_{\vartheta_2\vartheta_2}^{(y)}, l_{\psi\psi} = l_{\psi\psi}^{(x)} + l_{\psi\psi}^{(y)}$$

$$l_{\theta\phi} = l_{\vartheta_1\vartheta_1}^{(x)} - l_{\vartheta_2\vartheta_2}^{(y)}, l_{\theta\psi} = l_{\vartheta_1\psi}^{(x)} - l_{\vartheta_2\psi}^{(y)}, l_{\phi\psi} = l_{\vartheta_1\psi}^{(x)} + l_{\vartheta_2\psi}^{(y)}.$$

Then the estimates $\hat{\varphi}; \hat{\psi}$ are the solutions of the equations

$$\begin{aligned} l_{\vartheta_1}^{(x)}(\hat{\varphi};\hat{\psi}) + l_{\vartheta_2}^{(y)}(\hat{\varphi};\hat{\psi}) &= 0 \\ l_{\psi}^{(x)}(\hat{\varphi};\hat{\psi}) + l_{\psi}^{(y)}(\hat{\varphi};\hat{\psi}) &= 0 \end{aligned}$$

Now we can calculate the z- and v-values in dependence on the assumed distribution (likelihood function) from (5.22) and (5.23), respectively.

Example 5.9 We observe a binary variable with probability p_1 for a certain event (called success) in population 1 and probability p_2 for the same event in population 2, and we like to test the hypothesis

$H_0 : p_1 = p_2$ against
$H_0 : p_1 = p_2 + \delta; 0 < \delta < 1.$

Our test parameter is in this case $\theta = \ln \frac{p_1(1-p_2)}{p_2(1-p_1)}$ and we obtain after n_1 observations from population 1 with y_1 successes and n_2 independent observations from population 2 with y_2 successes and $N = n_1 + n_2$:

$$z_N = \frac{n_2 y_1 - n_1 y_2}{N} \text{ and } v_N = \frac{n_1 n_2 (y_1 + y_2)(N - y_1 - y_2)}{N^3}$$

Example 5.10 We have two normal populations with location parameters μ_x, μ_y and a common variance σ^2. Our purpose is to take sequentially two independent random samples $(\boldsymbol{x}_{11}, \ldots, \boldsymbol{x}_{1n_x})$ and $(\boldsymbol{y}_{21}, \ldots, \boldsymbol{y}_{2n_y})$ of sizes n_x and n_y from the two populations in order to test the null hypothesis

$H_0 : \mu_x = \mu_y$ against the following one-sided alternative hypothesis

$H_A : \mu_x > \mu_y + \delta, \delta > 0.$

Our test parameter in this case is $\theta = \frac{\mu_x - \mu_y}{\sigma} = \frac{\delta}{\sigma}$ and we obtain after n_x observations from population 1 and n_y independent observations from population 2 and $N = n_x + n_y$; sample means $\overline{x}_{n_x}; \overline{y}_{n_y}$: pooled sample variance from Section 2.2.1.5, $s^2 = \frac{(n_x-1)s_x^2 + (n_y-1)s_y^2}{n_x+n_y-2}$ and the maximum likelihood estimate

$\tilde{\sigma}^2 = \frac{s^2(n_x+n_y-2)}{n_x+n_y}$:

$$\begin{aligned} z_n &= \frac{n_x \cdot n_y}{N} \frac{\overline{x}_{n_x} - \overline{y}_{n_y}}{\tilde{\sigma}} \\ v_N &= \frac{n_x n_y}{N} - \frac{z_N^2}{2N} \end{aligned}$$

Example 5.11 We have two normal populations with location parameters μ_x, μ_y and variances σ_x^2, σ_y^2, respectively. Our purpose is to take sequentially two independent random samples $(\boldsymbol{x}_{11}, \ldots, \boldsymbol{x}_{1n_x})$ and $(\boldsymbol{y}_{21} \ldots, \boldsymbol{y}_{2n_y})$ of sizes n_x and n_y from the two populations in order to test the null hypothesis

$H_0 : \mu_x = \mu_y$ against one of the following one-sided alternative hypotheses

$H_A : \mu_x > \mu_y + \delta, \delta > 0$

$H_A : \mu_x < \mu_y + \delta, \delta > 0$.

We put $\theta = \frac{\mu_x}{\sigma_x} - \frac{\mu_y}{\sigma_y}; \varphi = \frac{\mu_x}{\sigma_x} + \frac{\mu_y}{\sigma_y}$ and the nuisance parameter is $\psi^T = (\sigma_x^2, \sigma_y^2)$. We obtain after n_x observations from population 1 and n_y independent observations from population 2 and $N = n_x + n_y$ the log-likelihood function

$$\begin{aligned} l(\theta, \varphi, \sigma_x^2, \sigma_y^2) = & -(n_x + n_y)\ln\sqrt{2\pi} - n_x \ln\sigma_x - n_y \ln\sigma_y \\ & -\frac{1}{2}\left[\sum_{i=1}^{n_x}\left(\frac{x_i}{\sigma_x} - \frac{\mu_x}{\sigma_x}\right)^2 + \sum_{i=1}^{n_y}\left(\frac{y_i}{\sigma_y} - \frac{\mu_y}{\sigma_y}\right)^2\right] \end{aligned}$$

Therefore the maximum likelihood estimates are

$$\tilde{\mu}_x = \overline{x}; \tilde{\mu}_y = \overline{y}; \tilde{\sigma}_x^2 = \frac{\sum_{i=1}^{n_x}(x_i - \overline{x})^2}{n_x}; \tilde{\sigma}_y^2 = \frac{\sum_{i=1}^{n_y}(y_i - \overline{y})^2}{n_y}$$

Putting also $l_{\theta\phi} = -\frac{n_x}{\sigma_x^2} + \frac{n_y}{\sigma_y^2}$ equal to zero shows us that we should try to obtain $\frac{n_x}{n_y} \approx \frac{\tilde{\sigma}_x^2}{\tilde{\sigma}_y^2}$. Thus in case of continuation we sample from population 1 if $\frac{n_x}{n_y} > \frac{\tilde{\sigma}_x^2}{\tilde{\sigma}_y^2}$ and from population 2 if $\frac{n_x}{n_y} < \frac{\tilde{\sigma}_x^2}{\tilde{\sigma}_y^2}$. In the improbable case that an equality sign occurs, we can sample in the next step from any population. The corresponding test is the sequential analogue to the Welch test from Section 2.4.1.4. Thus we obtain from (5.21) and (5.22)

$$\begin{aligned} z_N &= \frac{n_x \cdot n_y}{N}\left(\frac{\overline{x}}{\sigma_x} - \frac{\overline{y}}{\sigma_y}\right) \\ v_N &= \frac{n_x n_y}{N} - \frac{z_N^2}{2N} \end{aligned}$$

5.4 Triangular designs

In triangular designs the continuation region is closed and in triangular form. This has the advantage that the maximum number of observations is known in advance. Even if this number is larger than in the fixed sample size cases discussed in Chapter two, the expected number of observations needed is smaller than in the fixed sample cases. The authors have much experience in applying triangular designs and have found until now no case where the size of a triangular design exceeded the corresponding fixed sample case.

In all sequential procedures data collecting must be stopped as soon as the continuation region is left. Otherwise the risks fixed in advance cannot be guaranteed. This means that after each observation the data collected so far must be analysed. In group sequential procedures—not discussed in this book—special techniques have been developed to solve such problems. If the continuation region is left and sampling is continued, we have a so-called overshooting. In this case the requested risks α and β cannot be guaranteed.

In this section we restrict ourselves on tests about means and probabilities. An extended introduction to triangular designs and further tests for ordinary variables and censored samples can be found in Whitehead (1997) who is one of the first who gave a description of those designs and FORTRAN programs.

5.4.1 Basic principles

A triangular sequential design is based on the asymptotic test described in Section 5.3.

The tests were first developed for the special (symmetric)) case $\alpha = \beta$. This case, however, is of no practical interest because an experimenter very seldom chooses $\alpha = \beta$. We describe this special case only to show how it can be generalised to the more interesting one of arbitrary α and β (which of course includes the case $\alpha = \beta$ as a special case). For the moment we assume $\alpha = \beta$.

In the one-sample problem, we test the hypothesis
$H_0 : \theta = \theta_0$ against the alternative hypothesis
$H_A : \theta = \theta_1$.

With the sequence of statistics $z_n; v_n$ defined for the approximate test in Section 5.3 the continuation region is given by

$-a + 3bv_n < z_n < a + bv_n$; if $\theta > \theta_0$
$-a + bv_n < z_n < a + 3bv_n$; if $\theta < \theta_0$

$H_0 : \theta = \theta_0$ is accepted, if $z_n \geq a + bv_n$ for $\theta > \theta_0$; and if $z_n \leq -a + bv_n$ for $\theta < \theta_0$. The other cases, when z_n leaves the continuation region or meets the corresponding bound, lead to the acceptation of $H_A : \theta = \theta_1$. The two constants a and b have to be chosen as follows:

$$a = 2\ln\left(\frac{1}{2\alpha}\right)/\theta_1; \quad b = \frac{\theta_1}{4}$$

The two boundary lines meet in the point $(v_{\max}; z_{\max}) = \left(\frac{a}{b}; 2a\right)$.

If just this point is met, then accept $H_A : \theta = \theta_1$. This point defines the maximum sample size. It is larger than the size needed for the fixed sample size problem with the same precision.

In the case of a two-sided alternative the continuation region consists of two overlapping triangles. To test the hypothesis

$H_0 : \theta = \theta_0$ against

$H_A : \theta = \theta_1 < \theta_0$ or $\theta = \theta_2 > \theta_0$,

we act as demonstrated in the Sections 5.4.2 and 5.4.3.

$$\begin{array}{ccccc} \frac{2\sigma}{\theta_2-\theta_0}\ln(2\alpha) + \frac{3}{4}(\theta_2-\theta_0)\theta_n & < & z_n & < & -\frac{2\sigma}{\theta_2-\theta_0}\ln(2\alpha) + \frac{1}{4}(\theta_2-\theta_0)\theta_n \\ \frac{2\sigma}{\theta_0-\theta_1}\ln(2\alpha) + \frac{1}{4}(\theta_0-\theta_1)\theta_n & < & z_n & < & -\frac{2\sigma}{\theta_0-\theta_1}\ln(2\alpha) - \frac{3}{4}(\theta_0-\theta_1)\theta_n \end{array}$$

If $\theta_0 - \theta_1 = \theta_2 - \theta_0$, the two triangles are symmetric. In the next two sections we consider the one- and two-sample problem for means and probabilities for the general case with arbitrary α and β.

In the non-symmetric case $\alpha \neq \beta$, we proceed as follows. We replace θ by a transformed parameter θ^* and perform a triangular test with the given $\alpha(=\beta^*)$ which is chosen in such a way that the fixed sample size from Chapter two for the triple $(\alpha, \beta^*, \theta)$ is the same as for the triple $(\alpha, \beta, \theta^*)$. This leads to $\theta_1^* = 2\theta_1/(1 + u_{1-\beta}/u_{1-\alpha})$.

Therefore, in the general case, the constants in the boundary lines are given by

$$a = (1 + u_{1-\beta}/u_{1-\alpha})\ln(1/2\alpha)/\theta_1 \tag{5.26}$$

$$b = \theta_1/[2(1 + u_{1-\beta}/u_{1-\alpha})] \tag{5.27}$$

5.4.2 Testing hypotheses about means of normal distributions

5.4.2.1 One-sample problem

We test the hypothesis $H_0 : \mu = \mu_0$ against the alternative $H_1 : \mu = \mu_1$ for a normally distributed variable with mean μ and unknown variance σ^2. Here we have

$$z_n = \sum_{i=1}^{n} x_i \Big/ \sqrt{\frac{1}{n}\sum_{i=1}^{n} x_i^2} \tag{5.28}$$

$$\text{and } v_n = n - z_n^2/2n \tag{5.29}$$

The slope and intercept values for the boundary lines are given by (5.26) and (5.28), respectively, with $\theta_1 = (\mu_1 - \mu_0)/\sigma$.

Example 5.12 We wish to test the null hypothesis that male students have a mean body height of 180 cm against the one-sided alternative that the mean body height is 185 cm or larger.

The data collected read as follows: 183, 187, 179, 190, 184, 192, 198, 182, 188, 186.

We use $\alpha = 0.01$ and $\beta = 0.1$ and assume prior knowledge on the variance: $\sigma^2 = 16$.

Setting $\theta_1 = (185 - 180)/\sigma = 5/\sigma = 1.25$, we obtain $a = 4.8537$ and $b = 0.4030$. We see that the region of continuation is left already for $n = 10$ since $z_{10} = 7.983 < -a + 3bv_{10} = 7.599$, whereas for $n = 9$, we are within the bounds: $-a + 3bv_9 = 2.653 < z_9 = 7.088 < a + bv_9 = 7.356$. Since $\theta_1 > 0$ and z_{10} exceeds the upper boundary line, we have to accept the alternative hypothesis H_A. In the fixed sample case we would have needed $N_{fix} = 12$ samples. For an illustration of the acceptance regions and the continuation region, see Figure 5.1.

5.4.2.2 Two-sample problem

Let us now consider the problem of testing $H_0 : \mu_1 = \mu_2$ against $H_A : \mu_1 > \mu_2$ for two normal distributions with equal variance σ^2. The test parameter is then

$$\theta = \frac{\mu_1 - \mu_2}{\sigma}$$

For given samples $x_{11}, \ldots, x_{1n_1}$ and $x_{21}, \ldots, x_{2n_2}$ we calculate the sample means $\bar{x_1}$ and $\bar{x_2}$, respectively, and the maximum likelihood estimator of σ^2 for $\theta = 0$:

$$S_n^2 = \frac{1}{n_1 + n_2}\left\{\sum_{i=1}^{n_1} x_{1i}^2 + \sum_{i=1}^{n_2} x_{2i}^2 - \left(\sum_{i=1}^{n_1} x_{1i} + \sum_{i=1}^{n_2} x_{2i}\right)^2 /(n_1 + n_2)\right\}$$

The statistics z_n and θ_n are given by

$$\begin{aligned} z_n &= \frac{n_1 n_2}{n_1 + n_2} \cdot \frac{\bar{x_1} - \bar{x}_2}{S_n} \\ v_n &= \frac{n_1 n_2}{n_1 + n_2} \cdot \frac{z_n^2}{2(n_1 + n_2)} \end{aligned}$$

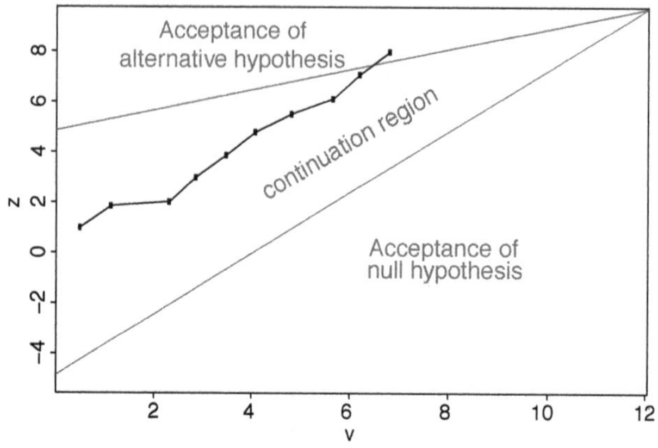

FIGURE 5.1
Triangular design for Example 5.12

The constants a and b are as indicated in (5.26) and (5.27), respectively.

Example 5.13 In this example we compare the mean body heights of male and female students which we collected in a classroom experiment. We test the null hypothesis $H_0 : \mu_1 = \mu_2$ that male and female students have equal body heights against the alternative hypothesis $H_1 : \mu_1 < \mu_2$ that female students are at least 8 cm smaller; $\mu_1 = 170$, $\mu_2 = 178$. Data are collected pairwise, starting with a female student followed by a male student, these reads as follows:

```
> heights.female = c(165, 168, 168, 173, 167, 169, 162)
> heights.male = c(179, 180, 188, 174, 185, 183, 179)
```

We chose $\alpha = 0.05$, $\beta = 0.2$ and assume, as before, prior knowledge on the variance: $\sigma^2 = 49$. This leads to $\theta_1 = -\frac{8}{\sigma} = -1.143$ and thus the constants a and b assume the values:

$$a = (1 + \frac{u(0.8)}{u(0.95)}) \log(\frac{1}{0.1}) \frac{1}{\theta_1} = -3.046$$

$$b = \frac{\theta_1}{2(1 + \frac{u(0.8)}{u(0.95)})} = -0.378$$

With the last observation of male students, i.e., $n_1 = n_2 = 7$, we leave the continuation region and accept the alternative hypothesis, see Figure 5.2.

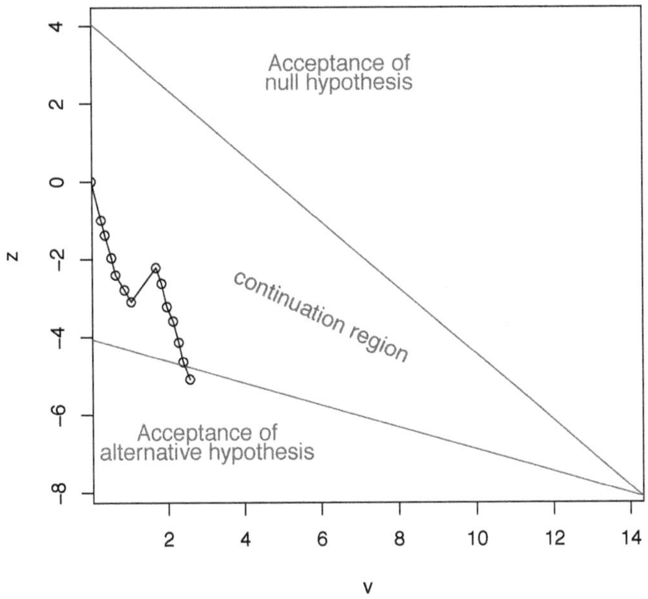

FIGURE 5.2
Triangular design for Example 5.13

5.4.3 Testing hypotheses about probabilities

5.4.3.1 One-sample problem

We test the hypothesis $H_0 : p = p_0$ against the alternative $H_A : p = p_1$ for a Bernoulli variable (binary variable) with unknown probability p. In this case we have

$$z_n = r_n - np_0, \quad v_n = np_0(1 - p_0) \tag{5.30}$$

where r_n denotes the number of successes in n trials. The slope and intercept values for the boundary lines are defined as in (5.26) and (5.27), respectively, with

$$\theta_1 = \ln \frac{p_1(1 - p_0)}{p_0(1 - p_1)} \tag{5.31}$$

being the log-adds ratio.

Example 5.14 In a public opinion poll, 842 out of $n = 2000$ people expressed their intention to vote for Party A in the upcoming election. The opinion leaders are interested to know whether or not this party can achieve the absolute majority of 50.1%. This means we test $H_0 : p = p_0 = 0.501$ against $H_1 : p = p_1 < p_0$, where we put $p_1 = 0.45$, and we choose $\alpha = 0.05, \beta = 0.1$.

Clearly, in the above fixed case we have to reject the null hypothesis, observing that with $r_n = 842$ the relative frequency is $\hat{p} = 842/2000 = 0.421$ and the (approximate) test-statistic leads to $u = (0.421 - 0.501)/\sqrt{0.421 \cdot 0.579/2000} = -7.246 < -u(1 - 0.05) = -1.645$.

Now, let us consider the triangular test case and find out how many people we would have to ask for their opinion to come to a sequential decision in favour of or against H_0. Observing that $\theta_1 = \ln(0.45 \cdot 0.499/0.501 \cdot 0.55) = -0.20467$, we find $a = -20.0155$ and $b = -0.0575$. The data recorded are stored as a binary vector named data.poll, the first 20 entries read as follows: 0 1 0 0 1 1 0 1 0 1 0 1 0 0 1 0 0 1 0 0, where "1" means a vote for Party A. Computing the z- and v-values with our OPDOE-function `tri_test.prop.one_sample`, we see that we have to stop after 243 observations, since $Z_{243} = -23.743$ falls below the lower bound: $z_{243} < a + bv_{243} = -23.510$, whereas for $n = 242$ we are still within the bounds: $a + bv_{242} = -23.495 < z_{242} = -23.242 < -a + 3bv_{242} = 9.576$. Therefore, we have to reject the null hypothesis and accept H_A. For a graphical illustration, see Figure 5.3.

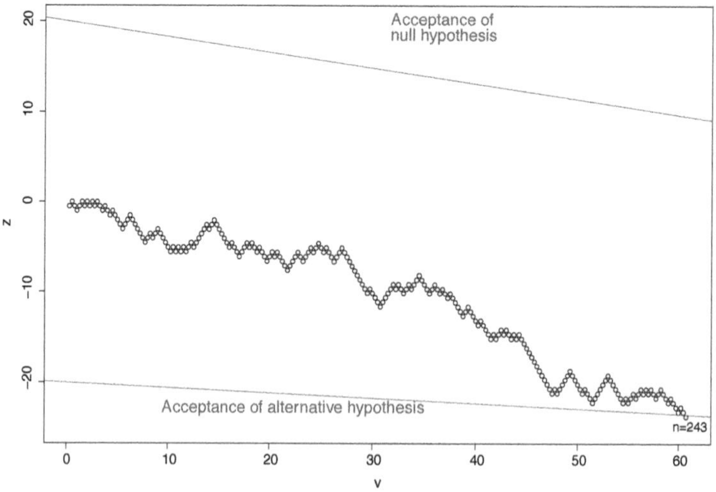

FIGURE 5.3
Triangular design for Example 5.14

5.4.3.2 Two-sample problem

We consider the problem of comparing the success probabilities of two binary variables; i.e., we test the hypothesis $H_0 : p_1 = p_2$ vs. $H_1 : p_1 = 0.6, p_2 = 0.5$. The triangular test is again based on the log-odds ratio

$$\theta_1 = \ln\left(\frac{p_1(1-p_2)}{p_2(1-p_1)}\right)$$

leading to

$$a = (1 + u_{1-\beta}/u_{1-\alpha})\ln(1/2\alpha)/\theta_1, b = \theta_1/[2(1 + u_{1-\beta}/u_{1-\alpha})]$$

The variables z and v now have to be modified such that

$$z_n = \frac{n_2 r_1 - n_1 r_2}{n_1 + n_2}, \quad n = n_1 + n_2$$

$$v_n = \frac{n_1 n_2 (r_1 + r_2)(n_1 + n_2 - (r_1 + r_2))}{(n_1 + n_2)^3}$$

Here n_1 and n_2 stand for the sample sizes and r_1 and r_2 for the number of successes in the first and second sample, respectively.

Example 5.15 Consider the problem of testing $H_0 : p_1 = p_2$ vs. $H_A : p_1 = 0.6, p_2 = 0.5$. The log-odds ratio then becomes $\theta_1 = \ln(0.6 \cdot 0.5/0.4 \cdot 0.5) = 0.4055$. Putting here $\alpha = 0.05$ and $\beta = 0.2$, we further obtain $a = (1 + u(0.8)/u(0.95))\ln(1/0.1)/0.4055 = 8.58$ and $b = 0.4055/[2(1 + u(0.8)/u(0.95))] = 0.1341$. This, in turn, implies the following sample sizes: $N_{max} = 1036, N_{fix} = 608$, $N_{exp} = 325$ for $\theta = 0$ or $\theta = \theta_1$ and $N_{exp} = 434$ for $\theta = \theta_1/2$.

For the triangular test we can use our OPDOE-function `tri_test.prop.two_sample`. The use of this function is demonstrated in the manual accompanying our OPDOE package.

Changing to standardized coordinates

$$z_s = \theta_1 z \text{ and } v_s = \theta_1^2 v$$

we obtain bounds which are independent of θ_1 and only depend on α and β. The constants to be used in the boundary lines read

$$a_s = (1 + u_{1-\beta}/u_{1-\alpha})\ln(1/2\alpha)$$

$$b_s = 1/[2(1 + u_{1-\beta}/u_{1-\alpha})]$$

The boundary lines of the continuation region in the standardized z_s, v_s-coordinate system are then given by

$$\text{upper bound: } z_s = a_s + b_s v_s$$
$$\text{lower bound: } z_s = -a_s + 3b_s v_s$$

5.5 A sequential selection procedure

As in Section 2.3 we have to select the population with the largest expectation from $a > 1$ normal distributions, when all populations have the same unknown variance σ^2, but now in a sequential procedure (and not only in a two-stage approach as in Example 5.3).

As in sequential testing the average sample size is smaller than in a non-sequential approach, we can especially save observations if one expectation is clearly larger than the others.

As in Section 2.3 the precision requirement in Bechhofer's indifference zone formulation is:

In the case $\mu_{[a-t+1]} - \mu_{[a-t]} \geq \delta$ the probability of an incorrect selection should be at most $\beta = 1 - P$.

As the *sampling rule* we use the approach to select in a first step two observations from each of the a populations, and in each further step we select one observation from each population. If we have received $n > 2$ observations from each population, we can arrange the data as in Table 5.3, where also the sample variances are given in the last row.

TABLE 5.3
Data and Estimated Variances after $n > 2$ Stages.

	1	2	...	a
1	y_{11}	y_{21}	...	y_{a1}
.	.	.	.	.
.	.	.	.	.
.	.	.	.	.
n	y_{1n}	y_{2n}	...	y_{an}
Sample sums	$Y_{1.}^{(n)}$	$Y_{2.}^{(n)}$	...	$Y_{a.}^{(n)}$
Sample variances	s_{1n}^2	s_{2n}^2	...	s_{an}^2

The *stopping rule* depends on the a sample sums and on the pooled estimates of the a estimated variances with $a(n-1)$ degrees of freedom which is the residual mean squares of a one-way ANOVA with the a populations as levels of a factor A or simply $s_n^2 = \frac{\sum_{i=1}^{a} s_{in}^2}{a}$. The sample sums are as usual in the ANOVA notation given by $Y_{i.}^{(n)} = \sum_{j=1}^{n} y_{ij}$. From these sums we calculate some auxiliary statistics as follows:

At first we rank the sample sums as $Y_{[1].}^{(n)} \leq Y_{[2].}^{(n)} \leq \ldots \leq Y_{[a].}^{(n)}$. If there are ties the tied sums are ranked arbitrarily. Next we calculate signed differences of the ranked sample sums $D_{i,k}^{(n)} = Y_{[i].}^{(n)} - Y_{[k].}^{(n)}$. From these signed differences positive quadratic forms of rank $a-1$ are calculated as follows:

$$Q_{a-i+1}^{(n)} = \sum_{r=1}^{a}\sum_{s=1}^{a} a_{r,s}(D_{i,r}^{(n)} - n\delta)(D_{i,s}^{(n)} - n\delta); \quad r, s \neq i$$

The coefficients are defined as: $a_{rs} = \begin{cases} \frac{2(a-1)}{a} & \text{if } r = s \\ -\frac{2}{a} & \text{if } r \neq s \end{cases}$

$r, s = 1, \ldots, a$.

It can be shown that $Q_1^{(n)} \leq Q_2^{(n)} \leq \ldots \leq Q_a^{(n)}$. We use these quadratic forms and the pooled estimate of the variance and calculate

$$M_i^{(n)} = \left[1 + \frac{Q_{a-i+1}^{(n)}}{2an(n-1)s_n^2}\right]^{-\frac{an-1}{2}} ; i = 1, \ldots, a \text{ and } z_n = \frac{1}{M_a^{(n)}} \sum_{i=1}^{a-1} M_i^{(n)}$$

It can be shown that $M_1^{(n)} \leq M_2^{(n)} \leq \ldots \leq M_a^{(n)}$. The *stopping rule* is as follows:

$$\text{If} \begin{cases} z_n \leq \frac{1-P}{P} = \frac{\beta}{1-\beta} & \text{stop sampling} \\ z_n > \frac{1-P}{P} = \frac{\beta}{1-\beta} & \text{continue sampling} \end{cases}$$

The *decision rule* is to select that population as the best one which gives rise to $Y_{[a].}^{(n)}$ at the final stage.

Example 5.16 We consider the milk fat performance (in kg per lactation) of heifers of three sires from Holstein Frisian cattle to select the sire with the highest breeding value for milk fat performance. This sire will be used in artificial insemination in the herd.

The precision requirements may be fixed by $P = 1 - \beta = 0.95$ and $\delta = 10$ kg. This leads to the stopping rule

$$\text{If} \begin{cases} z_n \leq \frac{0.05}{0.95} = 0.05263 & \text{stop sampling} \\ z_n > 0.05263 & \text{continue sampling} \end{cases}$$

In the case of continuation we collect fat performances of an additional heifer from each of the three sires. The first observations for five heifers per sires are given in Table 5.4.

Of course we select, in accordance with the sampling rule given above, in a first step, two observations from each sire.

TABLE 5.4
Data and Derived Statistics from Example 5.16 after $n = 2$ Stages.

Observation	Sire		
	1	2	3
1	$y_{1,1} = 132$	$y_{2,1} = 173$	$y_{3,1} = 147$
2	$y_{1,2} = 128$	$y_{2,2} = 166$	$y_{3,2} = 152$
3	$y_{1,3} = 135$	$y_{2,3} = 172$	$y_{3,3} = 150$
4	$y_{1,4} = 121$	$y_{2,4} = 176$	$y_{3,4} = 152$
5	$y_{1,5} = 138$	$y_{2,5} = 169$	$y_{3,5} = 146$
Sample sums	$Y_{1.}^{(2)} = 260$	$Y_{2.}^{(2)} = 339$	$Y_{3.}^{(2)} = 299$
	$Y_{[1].}^{(2)} = 260$	$Y_{[3].}^{(2)} = 339$	$Y_{[2].}^{(2)} = 299$
Sample variances	$s_{1,2}^2 = 8$	$s_{2,2}^2 = 24.5$	$s_{3,2}^2 = 12.5$
$D_{1,k}^{(2)}$	...	$D_{1,2}^{(2)} = -39$	$D_{1,3}^{(2)} = -79$
$D_{2,k}^{(2)}$	$D_{2,1}^{(2)} = 39$	...	$D_{2,3}^{(2)} = -40$
$D_{3,k}^{(2)}$	$D_{3,1}^{(2)} = 39$	$D_{3,2}^{(2)} = -40$	...
$Q_{3-i+1}^{(2)}$	$Q_3^{(2)} = 9921.33$	$Q_2^{(2)} = 6801.33$	$Q_1^{(2)} = 3601.33$
$M_i^{(2)}$	$10^5 \cdot M_1^{(2)} = 4.24$	$10^5 \cdot M_2^{(2)} = 10.67$	$10^5 \cdot M_3^{(2)} = 49.44$

We now calculate the quadratic forms after two observations (i.e., after the first stage) and denote them by $Q_i^{(2)}$.

We have $a_{rs} = \begin{cases} \frac{2(3-1)}{3} \text{ if } r = s \\ -\frac{2}{3} \text{ if } r \neq s \end{cases} ; r, s = 1, \ldots, a.$

The calculation of $Q_3^{(2)} = \sum_{r=1}^{3} \sum_{s=1}^{3} a_{r,s}(D_{1,r}^{(2)} - 20)(D_{1,s}^{(2)} - 20); r, s \neq 1$ leads to

$$Q_3^{(2)} = \tfrac{4}{3}\{(-59) \cdot (-59) + (-99) \cdot (-99)\}$$
$$-\tfrac{2}{3}\{(-59) \cdot (-99) + (-99) \cdot (-59)\} = 9921.33$$

In Table 5.4 the other Q-values and the $M_i^{(2)}$ are shown. Finally, we calculate the z-value and obtain $z_2 = \frac{4.24+10.67}{49.44} = 0.3016$.

This is larger than 0.05263 and we have to continue sampling. The z-values in the following steps are $z_3 = 0.0887; z_4 = 0.0581; z_5 = 0.016$.

Therefore we have to stop sampling after the fifth observation from each sire, and select the second sire as the one with the highest breeding value for milk fat production.

In our R-package OPDOE we have the function `size.seq_select.mean` available for doing all the necessary calculations. For the data given in Table 5.4 we obtain

```
> data.cattle = matrix(c(132, 128, 135, 121, 138, 173, 166, 172,
+      176, 169, 147, 152, 150, 152, 146), 5, 3)
> data.cattle

     [,1] [,2] [,3]
[1,]  132  173  147
[2,]  128  166  152
[3,]  135  172  150
[4,]  121  176  152
[5,]  138  169  146

> size.seq_select.mean(data = data.cattle, delta = 10, P = 0.95)

$Q
          [,1]      [,2]      [,3]
[1,]  9921.333  6801.333  3601.333
[2,] 21477.333 14997.333  7557.333
[3,] 45028.000 31428.000 17668.000
[4,] 63889.333 45289.333 23489.333

$M
             [,1]         [,2]         [,3]
[1,] 4.238718e-05 1.067420e-04 4.943891e-04
[2,] 1.074288e-07 4.379757e-07 6.145879e-06
[3,] 4.956877e-09 3.329562e-08 6.579088e-07
[4,] 1.581132e-10 1.569310e-09 1.095461e-07

$z
[1] 0.30164336 0.08874313 0.05814255 0.01576892

$b
[1] "CONTINUE" "CONTINUE" "CONTINUE" "STOP"

$stage
[1] 5

$BestPopulation
[1] 2
```

We have thus reproduced the above-mentioned results: stopping at stage four (after five observations) and declaring the second population as the best one.

Part II

Construction of Optimal Designs

In this part, special methods of constructing often-used experimental designs are given. First we discuss in Chapter six the construction of completely balanced incomplete block designs. Partially balanced incomplete block designs are less important in many fields of applications—except agricultural field experiments. In Chapter seven we describe methods of constructing fractional factorial experiments with factors all having two levels and with factors all having three levels. Finally in Chapter eight we show how optimal experimental designs for regression problems can be calculated.

6

Constructing Balanced Incomplete Block Designs

6.1 Introduction

In this book we restrict ourselves to eliminating one noise factor by blocking or stratification. One method of eliminating two noise factors is the use of so-called row-and-column designs, where the rows correspond to different levels of one blocking factor and the columns to levels of the other blocking factor (for details see 1/21/4200-1/21/4250 in Rasch et al. 2008). Special cases are Latin Squares and Latin Rectangles or change over designs.

Blocking is especially useful when the noise factor is qualitative, since analysis of covariance is not then available, but blocking is also possible when the noise (blocking) factor is quantitative. In this case we can group its values into classes and use these classes to define the levels of the noise factor.

Of course we need a priori information in order to choose the size and number of the blocks before the experiment begins (which should be the rule).

General principles of blocking have been already described in Section 1.3. We only repeat here some notations.

The number b of the blocks is determined as part of the experimental design process as described in Chapter two. In this book we assume that all b blocks have the same block size k and further that each of the v levels of the treatment factor, which will be called treatments, occur exactly r times. In place of b we can also determine r by the methods of Chapter two, due to the relation (6.4) below.

We assume in this chapter that we are faced with the following situation: the block size k as well as the number v of treatments is given in advance. We write R-programs which help us to find the smallest design which exists for this (v, k) pair especially in the case when $k < v$ which means that not all treatments can occur in any block together. Block designs in this case have been called incomplete block designs. An early reference on incomplete block designs is Yates (1936). Methods of construction are discussed in many articles and books, for instance, Bailey (2008), Hall (1986), Hanani (1961, 1975), Hedayat and John (1974), John (1987), Kageyama (1983), Kirkman

(1947), Linek and Wantland (1998), Mathon and Rosa (1990), Raghavarao et al. (1985), Street and Street (1987), Vanstone (1975), Williams (2002), and many others. Further, we assume that some kind of balance is fulfilled. To explain the need for those balances, we first look at the model of a block design.

$$\boldsymbol{y}_{ij} = \mu + a_i + b_j + \boldsymbol{e}_{ij} \tag{6.1}$$

where the value $\boldsymbol{y}_{ij}$ of the character to be measured depends on a general mean μ of the experiment, the treatment effect a_i of the i-th treatment with $\sum a_i = 0$, the effect b_j of the j-th block with $\sum b_j = 0$ and of error terms $\boldsymbol{e}_{ij}$ with expectation zero and a common variance. Remember that random variables are bold print. Often the block effects are assumed to be random; an alternative model for a block design is therefore

$$\boldsymbol{y}_{ij} = \mu + a_i + \boldsymbol{b}_j + \boldsymbol{e}_{ij} \tag{6.2}$$

without $\Sigma \boldsymbol{b}_j = 0$ but $E\boldsymbol{b}_j = 0$ and $var\boldsymbol{b}_j = \sigma_b^2$.

Both models are models of the two-way analysis of variance (ANOVA) without interaction effects and with maximum one observation in the subclasses. Model (6.1) is called a model with fixed effects or a model I, and model (6.2) is called a mixed model (see Chapter three).

The least squares estimator of the treatment effects can be easily written down if each treatment occurs exactly r times and each treatment pair occurs exactly λ times together in the block experiment. In this case we get $\lambda(v-1) = r(k-1)$ which is shown in (6.5). Then when T_i is the sum of all measurements of treatment i and B_i is the sum of all block totals that include the i-th treatment, we have $\hat{a}_i = \frac{k}{r(k-1)+\lambda}\left(T_i - \frac{B_i}{k}\right) = \frac{k}{\lambda v}\left(T_i - \frac{B_i}{k}\right)$ (by virtue of (6.5)). The differences between treatment i' and treatment i are given by $\hat{a}_i - \hat{a}_{i'}$.

The reason to assume that all treatments occur r times in one of the b blocks and that all treatment pairs occur equally often in the blocks is that all estimators of the treatment differences have the same variance, i.e., all treatments are estimated with the same precision. Further, all the treatment differences can be estimated with the same precision.

Example 6.1 Let us assume we construct a block design for $v = 11$ varieties to be grown in blocks with size $k = 5$ (plots). What is the smallest number b of blocks from which a balanced design with constant r and λ can be applied? With our R-program this question can easily be solved.

We create a design with the parameters $v = 11; k = 5; b = 11; r = 5$ and $\lambda = 2$.

We can easily see from this design that each of the numbers 1 to 11 (which represent the treatments) occurs exactly 5 times in each block, and that each pair of any of the numbers 1 to 11 occurs in exactly 2 blocks together.

If one tries to find such a design by hand, one soon meets with difficulties if v becomes large. Up to now there are still cases of pairs (v, k) where the smallest design is either not known or cannot be constructed by methods developed so far. There further exists no unique method for the construction of such designs.

That is the reason that we have to use the methods below for different (v, k) pairs.

The construction of block designs is **one** important step in the process of designing an experiment. Also very important is the **randomisation** for a practical layout of the experiment. That means that the abstract blocks given in the output of the R-programs must be randomly assigned to the actual blocks used in the experiment. Further the treatments in a pair of parentheses of any abstract block have to be randomly assigned to the actual experimental units. On the other hand, the minimum number of replications of a treatment (the sample size in the language of Chapter two) has to be determined in advance for the analysis planned and the precision of the statistical procedure demanded by the methods of Chapter two. If the number of replications needed is larger than that of the smallest design for given v and k, we must use the block design offered by the R-program more than once. How to do all this is shown in a practical example (Example 6.38).

The problems of the analysis of block designs are beyond the scope of this book. An excellent overview for this is given by Calinski and Kageyama (2000) and Hinkelmann and Kempthorne (1994, 2005). Incomplete block designs discussed in this chapter are optimal in some sense. But they may lose this property if, due to experimental conditions in the practical experiments (especially in experiments in open air, such as field experiments, or in experiments under industrial conditions), some experimental units contribute no information because they are destroyed or otherwise lost. Calinski and Kageyama (2003) give some results about the robustness of the designs against missing observations and some references to this topic.

6.2 Basic definitions

Let us first define some terminology before we give the R-programs for the construction.

Block designs are special experimental designs for reducing the influence of noise factors by blocking. A relatively homogeneous group of experimental units is called a **block**. We say that a block is ***complete*** if each of the v treatments occurs at least once in this block, otherwise the block is called ***incomplete***. A block design is called *complete* if all its blocks are complete and is called *incomplete* if at least one of its blocks is incomplete.

Complete block designs are easy to handle in the design phase as well as in the analysis phase and are not discussed further in this chapter.

An ***incidence matrix*** $N = (n_{ij})$ of a block design is a rectangular array with v rows and b columns with integer entries n_{ij}, where n_{ij} represents how many times the i-th treatment (represented by the i-th row) appears in the j-th block (represented by the *j-th* column). If all n_{ij} are either 0 or 1 then the incidence matrix and the corresponding block design are called **binary**.

The elements of incidence matrices of complete block designs are all positive ($n_{ij} \geq 1$). Incomplete block designs have incidence matrices with at least one element equal to zero. Mainly incomplete block designs are characterised in **compact form** in place of an incidence matrix in writing a block as a pair of parentheses which includes the numbers of the treatments in the block in lexicographical order.

Example 6.2 A block design with 4 treatments and 6 blocks is defined by the incidence matrix:

$$\begin{pmatrix} 1 & 0 & 1 & 0 & 0 & 0 \\ 0 & 1 & 0 & 1 & 1 & 1 \\ 1 & 0 & 1 & 0 & 0 & 0 \\ 0 & 1 & 0 & 1 & 1 & 1 \end{pmatrix}$$

Since this matrix contains zeros, it represents an incomplete block design. In compact form it may be written in the form (1,3), (2,4), (1,3), (2,4), (2,4), (2,4). In this representation, for example, the first set of parantheses represents block 1 to which treatments 1 and 3 are assigned, and this corresponds to the first column of the matrix (representing block 1) having a 1 in rows 1 and 3 (representing the treatments).

A block design with a symmetrical incidence matrix is called a ***symmetrical block design***. If in a block design all row sums of the incidence matrix are equal, i.e., the treatments occur equally often (the number of replications $r_i = r$), then we call this design ***equireplicated***. If in a block design all the column sums of the incidence matrix are equal, i.e., the number of experimental units per block is the same for all blocks, we call this design ***proper***. The number of experimental units in the j-th block will be denoted by k_j; thus for proper designs we have $k_j = k$.

It can easily be seen that both the sum of the replications r_i and the sum of the block sizes k_j have to be equal to the total number of experimental units of a block design. We therefore have for any block design

$$\sum_{i=1}^{v} r_i = \sum_{j=1}^{b} k_j = N \tag{6.3}$$

In the case of equireplicated and proper block designs ($r_i = r$ and $k_j = k$), we get

$$vr = bk \tag{6.4}$$

In symmetrical block designs we have $b = v$ and $r_i = k_i \quad (i = 1, \ldots, v)$.

An incomplete block design is ***connected*** if for any pair (A_k, A_l) of the treatments $A_1, \ldots, A_v$ a chain of treatments starting with A_k and ending with A_l can be found in such a way that any consecutive treatments in the chain occur together at least once in a block. Otherwise the block design is called ***disconnected***. The incidence matrix of a disconnected block design can (by interchanging rows and/or columns) be written as a block diagonal matrix of at least two sub-matrices.

Disconnected block designs can be considered and analysed as two (or perhaps more) independent experimental designs.

In the design of Example 6.2 the first and the second treatments do not jointly occur in any of the 6 blocks. But also there is no chain in the sense of the definition above to ensure that the design is connected. Hence it is a disconnected block design. What this means can be seen by rearranging the blocks and the treatments or, equivalently, by rearranging the columns and the rows of the incidence matrix. We interchange blocks 2 and 3 (the second block becomes the third one and vice versa) and the treatments 1 and 4. In the incidence matrix these interchanges result in the following matrix:

$$\boldsymbol{N} = \begin{pmatrix} 0 & 0 & 1 & 1 & 1 & 1 \\ 0 & 0 & 1 & 1 & 1 & 1 \\ 1 & 1 & 0 & 0 & 0 & 0 \\ 1 & 1 & 0 & 0 & 0 & 0 \end{pmatrix}$$

It can now be seen that this experimental design is composed of two designs with two disjoint subsets of the treatments. In the first experiment, 2 treatments (4 and 2—old numeration) occur in 4 blocks. In the second experiment, 2 further treatments (3 and 1) occur in 2 different blocks. Therefore, it is not possible to analyse the total experiment by one analysis of variance; this can only be done for the two parts separately, and by eventually pooling the residual sum of squares to increase the degrees of freedom, if the variances are assumed to be equal. Comparisons between treatments of the first group (4,2) and the second group (3,1) cannot be handled.

A (completely) ***balanced incomplete block design (BIBD)*** is a proper and equireplicated incomplete block design with the additional property that all treatment pairs occur in the same number λ of blocks. If a BIBD has v treatments occurring with r replications in b blocks of size $k < v$, it will be denoted by $B(v, k, \lambda)$.

A ***balanced incomplete block design*** is called **resolvable** or a **RBIBD** if the set of blocks can be partitioned into parallel classes (resolution classes) so that every treatment occurs once in each parallel class. We write $RB(v, k, \lambda)$. An RBIBD is called **affine α-resolvable**, if two non-parallel blocks have exactly α treatments in common, where α is a positive integer.

For an affine $RB(v, k, \lambda)$ we have $b = v + r - 1$ and $\alpha = \frac{k^2}{v}$.

Example 6.3 The BIBD with $v = 9, k = 3$ and $b = 12$ is resolvable in 4 classes (the columns below).

$$\left\{ \begin{array}{cccc} (1,2,3) & (1,4,7) & (1,5,9) & (1,6,8) \\ (4,5,6) & (2,5,8) & (2,6,7) & (3,5,7) \\ (7,8,9) & (3,6,9) & (3,4,8) & (2,4,9) \end{array} \right\}$$

In the notation $B(v, k, \lambda)$ only three of the five parameters v, b, k, r, λ of a BIBD appear. This is sufficient because only three of these five parameters can be chosen independently; the others are fixed. To show this we first determine the number of possible treatment pairs in the design. This is given by $\binom{v}{2} = \frac{v(v-1)}{2}$. On the other hand, there are in each of the b blocks exactly $\binom{k}{2} = \frac{k(k-1)}{2}$ treatment pairs, and therefore we obtain

$$\lambda v(v-1) = bk(k-1)$$

because each of the $\binom{v}{2}$ treatment pairs occurs λ times in the experiment. Using formula (6.4) we replace bk by vr in the last equation and get (after division by v)

$$\lambda(v-1) = r(k-1) \tag{6.5}$$

Equations (6.4) and (6.5) are ***necessary*** **conditions** for the existence of a BIBD. These necessary conditions reduce the set of quintuples of five positive integers v, b, r, k, λ to the subset of those integers for which a BIBD could be constructed. The latter subset corresponds to a set of three positive integers, for instance, $\{v, k, \lambda\}$. From the three elements v, k, λ, the remaining two parameters of the BIBD can be calculated using (6.4) and (6.5).

It should be noted that the necessary conditions are not always sufficient; it may happen that for five positive integers fulfilling (6.4) and (6.5), no BIBD exists. To show this, a single example will be sufficient.

Example 6.4 In this example it can be seen that the conditions necessary for the existence of a BIBD need not always be sufficient. Consider the values $v = 16, r = 3, b = 8, k = 6, \lambda = 1$. Using (6.4) $16 \times 3 = 8 \times 6$ and (6.5) $1 \times 15 = 3 \times 5$, the necessary conditions for the existence of a BIBD are met. Nevertheless no BIBD exists for this combination of parameters.

Besides (6.4) and (6.5) there is another necessary condition given by Fisher's inequality (Fisher, 1940), which states that the following inequality must always be satisfied:

$$b \geq v \tag{6.6}$$

This inequality does not hold in Example 6.4. But there are still examples where (6.4), (6.5) and (6.6) hold and yet no BIBD exists.

A BIBD (a so-called unreduced or trivial BIBD) can always be constructed for any positive integer v and k by forming all possible combinations of the v numbers in groups of size $k < v$.

Hence $b = \binom{v}{k}, r = \binom{v-1}{k-1}$ and $\lambda = \binom{v-2}{k-2}$.

Often a BIBD can be found as a part of such an unreduced BIBD and this is a reduced BIBD. One case for which such a reduction is not possible is when $v = 8$ and $k = 3$. There is no other case for $v \leq 25$ and $2 < k < v - 1$ where no unreduced BIBD exists.

A BIBD is **elementary** if there exists no subset of blocks which is also a BIBD. Some theorems on elementary BIBD can be found in Rasch and Herrendörfer (1985; pp. 74-77, 113-121).

From a BIBD with parameters v, b, r, k, λ a so-called ***complementary*** **BIBD** with parameters $v^* = v, b^* = b, r^* = b - r, k^* = v - k, \lambda^* = b - 2r + \lambda$ can be obtained if we interchange in the incidence matrix zeros and ones.

In place of a BIBD with parameters v, b, r, k, λ, we sometimes write BIBD (v, b, r, k, λ).

We are interested in constructing BIBDs with given v and k. Such designs are called $\boldsymbol{B(v, k)}$**-designs**. In practical applications usually the number k of elements in a block is fixed by the experimental conditions, and the number v of treatments is usually also given by the problem formulation. So, for instance, in an experiment with twins, the block size is equal to 2. Amongst the $B(v, k)$-designs we are looking for the smallest existing ones. We call a $B(v, k)$-design a minimum $B(v, k)$-design if either r, b or λ is minimum and the design is existing. In Table 6.1 the BIBD listed are minimum. We try to find as many as possible minimum $B(v, k)$-designs which will be constructed by our R-programs.

TABLE 6.1: Smallest (v, k)-BIBD for $v \leq 25$ and Methods of Construction.

Number		**Method**	**Design**				**Complementary Design Constructed by Method 2**		
	v		k	b	r	λ	k^*	r^*	λ^*
1	6	3,9	**3**	10	5	2	3	5	2
2	7	3,9,16	**3**	7	3	1	4	4	2
3	**8**	6,9	**3**	56	21	6	5	35	20
4		4,5,6	**4**	14	7	3	4	7	3
5	9	4,6,9,18	**3**	12	4	1	6	8	5
6		6,7,19	**4**	18	8	3	5	10	5
7	10	9,10	**3**	30	9	2	7	21	14
8		7,18	**4**	15	6	2	6	9	5
9		18	**5**	18	9	4	5	9	4
10	11	6,7,9	**3**	55	15	3	8	40	28
11		6,7	**4**	55	20	6	7	35	21
12		6,7,15	**5**	11	5	2	6	6	3
13	12	9	**3**	44	11	2	9	33	24
14		8	**4**	33	11	3	8	22	14
15		15	**5**	132	55	20	7	77	42
16		5,8,18	**6**	22	11	5	6	11	5
17	13	6,7,9	**3**	26	6	1	10	20	15
18		3,6,7,12	**4**	13	4	1	9	9	6
19		6,7,17	**5**	39	15	5	8	24	14
20		6,7	**6**	26	12	5	7	14	7
21	14	9	**3**	182	39	6	11	143	110
22		7	**4**	91	26	6	10	65	45
23		21	**5**	182	65	20	9	117	72
24		21	**6**	91	39	15	8	52	28
25		8	**7**	26	13	6	7	13	6
26	15	3,9	**3**	35	7	1	12	28	22
27		21	**4**	105	28	6	11	77	55
28		3,8	**5**	42	14	4	10	28	18
29		8	**6**	35	14	5	9	21	12
30		3,7,15,16	**7**	15	7	3	8	8	4
31	16	6,7,9	**3**	80	15	2	13	65	52
32		4,6,13,18	**4**	20	5	1	12	15	11
33		6,7	**5**	48	15	4	11	33	22
34		6	**6**	16	6	2	10	10	6
35		6	**7**	80	35	14	9	45	24
36		4,5,6,8	**8**	30	15	7	8	15	7
37	17	6,9	**3**	136	24	3	14	112	91
38		6,7,10	**4**	68	16	3	13	52	39
39		6,7,17	**5**	68	20	5	12	48	33
40		6	**6**	136	48	15	11	88	55
41		6	**7**	136	56	21	10	80	45

TABLE 6.1: continued.

Number		Method	Design				Complementary Design Constructed by Method 2		
	v		k	b	r	λ	k^*	r^*	λ^*
42		6,7	**8**	34	16	7	9	18	9
43	18	9	**3**	102	17	2	15	85	70
44		21	**4**	153	34	6	14	119	91
45		21	**5**	306	85	20	13	221	156
46		8	**6**	51	17	5	12	34	22
47		21	**7**	306	119	42	11	187	110
48		21	**8**	153	68	28	10	85	45
49		8	**9**	34	17	8	9	17	8
50	19	6,9	**3**	57	9	1	16	48	40
51		6,7	**4**	57	12	2	15	45	35
52		6	**5**	171	45	10	14	126	91
53		6,7	**6**	57	18	5	13	39	26
54		6,17	**7**	57	21	7	12	36	22
55		6	**8**	171	72	28	11	99	55
56		6,7,15,16	**9**	19	9	4	10	10	5
57	20	9	**3**	380	57	6	17	323	272
58		8	**4**	95	19	3	16	76	60
59		8	**5**	76	19	4	15	57	42
60		21	**6**	190	57	15	14	133	91
61		21	**7**	380	133	42	13	247	156
62		21	**8**	95	38	14	12	57	33
63		21	**9**	380	171	72	11	209	110
64		5,8	**10**	38	19	9	10	19	9
65	21	9	**3**	70	10	1	18	60	51
66		7	**4**	105	20	3	17	85	68
67		7,14	**5**	21	5	1	16	16	12
68		7	**6**	42	12	3	15	30	21
69		21	**7**	30	10	3	14	20	13
70		21	**8**	105	40	14	13	65	39
71		21	**9**	35	15	6	12	20	11
72		21	**10**	42	20	9	11	22	11
73	22	9	**3**	154	21	2	19	133	114
74		7	**4**	77	14	2	18	63	51
75		21	**5**	462	105	20	17	357	272
76		7	**6**	77	21	5	16	56	40
77		7	**7**	44	14	4	15	30	20
78		7	**8**	66	24	8	14	42	26
79		21	**9**	154	63	24	13	91	52
80		21	**10**	77	35	15	12	42	22
81		7	**11**	42	21	10	11	21	10
82	23	6,9	**3**	253	33	3	20	220	190
83		6	**4**	253	44	6	19	209	171

TABLE 6.1: continued.

Number		Method	Design				Complementary Design Constructed by Method 2		
	v		k	b	r	λ	k^*	r^*	λ^*
84		6	**5**	253	55	10	18	198	153
85		6	**6**	253	66	15	17	187	136
86		6	**7**	253	77	21	16	176	120
87		6	**8**	253	88	28	15	165	105
88		6	**9**	253	99	36	14	154	91
89		6	**10**	253	110	45	13	143	78
90		6,15,16	**11**	23	11	5	12	12	6
91	24	9	**3**	184	23	2	21	161	140
92		21	**4**	138	23	3	20	115	95
93		21	**5**	552	115	20	19	437	342
94		21	**6**	92	23	5	18	69	51
95		21	**7**	552	161	42	17	391	272
96		21	**8**	69	23	7	16	46	30
97		21	**9**	184	69	24	15	115	70
98		21	**10**	276	115	45	14	161	91
99		11	**11**	552	253	110	13	299	156
100		5	**12**	46	23	11	12	23	11
101	25	6,9	**3**	100	12	1	22	88	77
102		6	**4**	50	8	1	21	42	35
103		6	**5**	30	6	1	20	24	19
104		6	**6**	100	24	5	19	76	57
105		6,17	**7**	100	28	7	18	72	51
106		6	**8**	75	24	7	17	51	34
107		6	**9**	25	9	3	16	16	10
108		6	**10**	40	16	6	15	24	14
109		6	**11**	300	132	55	14	168	91
110		6	**12**	50	24	11	13	26	13

Note: For $k = 2$ Method 1 is always used and for $k > \frac{v}{2}$ Method 2 is always used.

Occasionally so-called partially balanced incomplete block designs (PBIBDs) are used which are designed analogously to BIBDs, but which allow two or even more (t) classes of treatment groups, and members of the group jointly occur in $\lambda_1, \lambda_2, \ldots, \lambda_t$ blocks. We will not discuss methods of constructing PBIBD. For the case $t = 2$ methods of construction can be found in Raghavarao (1971) and Calinski and Kageyama (2003). We do not include these because they are seldom in use in practise. For special cases of incomplete block designs with a small number of replications r, there exist cyclic designs and α-designs (also written as alpha-designs) described in John and Williams (1995). For those designs a C-package "CycDesigN" exists but is not free of charge.

6.3 Construction of BIBD

Example 6.5 Let us consider the case $v = 7, k = 3$. The trivial design is given by the following 35 blocks:

(1, 2, 3)	(1, 3, 6)	(1, 6, 7)	(2, 4, 7)	*(3, 5, 6)*
(1, 2, 4)	**(1, 3, 7)**	(2, 3, 4)	(2, 5, 6)	(3, 5, 7)
(1, 2, 5)	(1, 4, 5)	**(2, 3, 5)**	(2, 5, 7)	(3, 6, 7)
(1, 2, 6)	(1, 4, 6)	(2, 3, 6)	**(2, 6, 7)**	(4, 5, 6)
(1, 2, 7)	(1, 4, 7)	*(2, 3, 7)*	(3, 4, 5)	**(4, 5, 7)**
(1, 3, 4)	**(1, 5, 6)**	*(2, 4, 5)*	**(3, 4, 6)**	*(4, 6, 7)*
(1, 3, 5)	*(1, 5, 7)*	(2, 4, 6)	(3, 4, 7)	(5, 6, 7)

Rasch and Herrendörfer (1982, 1985) have shown that this trivial BIBD can be split into 3 smaller ones—one with 21 blocks, which is elementary, and two with 7 blocks each, which are elementary and isomorphic (i.e., can be transformed to each other by permutation). The designs are shown by different printing of the blocks; bold; italics and normal.

6.3.1 Specific methods

In this section we list only methods giving the smallest (or at least small) BIBD for given v and k. The smallest BIBD for given $v > 2$ and $k > 1$ is that with minimum r or b or λ. There is no problem to construct the trivial BIBD with $b = \binom{v}{k}$ blocks by simply writing down all the combinations of k elements out of r. This, and nothing else, is done in the R-function `design.bib` in the library "agricolae" mentioned in Section 1.5. In the methods used below several techniques from discrete mathematics are used. These are briefly described in Appendix A. That the methods really provide the construction of BIBDs with the parameters mentioned is proved by theorems which can be found in Rasch and Herrendörfer (1985) or in Dey (1986).

All methods are implemented in the R-function `bibd`.

Examples are included in the text. See Table 6.1 for an overview of known minimal designs and methods to construct them. The method is not given in the R-syntax because not all methods work for all (v, k)-pairs. For the user the method of construction possible can be found in Table 6.1.

```
> bibd(v=x, k=y, method=z)

[1] ... result
```

Here, z stands for any number indicated in the column "Method" in Table 6.1.

In general, to find minimal BIBD for v and k of interest ($v < 26$) use

```
> bibd(v=x, k=y)

[1] ... result
```

i.e., it is not necessary to indicate a specific method; it is just an option.

The BIBD with the smallest possible number of blocks b will be returned.

Those who like to use one of the methods given below to construct a small BIBD by hand should know that some of the methods number the v treatments from $0, 1, \ldots, v-1$. The BIBD is then obtained by increasing all values by 1. Some methods number the v treatments from $0, 1, \ldots, v-2$ or from $1, 2, \ldots, v-1$ and have a further element ∞. This element will remain constant by any arithmetic operation (like addition (mod p)) and after the construction has to be replaced by v.

Method 1 Writing all combinations of $v > 2$ elements $1, 2, \ldots, v$ taking k at a time (i.e., all subgroups of size k). This gives a BIBD with parameters $v, b = \begin{pmatrix} v \\ k \end{pmatrix}, r = \begin{pmatrix} v-1 \\ k-1 \end{pmatrix}, k, \lambda = \begin{pmatrix} v-2 \\ k-2 \end{pmatrix}$.

This can be the smallest design as in the case $v = 8, k = 3$.

Method 1 gives smallest BIBDs for $k = 2$ and (due to Method 2) also for $k = v - 2$ and further for $k = v - 1$ starting from a design with each of the v elements as the v blocks. For $k = 2$ the other parameters are $b = \frac{v(v-1)}{2}; r = v - 1,\ \lambda = 1$, and for $k = v - 2$ these are $b = \frac{v(v-1)}{2};\ r = \frac{(v-1)(v-2)}{2};\ \lambda = \frac{(v-2)(v-3)}{2}$.

Example 6.6 We use the R-program `bibd(v = 5, k = 2)` to construct the BIBD $v = 5; k = 2$ by Method 1.

The result is

```
 block  treatments

  1      (1, 2)
  2      (1, 3)
  3      (1, 4)
  4      (1, 5)
  5      (2, 3)
  6      (2, 4)
  7      (2, 5)
  8      (3, 4)
  9      (3, 5)
 10      (4, 5)
```

```
v = 5    k = 2    b = 10    r = 4    lambda = 1
```

Method 2 From a BIBD with parameters v, b, r, k, λ we obtain a BIBD with parameters $v^* = v, b^* = b, r^* = b - r, k^* = v - k, \lambda^* = b - 2r + \lambda$ by building up the complementary BIBD. By this method we can construct all designs with $v > k > \frac{v}{2}$ if those with $k \leq \frac{v}{2}$ are known.

Due to Methods 1 and 2, for any $v \geq 6$, we therefore only have to find methods for constructing BIBDs for $3 \leq k \leq \frac{v}{2}$.

Example 6.7 We use the R-program `bibd(v = 5, k = 3)` to construct the BIBD $v = 5; k = 3$ by Method 2.

The result is

```
 block  treatments


   1     (1,  2,  3)
   2     (1,  2,  4)
   3     (1,  2,  5)
   4     (1,  3,  4)
   5     (1,  3,  5)
   6     (1,  4,  5)
   7     (2,  3,  4)
   8     (2,  3,  5)
   9     (2,  4,  5)
  10     (3,  4,  5)


v = 5    k= 3    b = 10    r = 6    lambda = 3
```

Method 3 Construct a projective geometry PG(n, s) (Definition A.10) and regard its points as v treatments and, for each m, the m-dimensional subspaces as blocks. This leads to a BIBD with

$$
\begin{aligned}
v &= \frac{s^{n+1} - 1}{s - 1} \\
b &= \varphi(n, m, s) \\
r &= \frac{s^{m+1} - 1}{s^{n+1} - 1}\varphi(n, m, s) \\
k &= \frac{s^{m+1} - 1}{s - 1} \\
\lambda &= \frac{(s^{m+1} - 1) \cdot (s^m - 1)}{(s^{n+1}) \cdot (s^n - 1)}\varphi(n, m, s)
\end{aligned}
$$

The term $\phi(n, m, s)$ is described in Appendix A after Definition A.10.

Example 6.8 We construct a PG(3,2) with $s = p = 2; h = 1$ (see Definition A.10), $n = 3$. The GF(2) (see Definition A.7) is $\{0, 1\}$. We need no minimum function because $h = 1$ and we determine the 15 elements (treatments) of the PG(3,2) by using all combinations of (0;1)-values in $x = (x_0, \ldots, x_3)$ besides $(0,0,0,0)$: $\{(1,0,0,0), (0,1,0,0), (0,0,1,0), (0,0,0,1), (1,1,0,0), (1,0,1,0), (1,0,0,1), (0,1,1,0), (0,1,0,1), (0,0,1,1), (1,1,1,0), (1,1,0,1), (1,0,1,1), (0,1,1,1), (1,1,1,1)\}$. With $m = 2$ the equation $(n - m = 1)$ for the 2-dimensional subspaces is $a_0 + a_1x_1 + a_2x_2 + a_3x_3 = 0$ with all combinations of coefficients from GF(2) (besides (0,0,0,0)). These are just the same quadruples as given above for the 15 points. We now write the 15×15-matrix with rows defined by the treatments and columns defined by the subspaces (blocks). We write in each cell a 1 if the point lies in the block and a 0 otherwise. Let us consider the first block defined by $a_0 = 0$. All points with a 0 in the first place are situated in this block. These are the points 2, 3, 4, 8, 9, 10 and 14. The second equation is $x_1 = 0$. In this block we find all points with a 0 as a second entry. These are the points 1, 3, 4, 6, 7, 10, 13. In this way we continue until all 15 blocks are obtained. The design is number 30 in Table 6.1.

We use the R-program `bibd(v = 15, k = 7)` to construct the BIBD $v = 15; k = 7$ by Method 3.

The result is

```
 block   treatments

   1    (1,  2,  4,  5,  8, 10, 15)
   2    (2,  3,  5,  6,  9, 11,  1)
   3    (3,  4,  6,  7, 10, 12,  2)
   4    (4,  5,  7,  8, 11, 13,  3)
   5    (5,  6,  8,  9, 12, 14,  4)
   6    (6,  7,  9, 10, 13, 15,  5)
   7    (7,  8, 10, 11, 14,  1,  6)
   8    (8,  9, 11, 12, 15,  2,  7)
   9    (9, 10, 12, 13,  1,  3,  8)
  10    (10, 11, 13, 14,  2,  4,  9)
  11    (11, 12, 14, 15,  3,  5, 10)
  12    (12, 13, 15,  1,  4,  6, 11)
  13    (13, 14,  1,  2,  5,  7, 12)
  14    (14, 15,  2,  3,  6,  8, 13)
  15    (15,  1,  3,  4,  7,  9, 14)


v = 15    k = 7    b = 15    r = 7   lambda = 3
```

Method 4 Construct a Euclidean geometry EG(n, s) (Definition A.11) and regard its points as v treatments and, for each m, the m-dimensional subspaces as blocks. This leads to a BIBD with

$$\begin{aligned} v &= s^n \\ b &= \varphi(n,m,s) - \varphi(n-1,m,s) \\ r &= \frac{s^{m+1}-1}{s^{n+1}-1}\varphi(n,m,s) \\ k &= s^m \\ \lambda &= \frac{(s^{m+1}-1)\cdot(s^m-1)}{(s^{n+1})\cdot(s^n-1)}\varphi(n,m,s) \end{aligned}$$

Example 6.9 We construct an EG$(3,2)$ with $s = p = 2; h = 1$ (see Definition A.11), $n = 3$ and $m = 2$. The parameters of the block design are

$$\begin{aligned} v &= 2^3 = 8 \\ b &= \varphi(3,2,2) - \varphi(2,2,2) = 15 - 1 = 14 \\ r &= \frac{s^3-1}{s^4-1}\cdot 15 = 7 \\ k &= s^2 = 4 \\ \lambda &= \frac{(s^3-1)\cdot(s^2-1)}{(s^4-1)\cdot(s^3-1)}\cdot 15 = 3 \end{aligned}$$

Example 6.10 From the EG (3.2) in Example 6.9 we construct a BIBD with $v = 8;\ k = 4$. For this we use Theorem A.4 and take from Example 6.8 the elements without 0 in the first place. These are (1,0,0,0), (1,1,0,0), (1,0,1,0), (1,0,0,1), (1,1,1,0), (1,1,0,1), (1,0,1,1), (1,1,1,1). The 2-dimensional subspaces are defined by $a_0x_0 + a_1x_1 + a_2x_2 + a_3x_3$. We now write the 8×14-matrix with rows defined by the treatments and columns defined by the subspaces (blocks) (1,0,0,0), (0,1,0,0), (0,0,1,0), (0,0,0,1), (1,1,0,0), (1,0,1,0), (1,0,0,1), (0,1,1,0), (0,1,0,1), (0,0,1,1), (1,1,1,0), (1,1,0,1), (1,0,1,1), (0,1,1,1). We write in each cell a 1 if the point lies in the block and a 0 otherwise analogous to Example 6.8. The design is number 4 in Table 6.1.

Example 6.11 We use the R-program `bibd(v = 8, k = 4, method = 4)` to construct the BIBD $v = 8; k = 4$ by Method 4.

The result is

```
block   treatments

  1     (1,  3,  5,  7)
  2     (1,  2,  5,  6)
```

```
 3    (1,  4,  5,  8)
 4    (1,  2,  3,  4)
 5    (1,  3,  6,  8)
 6    (1,  2,  7,  8)
 7    (1,  4,  6,  7)
 8    (2,  4,  6,  8)
 9    (3,  4,  7,  8)
10    (2,  3,  6,  7)
11    (5,  6,  7,  8)
12    (2,  4,  5,  7)
13    (3,  4,  5,  6)
14    (2,  3,  5,  8)


v = 8    k = 4    b = 14    r = 7   lambda = 3
```

Method 5 If N is the incidence matrix of a BIBD with parameters $v = b = 4l+3, r = k = 2l+1$ and $\lambda = l; (l = 1, 2, \ldots)$ and if $\tilde{N}$ is the incidence matrix of its complementary BIBD, then we obtain by the matrix $N^* = \begin{vmatrix} N & \tilde{N} \\ e'_v & 0'_v \end{vmatrix}$ the incidence matrix of a BIBD $(4l+4, 8l+6, 4l+3, 2l+2, 2l+1)$.

Example 6.12 We choose $l = 1$ and get

$$N = \begin{pmatrix} 1 & 1 & 1 & 0 & 0 & 0 & 0 \\ 1 & 0 & 0 & 1 & 1 & 0 & 0 \\ 0 & 1 & 0 & 1 & 0 & 1 & 0 \\ 1 & 0 & 0 & 0 & 0 & 1 & 1 \\ 0 & 0 & 1 & 1 & 0 & 0 & 1 \\ 0 & 0 & 1 & 0 & 1 & 1 & 0 \\ 0 & 1 & 0 & 0 & 1 & 0 & 1 \end{pmatrix} \quad \tilde{N} = \begin{pmatrix} 0 & 0 & 0 & 1 & 1 & 1 & 1 \\ 0 & 1 & 1 & 0 & 0 & 1 & 1 \\ 1 & 0 & 1 & 0 & 1 & 0 & 1 \\ 0 & 1 & 1 & 1 & 1 & 0 & 0 \\ 1 & 1 & 0 & 0 & 1 & 1 & 0 \\ 1 & 1 & 0 & 1 & 0 & 0 & 1 \\ 1 & 0 & 1 & 1 & 0 & 1 & 0 \end{pmatrix}$$

This leads to

$$N^* = \begin{pmatrix} 1 & 1 & 1 & 0 & 0 & 0 & 0 & 0 & 0 & 0 & 1 & 1 & 1 & 1 \\ 1 & 0 & 0 & 1 & 1 & 0 & 0 & 0 & 1 & 1 & 0 & 0 & 1 & 1 \\ 0 & 1 & 0 & 1 & 0 & 1 & 0 & 1 & 0 & 1 & 0 & 1 & 0 & 1 \\ 1 & 0 & 0 & 0 & 0 & 1 & 1 & 0 & 1 & 1 & 1 & 1 & 0 & 0 \\ 0 & 0 & 1 & 1 & 0 & 0 & 1 & 1 & 1 & 0 & 0 & 1 & 1 & 0 \\ 0 & 0 & 1 & 0 & 1 & 1 & 0 & 1 & 1 & 0 & 1 & 0 & 0 & 1 \\ 0 & 1 & 0 & 0 & 1 & 0 & 1 & 1 & 0 & 1 & 1 & 0 & 1 & 0 \\ 1 & 1 & 1 & 1 & 1 & 1 & 1 & 0 & 0 & 0 & 0 & 0 & 0 & 0 \end{pmatrix}$$

The BIBD 4 in Table 6.1.

Example 6.13 We use the R-program `bibd(v = 8, k = 4, method = 5)` to find another way to construct the BIBD number 4 in Table 6.1 with parameters $v = 8, b = 14, k = 4, r = 7, \lambda = 3$ again but now by Method 5.

The result is the same as in Example 6.11.

Method 6 Let $v = p^m$ and p a prime number; m a natural number. From the elements of a Galois field $\{a_0 = 0;\ a_1 = 1; \dots, a_{v-1}\}$ (Definition A.7) we construct $v-1$ Latin Squares $A_l = (a_{ij}^{(l1)});\ l = 1, \dots, v-1$ in the following way: $A_1 = (a_{ij}^{(1)})$ is the group addition table. The elements of $A_t = (a_{ij}^{(t)}); t = 2, \dots, v-1$ are $a_{ij}^t = a_{ij}^1 \cdot a_t$. Then we construct the $v(v-1)$ matrix $A = (A_1, \dots, A_{v-1})$. With the required block size k, we take k different elements from the Galois field. Then each column of A defines one block of a BIBD whose elements are just the row numbers A of the k selected elements of the field. If each block occurs $w \geq 2$ times, we delete it $w-1$ times. To find out whether blocks occur more than once, we should first order the elements in the blocks lexicographically by their size.The parameter of the original BIBD are $v = p^m$; $b = v(v-1)$; $r = k(v-1)$; k; $\lambda = k(k-1)$. The reduced BIBD has parameters: $v^* = v$, $b^* = \frac{b}{w}$, $r^* = \frac{r}{w}$, $k^* = k$, $\lambda^* = \frac{\lambda}{w}$.

Example 6.14 We construct the BIBD(9, 18, 8, 4, 3) (number 6 in Table 6.1) without and with the R-program. For $v = 9 = 3^2$ we have $p = 3; m = 2$. The minimum function is $x^2 + x + 2$ and $f(x) = \alpha_0 + \alpha_1 x$ with coefficients α_i; $i = 0, 1$ from $GF(3) = \{0, 1, 2\}$.

The function $F(X) \equiv f(x)$ (mod3; $x^2 + x + 2$) gives for all values of $f(x)$ the nine elements of GF(9):

α_0	α_1	$f(x) = F(x)$
0	0	$a_0 = 0$
0	1	$a_2 = x$
0	2	$a_3 = 2x$
1	0	$a_1 = 1$
1	1	$a_4 = 1 + x$
1	2	$a_5 = x^2 = 1 + 2x$
2	0	$a_6 = 2$
2	1	$a_7 = 2 + x$
2	2	$a_8 = 2 + 2x$

The group addition table of GF(9) is

$$\begin{pmatrix}
0 & 1 & x & 2x & 1+x & 1+2x & 2 & 2+x & 2+2x \\
1 & 2 & 1+x & 1+2x & 2+x & 2+2x & 0 & x & 2x \\
x & 1+x & 2x & 0 & 1+2x & 1 & 2+x & 2+2x & 2 \\
2x & 1+2x & 0 & x & 1 & 1+x & 2+2x & 2 & 2+x \\
1+x & 2+x & 1+2x & 1 & 2+2x & 2 & x & 2x & 0 \\
1+2x & 2+2x & 1 & 1+x & 2 & 2+x & 2x & 0 & x \\
2 & 0 & 2+x & 2+2x & x & 2x & 1 & 1+x & 1+2x \\
2+x & x & 2+2x & 2 & 2x & 0 & 1+x & 1+2x & 1 \\
2+2x & 2x & 2 & 2+x & 0 & x & 1+2x & 1 & 1+x
\end{pmatrix}$$

$$\begin{pmatrix}
0 & x & 1+2x & 2+x & 1 & 2+2x & 2x & 1+x & 2 \\
x & 2x & 1 & 2+2x & 1+x & 2 & 0 & 1+2x & 2+x \\
1+2x & 1 & 2+x & 0 & 2+2x & x & 1+x & 2 & 2x \\
2+x & 2+2x & 0 & 1+2x & x & 1 & 2 & 2x & 1+x \\
1 & 1+x & 2+2x & x & 2 & 2x & 1+2x & 2+x & 0 \\
2+2x & 2 & x & 1 & 2x & 1+x & 2+x & 0 & 1+2x \\
2x & 0 & 1+x & 2 & 1+2x & 2+x & x & 1 & 2+2x \\
1+x & 1+2x & 2 & 2x & 2+x & 0 & 1 & 2+2x & x \\
2 & 2+x & 2x & 1+x & 0 & 1+2x & 2+2x & x & 1
\end{pmatrix}$$

$$\begin{pmatrix}
0 & 2x & 2+x & 1+2x & 2 & 1+x & x & 2+2x & 1 \\
2x & x & 2 & 1+x & 2+2x & 1 & 0 & 2+x & 1+2x \\
2+x & 2 & 1+2x & 0 & 1+x & 2x & 2+2x & 1 & x \\
1+2x & 1+x & 0 & 2+x & 2x & 2 & 1 & x & 2+2x \\
2 & 2+2x & 1+x & 2x & 1 & x & 2+x & 1+2x & 0 \\
1+x & 1 & 2x & 2 & x & 2+2x & 1+2x & 0 & 2+x \\
x & 0 & 2+2x & 1 & 2+x & 1+2x & 2x & 2 & 1+x \\
2+2x & 2+x & 1 & x & 1+2x & 0 & 2 & 1+x & 2x \\
1 & 1+2x & x & 2+2x & 0 & 2+x & 1+x & 2x & 2
\end{pmatrix}$$

$$\begin{pmatrix}
0 & 1+x & 1 & 2 & 2+x & x & 2+2x & 2x & 1+2x \\
1+x & 2+2x & 2+x & x & 2x & 1+2x & 0 & 1 & 2 \\
1 & 2+x & 2 & 0 & x & 1+x & 2x & 1+2x & 2+2x \\
2 & x & 0 & 1 & 1+x & 2+x & 1+2x & 2+2x & 2x \\
2+x & 2x & x & 1+x & 1+2x & 2+2x & 1 & 2 & 0 \\
x & 1+2x & 1+x & 2+x & 2+2x & 2x & 2 & 0 & 1 \\
2+2x & 0 & 2x & 1+2x & 1 & 2 & 1+x & 2+x & x \\
2x & 1 & 1+2x & 2+2x & 2 & 0 & 2+x & x & 1+x \\
1+2x & 2 & 2+2x & 2x & 0 & 1 & x & 1+x & 2+x
\end{pmatrix}$$

$$\begin{pmatrix}
0 & 1+2x & 2+2x & 1+x & x & 2 & 2+x & 1 & 2x \\
1+2x & 2+x & x & 2 & 1 & 2x & 0 & 2+2x & 1+x \\
2+2x & x & 1+x & 0 & 2 & 1+2x & 1 & 2x & 2+x \\
1+x & 2 & 0 & 2+2x & 1+2x & x & 2x & 2+x & 1 \\
x & 1 & 2 & 1+2x & 2x & 2+x & 2+2x & 1+x & 0 \\
2 & 2x & 1+2x & x & 2+x & 1 & 1+x & 0 & 2+2x \\
2+x & 0 & 1 & 2x & 2+2x & 1+x & 1+2x & x & 2 \\
1 & 2+2x & 2x & 2+x & 1+x & 0 & x & 2 & 1+2x \\
2x & 1+x & 2+x & 1 & 0 & 2+2x & 2 & 1+2x & x
\end{pmatrix}$$

$$\begin{pmatrix}
0 & 2 & 2x & x & 2+2x & 2+x & 1 & 1+2x & 1+x \\
2 & 1 & 2+2x & 2+x & 1+2x & 1+x & 0 & 2x & x \\
2x & 2+2x & x & 0 & 2+x & 2 & 1+2x & 1+x & 1 \\
x & 2+x & 0 & 2x & 2 & 2+2x & 1+x & 1 & 1+2x \\
2+2x & 1+2x & 2+x & 2 & 1+x & 1 & 2x & x & 0 \\
2+x & 1+x & 2 & 2+2x & 1 & 1+2x & x & 0 & 2x \\
1 & 0 & 1+2x & 1+x & 2x & x & 2 & 2+2x & 2+x \\
1+2x & 2x & 1+x & 1 & x & 0 & 2+2x & 2+x & 2 \\
1+x & x & 1 & 1+2x & 0 & 2x & 2+x & 2 & 2+2x
\end{pmatrix}$$

$$\begin{pmatrix}
0 & 2+x & 1+x & 2+2x & 2x & 1 & 1+2x & 2 & x \\
2+x & 1+2x & 2x & 1 & 2 & x & 0 & 1+x & 2+2x \\
1+x & 2x & 2+2x & 0 & 1 & 2+x & 2 & x & 1+2x \\
2+2x & 1 & 0 & 1+x & 2+x & 2x & x & 1+2x & 2 \\
2x & 2 & 1 & 2+x & x & 1+2x & 1+x & 2+2x & 0 \\
1 & x & 2+x & 2x & 1+2x & 2 & 2+2x & 0 & 1+x \\
1+2x & 0 & 2 & x & 1+x & 2+2x & 2+x & 2x & 1 \\
2 & 1+x & x & 1+2x & 2+2x & 0 & 2x & 1 & 2+x \\
x & 2+2x & 1+2x & 2 & 0 & 1+x & 1 & 2+x & 2x
\end{pmatrix}$$

$$\begin{pmatrix}
0 & 2+2x & 2 & 1 & 1+2x & 2x & 1+x & x & 2+x \\
2+2x & 1+x & 1+2x & 2x & x & 2+x & 0 & 2 & 1 \\
2 & 1+2x & 1 & 0 & 2x & 2+2x & x & 2+x & 1+x \\
1 & 2x & 0 & 2 & 2+2x & 1+2x & 2+x & 1+x & x \\
1+2x & x & 2x & 2+2x & 2+x & 1+x & 2 & 1 & 0 \\
2x & 2+x & 2+2x & 1+2x & 1+x & x & 1 & 0 & 2 \\
1+x & 0 & x & 2+x & 2 & 1 & 2+2x & 1+2x & 2x \\
x & 2 & 2+x & 1+x & 1 & 0 & 1+2x & 2x & 2+2x \\
2+x & 1 & 1+x & x & 0 & 2 & 2x & 2+2x & 1+2x
\end{pmatrix}$$

We select the four elements 0; 1; 2; x, and we receive the blocks

(1,2,3,7) [from the first row of the addition table], (1,2,7,8); (1,4,6,9); (3,4,5,8); (4,6,7,9); (3,5,8,9); (1,2,5,7); (2,4,6,9); (3,5,6,8).

From the next matrix we obtain (1,2,5,9); (1,3,6,7); (2,4,6,8); (3,5,6,7); (1,4,5,9); (2,3,4,8); (2,3,7,8); (3,6,7,9); (1,5,8,9).

We continue in this way and obtain

(1,5,7,9); (2,3,6,7); (2,4,8,9); (3,6,7,8); (1,5,6,9); (2,4,5,8); (1,2,4,8); (3,4,6,7); (1,3,5,9); (1,3,4,6); (4,7,8,9); (1,3,4,5); (1,2,3,4); (3,7,8,9); (1,7,8,9); (2,5,6,9); (2,5,6,8); (2,5.6.7); (1,5,6,8); (3,4,5,7); (2,4,5,7); (2,3,6,9); (1,2,3,9); (1,4,6,8); (2,3,8,9); (1,6,7,8); (4,5,7,9); (1,2,4,7); (1,2,7,9); (3,4,6,9); (1,3,5,8); (4,6,8,9); (3,5,7,8); (1,2,6,7); (4.5.6.9); (2.3.5.8); (1,6,8,9); (4,5,6,7); (4,5,7,8); (2,3,7,9); (2,3,5,9); (1,2,6,8); (2,3,4,9); (1,3,6,8); (1,4,5,7); (1,3,4,8); (5,7,8,9); (1,3,4,7); (1,3,4,9); (2,7,8,9); (6,7,8,9); (2,3,5,6); (1,2,5,6); (2,4,5,6).

Because all the blocks are different, we have $w = 1$ and $v = 9$; $b = 72$; $r = 32$; $k = 4$; $\lambda = 12$.

But we know from Theorem 6.1 that there exists a BIBD with $v = 9$; $b = 18$; $r = 8$; $k = 4$; $\lambda = 3$ and this shows that Method 6 even if $w = 1$ must not lead to the smallest BIBD. It is also not known whether the smallest BIBD always is a subset of the BIBD obtained by Method 6, it could be a larger elementary BIBD as in the case $v = 7$; $k = 3$; $b = 21$, than the design 2 in Table 6.1.

Example 6.15 To construct the BIBD mentioned above by Method 6 the following R-code may be used: `bibd(v = 9, k = 4)`

(1,2,3,4)	(1,2,3,5)	(1,2,3,6)	(1,2,3,7)	(1,2,3,8)	(1,2,3,9)	(1,2,4,7)	(1,2,4,9)
(1,2,5,8)	(1,2,5,9)	(1,2,6,7)	(1,2,6,8)	(1,3,4,7)	(1,3,4,8)	(1,3,5,7)	(1,3,5,9)
(1,3,6,8)	(1,3,6,9)	(1,4,5,6)	(1,4,5,7)	(1,4,5,9)	(1,4,6,7)	(1,4,6,8)	(1,4,7,8)
(1,4,7,9)	(1,5,6,8)	(1,5,6,9)	(1,5,7,9)	(1,5,8,9)	(1,6,7,8)	(1,6,8,9)	(1,7,8,9)
(2,3,4,8)	(2,3,4,9)	(2,3,5,7)	(2,3,5,8)	(2,3,6,7)	(2,3,6,9)	(2,4,5,6)	(2,4,5,8)
(2,4,5,9)	(2,4,6,7)	(2,4,6,9)	(2,4,7,9)	(2,4,8,9)	(2,5,6,7)	(2,5,6,8)	(2,5,7,8)
(2,5,8,9)	(2,6,7,8)	(2,6,7,9)	(2,7,8,9)	(3,4,5,6)	(3,4,5,7)	(3,4,5,8)	(3,4,6,8)
(3,4,6,9)	(3,4,7,8)	(3,4,8,9)	(3,5,6,7)	(3,5,6,9)	(3,5,7,8)	(3,5,7,9)	(3,6,7,9)
(3,6,8,9)	(3,7,8,9)	(4,5,6,7)	(4,5,6,8)	(4,5,6,9)	(4,7,8,9)	(5,7,8,9)	(6,7,8,9)

Remark to Method 6

The selection of k from the v elements of the Galois field is crucial. Which subset we select has an influence on the non-reducible number of blocks of the BIBD constructed by Method 6. In the case $v = 31$ we found for $k = 4$ a BIBD with $b = 465$ by selecting the first four elements (0, 1, 2, 3) of GF(31), and this could of course not be reduced to 310 blocks. By another choice (0,1,2,4), we found a BIBD with $b = 930$ (which is the maximum number 31*30) blocks which also cannot be reduced to 310 blocks. Our program was therefore improved by randomly selecting several k-tupels and, from the so-constructed BIBDs, selecting the smallest one. After several random selections we found the minimum block number 310 using the elements (0, 8, 17, 29) of GF(31).

We propose to apply Method 6 only in those cases where no other specific method is available and to compare the result with that of the general method described in Section 6.3.2. However when the reader is using our R-program with Method 6 and fixes the number of random selection of subsets of size k large enough, there is a good chance of coming out with the smallest possible design. In Table 6.1 in those cases where Method 6 is listed as the only one, the number of blocks is that which we reached as the smallest one in several runs of our R-program.

Example 6.15 – continued To construct a smaller BIBD as shown above we can try using the following R-code in which Nsim means the number of random selection of k elements from the Galois field demanded by the user.

```
> bibd(p=3,m=2,k=4,Nsim=20)
```

[1] (1,2,4,6) (1,2,5,7) (1,2,8,9) (1,3,4,9) (1,3,5,6) (1,3,7,8) (1,4,5,8) (1,6,7,9) (2,3,4,5) (2,3,6,8) (2,3,7,9) (2,4,7,8) (2,5,6,9) (3,4,6,7) (3,5,8,9) (4,5,7,9) (4,6,8,9) (5,6,7,8)

Theorem 6.1 (First Fundamental Theorem, Theorem I from Bose, 1939) From a (v, k, ρ, s, t)-difference set $M = \{M_1, \dots, M_t\}$ in Definition A.12 by the relation $x_j^d = x_j^g + a$; $x_j^g \in M_j$; $a \in A$, together with the sets in M, mt sets are defined which compose a BIBD with parameters $v = ms$; $b = mt$, r, k, λ.

Method 7 Let there be a (v, k, ρ, s, t)-difference set (Definition A.12) based on a module with $m = v/s$ elements. From each of the sets $M_l (l = 1, \dots, t)$ we establish m sets $M_{lh}(l = 1, \dots, t, h = 1, \dots, m)$ by setting $M_{l1} = M_l$ and obtain M_{li-1} from M_l by adding mod m the unity element from the appropriate class i times to each element of the set M_l. These mt sets M_{lh} form a BIBD with parameters $v = ms$, $b = mt, k = r = \frac{tk}{s}$, $\lambda = \frac{tk(k-1)}{s(ms-1)} = \frac{\rho}{s}$.

Example 6.16 We take the difference set: $\{0_1, 3_1, 0_2, 4_2\}$, $\{0_1, 2_1, 1_2, 3_2\}$, $\{1_1, 2_1, 3_1, 0_2\}$ (from the Example A.1) and construct the sets M_{lh}:

M_{11} to M_{15}	M_{21} to M_{25}	M_{31} to M_{35}
$(0_1, 3_1, 0_2, 4_2)$;	$(0_1, 0_2, 1_2, 3_2)$;	$(1_1, 2_1, 3_1, 0_2)$;
$(1_1, 4_1, 1_2, 0_2)$;	$(1_1, 1_2, 2_2, 4_2)$;	$(2_1, 3_1, 4_1, 1_2)$;
$(2_1, 0_1, 2_2, 1_2)$;	$(2_1, 2_2, 3_2, 0_2)$;	$(3_1, 4_1, 0_1, 2_2)$;
$(3_1, 1_1, 3_2, 2_2)$;	$(3_1, 3_2, 4_2, 1_2)$;	$(4_1, 0_1, 1_1, 3_2)$;
$(4_1, 2_1, 4_2, 2_2)$;	$(4_1, 4_2, 0_2, 2_2)$;	$(0_1, 1_1, 2_1, 4_2)$;

We use now the following coding: $0_1 \approx 1$, $0_2 \approx 2$, $1_1 \approx 3$, $1_2 \approx 4$, $2_1 \approx 5$, $2_2 \approx 6$, $3_1 \approx 7$, $3_2 \approx 8$, $4_1 \approx 9$, $4_2 \approx 10$.

In this way we have constructed the BIBD number 8 in Table 6.1. Tables of difference sets are available, for instances, in Takeuchi (1962), Raghavarao (1971), McFarland (1973), Rasch and Herrendörfer (1985) and Colbourn and Dinitz (2006).

Example 6.17 We use the R-program `bibd(v = 10, k = 4)` to construct the BIBD $v = 10; k = 4$ by Method 7.
The result is

```
 block   treatments

   1     (1,  4,  6, 10)
   2     (2,  5,  7,  6)
   3     (3,  1,  8,  7)
   4     (4,  2,  9,  8)
   5     (5,  3, 10,  9)
   6     (1,  6,  7,  9)
   7     (2,  7,  8, 10)
   8     (3,  8,  9,  6)
   9     (4,  9, 10,  7)
  10     (5, 10,  6,  8)
  11     (2,  3,  4,  6)
  12     (3,  4,  5,  7)
  13     (4,  5,  1,  8)
  14     (5,  1,  2,  9)
  15     (1,  2,  3, 10)

v = 10    k = 4    b = 15    r = 6   lambda = 2
```

Theorem 6.2 (Second Fundamental Theorem, Theorem II from Bose, 1939). From a (v, k, ρ, s, t, t')-difference set K (with the element ∞) in Definition A.14 by the relation $x_j^d = x_j^g \in M_j$ or N_j; $a \in A$, $\infty + a = \infty$ together with the sets in K, $m(t + t')$ sets are defined which compose a BIBD with parameters $v = ms + 1;\ m(t + t'), r = mt, k, \lambda$.

Method 8 Let there be a $(v; k; \rho; s; t; t')$-difference set (Definition A.15) based on a module containing $m = \frac{v-1}{s}$ elements and the element ∞ (a fixed element). $\{L_1, \ldots, L_{t+1'}\} = \{M_1, \ldots, L_{t+t'}\} = \{M_1, \ldots, M_t, N_1, \ldots, N_{t'}\}$ and take from each L_h exactly m sets $L_{ij}(h = 1, \ldots, m)$ by setting and obtain L_{iu+1} from L_i by adding (modulo m) u times the unit element of the corresponding class K_u to L_i while leaving element ∞ unchanged. These $m(t+t')$ sets L_{ih} form a BIBD with $v = ms + 1, b = m(t + t'), k, r = tm, \lambda = \frac{\rho}{s}$.

Example 6.18 We consider the module of Example A.1 with $s = t =$

t' and $K_\infty = \{(0,3,1);\ (\infty,0,1)\}$. From this we obtain (mod 5) the 10 = 5(1+1) blocks (0,3,1); (1,4,2); (2,0,3); (3,1,4); (4,2,0); (0,3,1); $(\infty,0,1)$; $(\infty,2,3)$; $(\infty,3,4)$; $(\infty,4,0)$. The $v = 5 \cdot 1 + 1 = 6$ treatments are $0,1,2,3,4,\infty$. Which can be renamed to $1,\ldots,6$.

Example 6.19 We use the R-program `bibd(v = 14, k= 7)` to construct the BIBD number 25 in Table 6.1 with parameters $v = 14, k = 7, b = 26, r = 13, \lambda = 6$.

The result is

block	*treatments*
1	(1, 2, 4, 5, 10, 11, 13)
2	(2, 3, 5, 6, 11, 12, 1)
3	(3, 4, 6, 7, 12, 13, 2)
4	(4, 5, 7, 8, 13, 1, 3)
5	(5, 6, 8, 9, 1, 2, 4)
6	(6, 7, 9, 10, 2, 3, 5)
7	(7, 8, 10, 11, 3, 4, 6)
8	(8, 9, 11, 12, 4, 5, 7)
9	(9, 10, 12, 13, 5, 6, 8)
10	(10, 11, 13, 1, 6, 7, 9)
11	(11, 12, 1, 2, 7, 8, 10)
12	(12, 13, 2, 3, 8, 9, 11)
13	(13, 1, 3, 4, 9, 10, 12)
14	(2, 4, 5, 10, 11, 13, 14)
15	(3, 5, 6, 11, 12, 1, 14)
16	(4, 6, 7, 12, 13, 2, 14)
17	(5, 7, 8, 13, 1, 3, 14)
18	(6, 8, 9, 1, 2, 4, 14)
19	(7, 9, 10, 2, 3, 5, 14)
20	(8, 10, 11, 3, 4, 6, 14)
21	(9, 11, 12, 4, 5, 7, 14)
22	(10, 12, 13, 5, 6, 8, 14)
23	(11, 13, 1, 6, 7, 9, 14)
24	(12, 1, 2, 7, 8, 10, 14)
25	(13, 2, 3, 8, 9, 11, 14)
26	(1, 3, 4, 9, 10, 12, 14)

v = 14   k = 7   b = 26   r = 13   lambda = 6

Method 9 (BIBDs with block size $k = 3$) It is sufficient (from a theorem in Hwang and Lin, 1974) to specify BIBDs with $k = 3$ and

$$\begin{aligned} \lambda &= 1, v \equiv 1 \text{ or } 3 \pmod 6 \\ \lambda &= 2, v \equiv 0 \text{ or } 4 \pmod 6 \\ \lambda &= 3, v \equiv 5 \pmod 6 \\ \lambda &= 6, v \equiv 2 \pmod 6 \end{aligned}$$

Take a set $D(v, \lambda)(\lambda = 1, 2, 3, 6)$ of admissible triples (Definition A.16) some with an index I, some without) containing each positive difference λ times. Construct the initial blocks of the form $(0, d_1, d_1 + d_2)$. Construct the blocks of the BIBD by adding to the initial block the numbers $0, 1, \ldots, v^* - 1$ (mod v^*) if the corresponding triple has no index. Construct the blocks of the BIBD by adding to the initial block the numbers $0, 1, \ldots, I - 1$ (mod v^*) if the corresponding triple has the index *I*.

For $\lambda = 1$ $v \neq 9, v = v^*$ and we obtain $D(v, 1) = Z(M \cup R)$ by means of the matrices M and R from Table 6.2.

$Z(A)$ with a 4x3 matrix $A = (a_{ij})$ is defined with $v = 6t + 1;\ t = 4h;\ h > 0$ as follows

$$Z(A) = \begin{pmatrix} a_{11} + 2i, 11h + a_{12} - i, 11h + a_{11} + a_{12} + i;\ i = 0, \ldots, a_{13} \\ a_{21} + 2i, 8h + a_{22} - i, 8h + a_{21} + a_{22} + i;\ i = 0, \ldots, a_{23} \\ 3h + a_{31} + 2i, 6h + a_{32} - i, 9h + a_{31} + a_{32} + i;\ i = 0, \ldots, a_{33} \\ 3h + a_{41} + 2i, 3h + a_{42} - i, 6h + a_{41} + a_{42} + i;\ i = 0, \ldots, a_{43} \end{pmatrix}$$

Note that triples with negative a_{i3} have been omitted.

For $v = 9$ take

$$D(9, 1) = \left\{ \begin{array}{c} (1, 2, 3) \\ (4, \infty, \infty) \end{array} \right\}$$

For $\lambda = 2$ and $v = 6t + 6$, we have $v = v*$ and take

$$D(6t + 6, 2) = \left\{ \begin{array}{c} (1 + 2i, t + 1 - i, t + 2 + i);\ i = 0, \ldots, t - 1 \\ (2 + 2i, 3t + 1 - i, 3t + 2 + i)^*;\ i = 0, \ldots, t - 1 \\ (3t + 2, \infty, \infty)^* \end{array} \right\}$$

$$D(6t + 4, 2) = \left\{ \begin{array}{c} (1 + 2i, t - 1 - i, t + 1 + i);\ i = 0, \ldots, t - 1 \\ (2 + 2i, 3t + 1 - i, 3t + 1 - i)^*;\ i = 0, \ldots, t - 1 \\ (3t + 1, \infty, \infty)^* \\ (2t + 1, 2t + 1, 2t + 1)^*_{2t+1} \end{array} \right\}$$

TABLE 6.2
Matrices M and R for Constructing BIBDs by Means of Method 9 for $\lambda = 1$.

$v = 6t + 1,\ \ t = 4h,\ \ h > 0$

$$M = \begin{pmatrix} 1 & 0 & h-1 \\ 2 & 0 & h-2 \\ 1 & 0 & h-1 \\ 2 & 0 & h-2 \end{pmatrix}, \qquad R = \{(2h, 6h+1, 8h+8)\};$$

$v = 6t + 1,\ \ t = 4h + 1$

$$M = \begin{pmatrix} 1 & 3 & h-1 \\ 2 & 2 & h-1 \\ 3 & 1 & h-1 \\ 2 & 1 & h-1 \end{pmatrix}, \qquad R = \{(2h+1, 6h+2, 8h+3)\};$$

$v = 6t + 1,\ \ t = 4h + 2$

$$M = \begin{pmatrix} 1 & 5 & h-1 \\ 2 & 3 & h-1 \\ 4 & 2 & h-1 \\ 3 & 1 & h-1 \end{pmatrix}, \qquad R = \left\{ \begin{array}{c} (2h+1, 6h+3, 8h+4) \\ (3h+2, 9h+5, 12h+6)^* \end{array} \right\};$$

$v = 6t + 1,\ \ t = 4h + 3$

$$M = \begin{pmatrix} 1 & 3 & h-1 \\ 2 & 2 & h-1 \\ 3 & 1 & h-1 \\ 2 & 1 & h-1 \end{pmatrix}, \qquad R = \left\{ \begin{array}{c} (2h+1, 6h+4, 8h+5) \\ (3h+2, 9h+6, 12h+8) \\ (3h+3, 9h+7, 12h+9)^* \end{array} \right\};$$

$v = 6t + 3,\ \ t = 4h$

$$M = \begin{pmatrix} 1 & 1 & h-1 \\ 2 & 0 & h-1 \\ 2 & 0 & h-1 \\ 1 & 0 & h-1 \end{pmatrix}, \qquad R = \{(8h+1, 8h+1, 8h+1)^*_{8h+1}\};$$

$v = 6t + 3,\ \ t = 4h + 1,\ \ h > 0$

$$M = \begin{pmatrix} 1 & -3 & h-2 \\ 2 & 2 & h-2 \\ 3 & 3 & h-1 \\ 5 & -1 & h-1 \end{pmatrix}, \qquad R = \left\{ \begin{array}{l} (2h-1, 3h+3, 5h+2) \\ (3h, 9h+3, 12h+3) \\ (3h+1, 9h+4, 12h+4)^* \\ (8h+3, 8h+3, 8h+3)^*_{8h+3} \end{array} \right\};$$

$v = 6t + 3,\ \ t = 4h + 2$

$$M = \begin{pmatrix} 1 & 6 & h-1 \\ 2 & 4 & h-2 \\ 4 & 3 & h-1 \\ 3 & 1 & h \end{pmatrix}, \qquad R = \left\{ \begin{array}{c} (3h+2, 9h+6, 12h+7)^* \\ (8h+5, 8h+5, 8h+5)^*_{8h+5} \end{array} \right\};$$

$v = 6t + 3,\ \ t = 4h + 3$

$$M = \begin{pmatrix} 1 & 9 & h \\ 2 & 6 & h-1 \\ 3 & 5 & h \\ 4 & 2 & h \end{pmatrix}, \qquad R = \{(8h+7, 8h+7, 8h+7)^*_{8h+7}\}$$

For $\lambda = 3$ we have $v = v^*$ take the $D(v, 3)$ from Table 6.3.

For $\lambda = 6$ we have $v^* = v - 1$ and take

$$D(6t{+}2, 2) = \left\{ \begin{array}{c} 5 \times D(6t+1,1);\ D(6t+1,1) \text{ reduced by any one triple } (d_1', d_2', d_3') \\ (d_1', \infty, \infty)^*, (d_2', \infty, \infty)^*, (d_3' \infty, \infty)^* \end{array} \right\}$$

Example 6.20 Let $v = 15 = 6t + 3$ with $t = 4h + 2, h = 0$; thus we have $\lambda = 1$. Since $h = 0$ only the fourth row of M from Table 6.2

TABLE 6.3
$D(v, 3)$ for Constructing BIBDs by Means of Method 9 for $\lambda = 3$ and $v = 6t{+}5$.

$t = 4h, h = 0$

$D(5, 3) = \{(1, 1, 2), (1, 2, , 2)^*\}$

$t = 4h, h > 0$

$$D(24h + 5, 3) = \left\{ \begin{array}{l} [D(24h + 7, 1) - (2h - 1, 10h + 4, 12h + 3)] \\ D^*(24h + 4, 2), (2h - 1, 10h + 4, 12h + 2)^* \end{array} \right\};$$

$t = 4h + 1, h > 1$

$$D(24h + 11, 3) = \left\{ \begin{array}{l} D(24h + 7, 1) \\ (1 + 2i, 6h + 2 - i, 6h + 3 + i),\ i = 0, \ldots, 2h - 1 \\ (2 + 2i, 10h + 4 - i, 10h + 6 + i)\ i = 0, \ldots, 2h - 1 \\ (1 + 2i, 10h + 4 - i, 10h + 5 - i)\ i = 0, \ldots, 2h - 1 \\ (2 + 2i, 6h + 2 - i, 6h + 4 + i)\ i = 0, \ldots, 2h - 1 \\ (4h + 2, 6h + 3, 10h + 5) \\ (4h + 1, 8h + 3, 12h + 4) \\ (4h + 2, 8h + 4, 12h + 5)^* \end{array} \right\};$$

$t = 4h + 2$

$$D(24h + 17, 3) = \left\{ \begin{array}{l} [D(24h + 19, 1) - (3h + 2, 9h + 6, 12h + 8) \\ -(3h + 3, 9h + 7, 12h + 9)^*] \\ D^*(24h + 16, 2) \\ (3h + 2, 9h + 7, 12h + 8)^* \\ (3h + 3, 9h + 6, 12h + 8)^* \end{array} \right\};$$

$t = 4h + 3$

$$D(24h + 23, 3) = \left\{ \begin{array}{l} (1 + 2i, 10h + 9i, 10h + 10 + 9),\ i = 0, \ldots, 2h + 1 \\ (2 + 2i, 6h + 5 - i, 6h + 7 + i),\ i = 0, \ldots, 2h \\ (1 + 2i, 6h + 5 - i, 6h + 6 + i),\ i = 0, \ldots, 2h + 1 \\ (2 + 2i, 10h + 9 - i, 10h + 11 + i),\ i = 0, \ldots, 2h \\ (1 + 2i, 6h + 6 - i, 6h + 7 + i),\ i = 0, \ldots, h \\ (2 + 2i, 10h + 9 - i, 10h + 11 + i),\ i = 0, \ldots, h - 1 \\ (2h + 3 + 2i, 9h + 8 - i, 11h + 11 + i),\ i = 0, \ldots, h \\ (2h + 4 + 2i, 5h + 4 - i, 7h + 8 + i),\ i = 0, \ldots, h - 1 \\ (4h + 4, 5h + 5, 9h + 9) \\ (2h + 2, 8h + 8, 10h + 10) \\ (4h + 4, 6h + 6, 10h + 10) \end{array} \right\}$$

gives a triple $Z(M)$, namely (3,1,4). We find $R = \left\{ \begin{array}{c} (2,6,7)^*) \\ (5,5,5)_5^* \end{array} \right\}$. Hence

$$D(15,1) = \left\{ \begin{array}{c} (1,3,4) \\ (2,7,6)^* \\ (5,5,5)_5^* \end{array} \right\}$$

This gives the initial blocks $(0,1,4); (0,2,8); (0,5,10)_5$. We obtain 15 blocks from each of the first two initial blocks, and from the third we obtain the blocks (0,5,10); (1,6,11); (2,7,12); (3,8,13); and (4,9,14). Adding a "1" to all elements gives us the usual notation from 1 to 15.

Example 6.21 We use the R-program `bibd(v = 14, k = 3)` to construct the BIBD number 21 from Table 6.1 with parameters $v = 14, k = 3, b = 182, r = 39, \lambda = 6$. Because the design has 182 blocks, we do not print it. The reader may run the program if he is interested in just this design.

Method 10 A BIBD with $v = s^2, b = s(s+1), k = s$ (for instance, constructed by Method 4 with $n = 2$ and $m = 1$) can be written in the form of $s+1$ groups with s blocks each. W.l.o.g. group 1 can be written as:

Block 1	1	2	...	s
Block 2	$s+1$	$s+2$	...	$2s$
...	...	...	...	...
Block s	$(s-1)s+1$	$(s-1)s+2$		s^2

We take now $s-1$ times groups 2 to s and group 1 once. From group 1 we build up additionally all $(s-1)$-tupels from each block and supplement them by the treatment $v+1 = s^2+1$. We obtain thus a BIBD with parameters $v = s^2+1$; $k = s$; $b = s(s^2+1); r = s^2; \lambda = s-1$. Contrary to other methods of course not all blocks are different; some blocks occur repeatedly.

Example 6.22 We start with the following BIBD with parameters $v = 9$; $k = 3$; $b = 12$; $r = 4$; $\lambda = 1$.

$$\text{Group 1: } \left\{ \begin{array}{c} (1,2,6) \\ (3,4,5) \\ (8,8,9) \end{array} \right\}; \text{ Group 2: } \left\{ \begin{array}{c} (1,3,7) \\ (2,4,9) \\ (5,6,8) \end{array} \right\}; \text{ Group 3: } \left\{ \begin{array}{c} (1,4,8) \\ (2,5,7) \\ (3,6,9) \end{array} \right\};$$

$$\text{Group 4: } \left\{ \begin{array}{c} (1,5,9) \\ (2,3,8) \\ (4,6,7) \end{array} \right\}$$

From this we obtain the BIBD with parameters $v = 10$; $k = 3$; $b = 30$; $r = 9; \lambda = 2$.

$$\left\{ \begin{array}{c} (1,2,6) \\ (3,4,5) \\ (7,8,9) \end{array} \right\}; \left\{ \begin{array}{c} (1,3,7) \\ (2,4,9) \\ (5,6,8) \end{array} \right\}; \left\{ \begin{array}{c} (1,3,7) \\ (2,4,9) \\ (5,6,8) \end{array} \right\}; \left\{ \begin{array}{c} (1,4,8) \\ (2,5,7) \\ (3,6,9) \end{array} \right\}; \left\{ \begin{array}{c} (1,4,8) \\ (2,5,7) \\ (3,6,9) \end{array} \right\};$$

$$\left\{ \begin{array}{c} (1,5,9) \\ (2,3,8) \\ (4,6,7) \end{array} \right\}; \left\{ \begin{array}{c} (1,5,9) \\ (2,3,8) \\ (4,6,7) \end{array} \right\}; \left\{ \begin{array}{c} (1,2,10) \\ (3,4,10) \\ (7,8,10) \end{array} \right\}; \left\{ \begin{array}{c} (1,10,6) \\ (3,10,5) \\ (7,10,9) \end{array} \right\}; \left\{ \begin{array}{c} (10,2,6) \\ (10,4,5) \\ (10,8,9) \end{array} \right\};$$

Example 6.23 We can use the R-program `bibd(v = 10, k = 3)` to construct the BIBD number 7 in Table 6.1 with $v = 10, k = 3$ by Method 10, the result is

$$\begin{array}{ccccc} (1,2,3) & (7,8,9) & (4,6,9) & (1,4,7) & (4,8,10) \\ (2,3,4) & (8,9,1) & (5,7,1) & (2,5,8) & (5,9,10) \\ (3,4,5) & (9,1,2) & (6,8,2) & (3,6,9) & (6,1,10) \\ (4,5,6) & (1,3,6) & (7,9,3) & (1,5,10) & (7,2,10) \\ (5,6,7) & (2,4,7) & (8,1,4) & (2,6,10) & (8,3,10) \\ (6,7,8) & (3,5,8) & (9,2,5) & (3,7,10) & (9,4,10) \end{array}$$

Method 11 (Abel, 1994) This method is described in Abel (1994); our program follows this method.

Example 6.24 We use the R-program `bibd(v = 24, k = 11)` to construct the BIBD number 99 in Table 6.1 with parameters $v = 24, k = 11, b = 552, r = 253, \lambda = 110$ by Method 11. Because the design has 552 blocks we do not print it, the reader may run the program if he is interested in just this design.

Method 12 If $12t + 1$ (t an integer) is a prime or a prime power and where t is odd then the initial blocks $(0, x^{2i}, x^{4t+2i}, x^{8t+2i})(i = 0, 1, \ldots, t-1)$ provide a construction of BIBDs with $v = 12t+1$; $k = 4, \lambda = 1, r = 4t, b = t(12t+1)$.

Example 6.25 We use the R-program `bibd(v = 13, k = 4)` to construct the BIBD 18 with parameters $v = 13, k = 4, b = 13, r = 4$ and $\lambda = 1$ by Method 12. The result is

```
block   treatments


  1      (1,  3,  9, 13)
  2      (2,  4, 10,  1)
  3      (3,  5, 11,  2)
  4      (4,  6, 12,  3)
  5      (5,  7, 13,  4)
  6      (6,  8,  1,  5)
  7      (7,  9,  2,  6)
  8      (8, 10,  3,  7)
  9      (9, 11,  4,  8)
 10     (10, 12,  5,  9)
 11     (11, 13,  6, 10)
 12     (12,  1,  7, 11)
 13     (13,  2,  8, 12)


v = 13    k = 4    b = 13    r = 4    lambda = 1
```

Method 13 If $4t+1$ (t an integer) is a prime or a prime power and x is a primitive root of GF$(4t+1)$, q is an odd integer and α is an odd integer such that $\frac{(x^\alpha+1)}{(x^\alpha-1)} = x^q$. The initial blocks:

$$(x_1^{2i}, x_1^{2t+2i}, x_2^{\alpha+2i}, x_2^{\alpha+2t+2i})\ (i=0,1,\dots,t-1),$$
$$(x_2^{2i}, x_2^{2t+2i}, x_3^{\alpha+2i}, x_3^{\alpha+2t+2i})\ (i=0,1,\dots,t-1),$$
$$(x_3^{2i}, x_3^{2t+2i}, x_1^{\alpha+2i}, x_1^{\alpha+2t+2i})\ (i=0,1,\dots,t-1),$$
$$\infty, 0_1, 0_2, 0_3$$

provide a construction of a RBIBDs with $v = 12t+4; k = 4, \lambda = 1, r = 4t+1, b = (4t+1)(3t+1)$.

Example 6.26 Using Method 13, let $t = 1$, then we have as the primitive element 2 of GF(5) and with $\alpha = 1 \frac{(2^\alpha+1)}{(2^\alpha-1)} = 3 = 2^3$ (mod5) which means that $q = 3$. The initial blocks are $(1_1, 4_1, 2_2, 3_2)$; $(1_2, 4_2, 2_3, 3_3)$; $(1_3, 4_3, 2_1, 3_1)$; $(\infty; 0_1, 0_2, 0_3)$ and can modulo 5 be developed to the complete BIBD, the 16 different symbols can finally be transformed into $1, 2, \dots, 16$.

Remark: The design with $v = 28$ is obtained by Method 13 via the minimum function x^2+x+2 from Table A.3.

Example 6.27 We use the R-program `bibd(v = 16, k = 4)` to construct the BIBD number 32 from Table 6.1 with parameters $v = 16, k = 4, b = 20, r = 5, \lambda = 1$ by Method 13.

The result is:

```
block   treatments

  1     (2,   5,  8,  9)
  2     (3,   1,  9, 10)
  3     (4,   2, 10,  6)
  4     (5,   3,  6,  7)
  5     (1,   4,  7,  8)
  6     (7,  10, 13, 14)
  7     (8,   6, 14, 15)
  8     (9,   7, 15, 11)
  9     (10,  8, 11, 12)
 10     (6,   9, 12, 13)
 11     (12, 15,  3,  4)
 12     (13, 11,  4,  5)
 13     (14, 12,  5,  1)
 14     (15, 13,  1,  2)
 15     (11, 14,  2,  3)
 16     (1,   6, 11, 16)
 17     (2,   7, 12, 16)
```

```
  18      (3,   8, 13, 16)
  19      (4,   9, 14, 16)
  20      (5,  10, 15, 16)

 v = 16     k = 4     b = 20     r = 5    lambda = 1
```

Method 14 If $20t + 1$ (t an positive integer) is a prime or a prime power and an odd power of a primitive root of GF$(20t + 1)$, then the initial blocks $(x^0, x^{4t}, x^{8t}, x^{12t}, x^{16t})$; $(x^2, x^{4t+2}, x^{8t+2}, x^{12t+2}, x^{16t+2})$; ...; $(x^{2t+2}, x^{6t+2}, x^{10t+2}, x^{14t+2}, x^{18t+2})$ provide a BIBD with $v = 20t + 1; k = 5, \lambda = 1, r = 5t, b = t(20t + 1)$.

Example 6.28 Using Method 14 and letting $t = 2$, then we have 7 as a primitive element of GF(41). Now $7^{4t} + 1 = 7^8 + 1 = 38 = 7^5 \pmod{41}$, which means that the initial blocks are (1, 37, 16, 18, 10) and (8, 9, 5, 21, 39).

Example 6.29 We use the R-program `bibd(v = 19, k = 9)` to construct the BIBD number 56 with parameters ($v = 19, k = 9, b = 19, r = 9, \lambda = 4$) by Method 14.

The result is:

```
  block   treatments

    1     (2,   5,  6,  7,  8, 10, 12, 17, 18)
    2     (3,   6,  7,  8,  9, 11, 13, 18, 19)
    3     (4,   7,  8,  9, 10, 12, 14, 19,  1)
    4     (5,   8,  9, 10, 11, 13, 15,  1,  2)
    5     (6,   9, 10, 11, 12, 14, 16,  2,  3)
    6     (7,  10, 11, 12, 13, 15, 17,  3,  4)
    7     (8,  11, 12, 13, 14, 16, 18,  4,  5)
    8     (9,  12, 13, 14, 15, 17, 19,  5,  6)
    9     (10, 13, 14, 15, 16, 18,  1,  6,  7)
   10     (11, 14, 15, 16, 17, 19,  2,  7,  8)
   11     (12, 15, 16, 17, 18,  1,  3,  8,  9)
   12     (13, 16, 17, 18, 19,  2,  4,  9, 10)
   13     (14, 17, 18, 19,  1,  3,  5, 10, 11)
   14     (15, 18, 19,  1,  2,  4,  6, 11, 12)
   15     (16, 19,  1,  2,  3,  5,  7, 12, 13)
   16     (17,  1,  2,  3,  4,  6,  8, 13, 14)
   17     (18,  2,  3,  4,  5,  7,  9, 14, 15)
   18     (19,  3,  4,  5,  6,  8, 10, 15, 16)
   19     (1,   4,  5,  6,  7,  9, 11, 16, 17)

 v = 19     k = 9     b = 19     r = 9    lambda = 4
```

Method 15 If $4t+1$ (t an positive integer) is a prime or a prime power and x is a primitive element of $GF(4t+1)$, then the initial block $x^0, x^2, x^4, \ldots, x^{4t-2}$ provides a BIBD with $v = b = 4t-1; r = k = 2t-1, \lambda = t-1$.

Example 6.30 We use the R-program `bibd(v = 11, k = 5)` to construct the BIBD number 12 with parameters $v = 11, k = 5, b = 11, r = 5, \lambda = 2$ by Method 15.

The result is

```
 block   treatments


  1      (2,  4,  5,  6, 10)
  2      (3,  5,  6,  7, 11)
  3      (4,  6,  7,  8,  1)
  4      (5,  7,  8,  9,  2)
  5      (6,  8,  9, 10,  3)
  6      (7,  9, 10, 11,  4)
  7      (8, 10, 11,  1,  5)
  8      (9, 11,  1,  2,  6)
  9      (10,  1,  2,  3,  7)
  10     (11,  2,  3,  4,  8)
  11     (1,  3,  4,  5,  9)

v = 11     k = 5     b = 11     r = 5    lambda = 2
```

Method 16 Let H be an Hadamard matrix of order $n = 4t$ and let B be a matrix obtained from H by deleting its first row and its first column. Then $N = \frac{1}{2}(B + e_n e_n^T)$ is the incidence matrix of a BIBD with $v = b = 4t-1; r = k = 2t-1, \lambda = t-1$.

Example 6.31 If $v = b = 15; r = k = 7, \lambda = 3$, we use the Hadamard matrix of order 16 ($t = 4$)

$$\begin{pmatrix}
1 & 1 & 1 & 1 & 1 & 1 & 1 & 1 & 1 & 1 & 1 & 1 & 1 & 1 & 1 & 1 \\
1 & -1 & 1 & -1 & 1 & -1 & 1 & -1 & 1 & -1 & 1 & -1 & 1 & -1 & 1 & -1 \\
1 & 1 & -1 & -1 & 1 & 1 & -1 & -1 & 1 & 1 & -1 & -1 & 1 & 1 & -1 & -1 \\
1 & -1 & -1 & 1 & 1 & -1 & -1 & 1 & 1 & -1 & -1 & 1 & 1 & -1 & -1 & 1 \\
1 & 1 & 1 & 1 & -1 & -1 & -1 & -1 & 1 & 1 & 1 & 1 & -1 & -1 & -1 & -1 \\
1 & -1 & 1 & -1 & -1 & 1 & -1 & 1 & 1 & -1 & 1 & -1 & -1 & 1 & -1 & 1 \\
1 & 1 & -1 & -1 & -1 & -1 & 1 & 1 & 1 & 1 & -1 & -1 & -1 & -1 & 1 & 1 \\
1 & -1 & -1 & 1 & -1 & 1 & 1 & -1 & 1 & -1 & -1 & 1 & -1 & 1 & 1 & -1 \\
1 & 1 & 1 & 1 & 1 & 1 & 1 & 1 & -1 & -1 & -1 & -1 & -1 & -1 & -1 & -1 \\
1 & -1 & 1 & -1 & 1 & -1 & 1 & -1 & -1 & 1 & -1 & 1 & -1 & 1 & -1 & 1 \\
1 & 1 & 1 & -1 & 1 & 1 & -1 & -1 & -1 & -1 & 1 & 1 & -1 & -1 & 1 & 1 \\
1 & -1 & -1 & 1 & 1 & -1 & -1 & 1 & -1 & 1 & 1 & -1 & -1 & 1 & 1 & -1 \\
1 & 1 & 1 & 1 & -1 & -1 & -1 & -1 & -1 & -1 & -1 & -1 & 1 & 1 & 1 & 1 \\
1 & -1 & 1 & -1 & -1 & 1 & -1 & 1 & -1 & 1 & -1 & 1 & 1 & -1 & 1 & -1 \\
1 & 1 & -1 & -1 & -1 & -1 & 1 & 1 & -1 & -1 & 1 & 1 & 1 & 1 & -1 & -1 \\
1 & -1 & -1 & 1 & -1 & 1 & 1 & -1 & -1 & 1 & 1 & -1 & 1 & -1 & -1 & 1
\end{pmatrix}$$

If we remove the first row and the first column we get a square matrix of order

15. This corresponds with the incidence matrix of a BIBD if we replace the entries -1 by 0.

We use the R-program `bibd(v = 15, k = 7)` to construct the BIBD number 30 in Table 6.1 with parameters $v = 15, k = 7, b = 15, r = 7, \lambda = 3$ by Method 16.

The result is

```
 block   treatments


   1    (1,   2,  4,  5,  8, 10, 15)
   2    (2,   3,  5,  6,  9, 11,  1)
   3    (3,   4,  6,  7, 10, 12,  2)
   4    (4,   5,  7,  8, 11, 13,  3)
   5    (5,   6,  8,  9, 12, 14,  4)
   6    (6,   7,  9, 10, 13, 15,  5)
   7    (7,   8, 10, 11, 14,  1,  6)
   8    (8,   9, 11, 12, 15,  2,  7)
   9    (9,  10, 12, 13,  1,  3,  8)
  10    (10, 11, 13, 14,  2,  4,  9)
  11    (11, 12, 14, 15,  3,  5, 10)
  12    (12, 13, 15,  1,  4,  6, 11)
  13    (13, 14,  1,  2,  5,  7, 12)
  14    (14, 15,  2,  3,  6,  8, 13)
  15    (15,  1,  3,  4,  7,  9, 14)

 v = 15    k = 7    b = 15    r = 7   lambda = 3
```

Balanced Incomplete Block Design: BIBD(15,15,7,7,3)

(1, 2, 3, 4, 5, 6, 7) (1, 2, 3, 8, 9,10,11) (1, 2, 3,12,13,14,15) (1, 4, 5, 8, 9,12,13) (1, 4, 5,10,11,14,15) (1, 6, 7, 8, 9,14,15) (1, 6, 7,10,11,12,13) (2, 4, 6, 8,10,12,14) (2, 4, 6, 9,11,13,15) (2, 5, 7, 8,10,13,15) (2, 5, 7, 9,11,12,14) (3, 4, 7, 8,11,12,15) (3, 4, 7, 9,10,13,14) (3, 5, 6, 8,11,13,14) (3, 5, 6, 9,10,12,15)

Method 17 Let $v = p^n = m(\lambda - 1) + 1$ where p is a prime and $n \geq 1$ and x be a primitive element of $GF(v)$; then the initial blocks $(0, x^i, x^{i+m}, x^{i+2m}, \ldots, x^{i(\lambda-2)m})$, $i = 0, \ldots, m-1$ provide a BIBD with v; $b = mv; k = \lambda, ; r = mk, \lambda$.

Example 6.32 We construct a design with $v = p = 29 = 7{*}4 + 1$, which means $m = 7, \lambda = 5$ by Method 17. The initial blocks are $(0, x^i, x^{i+7}, x^{i+14}, x^{i-21})$, $i = 0, \ldots 6$. A primitive element of GF(29) is $x = 2$ (see Table A.1).

The parameters of the resulting BIBD are $v = 29, b = 203, r = 35, k = 5, \lambda = 5$. Because the design has 552 blocks, we do not print it. The reader may run the program if he is interested in just this design.

Method 18 (Block section) From a symmetric BIBD with parameters $v = b, k = r, \lambda$ we delete one block and from all residual blocks all the elements contained in the deleted block. What remains is a BIBD with parameters $v^* = v - k, b^* = v - 1, k^* = k - \lambda, r^* = k, \lambda^* = \lambda$. If especially $v = b = 4t - 1; r = k = 2t - 1, \lambda = t - 1$, we have $v^* = v - k = 2t, b^* = v - 1 = 4t - 2, k^* = k - \lambda = t, r^* = k = 2t - 1, \lambda^* = \lambda = t - 1$. A BIBD constructed by Method 18 is called a **residual design** from the original BIBD.

Remark: Method 18 not always gives the smallest BIBD. For instance, the residual design from the BIBD 107 in Table 6.1 is a design with $v = 16$ and $k = 6$, $b = 24$ and $r = 9$ and this has more blocks than the BIBD 34 in Table 6.1.

Example 6.33 Consider the design with $v = b = 39 = 4{*}10 - 1, k = r = 19, \lambda = 9$ and $t = 10$. Method 18 provides the BIBD 64 with parameters $v^* = 20, b^* = 38, k^* = 10, r^* = 19, \lambda^* = 9$.

Example 6.34 We use the R-program `bibd(v = 10, k = 5)` to construct the BIBD number 9 with parameters $v = 10, k = 5, b = 18, r = 9, \lambda = 4$ by Method 18.

The result is

```
block   treatments

  1     (1,  2,  3,  5,  9)
  2     (2,  3,  4,  6,  1)
  3     (3,  4,  5,  7,  2)
  4     (4,  5,  6,  8,  3)
  5     (5,  6,  7,  9,  4)
  6     (6,  7,  8,  1,  5)
  7     (7,  8,  9,  2,  6)
  8     (8,  9,  1,  3,  7)
  9     (9,  1,  2,  4,  8)
 10     (1,  2,  5,  7, 10)
 11     (2,  3,  6,  8, 10)
 12     (3,  4,  7,  9, 10)
 13     (4,  5,  8,  1, 10)
 14     (5,  6,  9,  2, 10)
 15     (6,  7,  1,  3, 10)
 16     (7,  8,  2,  4, 10)
 17     (8,  9,  3,  5, 10)
 18     (9,  1,  4,  6, 10)
```

```
v = 10    k = 5    b = 18    r = 9   lambda = 4
```

Method 19 (Block intersection) From a symmetric BIBD with parameters $v = b, k = r, \lambda$ we delete one block and from all residual blocks all the elements not contained in the deleted block. What remains is a BIBD with parameters $v^* = k, b^* = v - 1, k^* = \lambda, r^* = k - 1, \lambda^* = \lambda - 1$. If especially $v = b = 4t - 1; r = k = 2t - 1, \lambda = t - 1$, we have $v^* = k = 2t - 1$, $b^* = v - 1 = 4t - 2$, $k^* = \lambda = t - 1$, $r^* = k - 1 = 2t - 2, \lambda^* = \lambda = t - 2$.

A BIBD constructed by Method 19 is called a **derived design** from the original BIBD.

Example 6.35 Consider the design with $v = b = 39, k = r = 19, \lambda = 9$ and $t = 10$. Method 19 provides the BIBD with parameters $v^* = 19, b^* = 38, k^* = 9, r^* = 18, \lambda^* = 8$. But we see that this method does not always give the smallest designs, because design 56 in Table 6.1 is half as large.

Theorem 6.3 A quasi-reduced (quasi-derived) BIBD is always a reduced (derived) BIBD if λ equals 1 or 2. For $\lambda > 2$ this is not true.

Bhattacharya (1945) gave an example of a BIBD which is quasi-reduced but not reduced. The BIBD has parameters $v = 16, b = 24, k = 6, r = 9, \lambda = 3$. Two of its blocks have 4 treatments in common and can therefore not stem from a symmetric BIBD with parameters $v = b = 25, k = r = 9, \lambda = 3$ due to Theorem A.6.

The following method is more academic, the designs that can be constructed have large v values. We therefore give no program.

Method 20 (Kronecker product designs) Let N_1, N_2 be the incidence matrices of two BIBD with parameters $v_i, b_i, k_i, r_i, \lambda_i, i = 1, 2$ with restriction $b_i = 4(r_i - \lambda_i)$ and $\overline{N}_i, \overline{N}_2$ be the incidence matrices of two complementary designs, respectively; then $N_1 \otimes N_2 + \overline{N}_i \otimes \overline{N}_2$ is the incidence matrix of a BIBD with parameters $v^* = v_1 v_2$, $b^* = b_1 b_2$, $k^* = k_1 k_2 + (v_1 - k_1)(v_2 - k_2)$, $r^* = r_1 r_2 - (b_1 - r_1)(b_2 - r_2)$ and $\lambda^* = r^* - \frac{b^*}{4}$.

Example 6.36 The designs number 34 and 108 from Table 6.1 fulfill the condition $b = 4(r - \lambda)$. From these designs we can construct a design with $v = 16 \cdot 16$ (using the same design twice) or design with $v = 16 \cdot 25$ or $v = 25 \cdot 25$.

There are more specific methods. Many of them like Method 20 give only a few solutions for $v < 51$ and are not mentioned here.

6.3.2 General method

Method 21 If for $v > 25$ a (v, k)-pair exists where none of the methods in Section 6.3.1 leads to a BIBD, we can proceed as follows.

A trivial solution is to take all $\binom{v}{k}$ k-tupels from the v elements. Then we can try by systematic search to find a subset of these $\binom{v}{k}$ k-tupels which is also a BIBD. For this it could be helpful to know the smallest r (or b, or λ) for which the necessary conditions (6.4) and (6.5) are fulfilled. For this we have to solve the diophantic equation $r = \frac{bk}{v} = \frac{\lambda(v-1)}{k-1}$ or $\frac{bk}{v} - \frac{v-1}{k-1}\lambda = 0$ which is of the form $mx + ny = c$ with $c = 0$, $m = \frac{k}{v}$, $n = -\frac{v-1}{k-1}$, $x = b$ and $y = \lambda$.

Example 6.37 Let us consider $v = 26, k = 4$. Then we have to solve $r = \frac{4}{26} \cdot b = \frac{25}{3}\lambda$. The smallest r is $r = 100$, $b = 650$, $\lambda = 12$. $b = 650$ is 1/23 of the number of blocks $(26 \cdot 25 \cdot 23)$ of the trivial BIBD. The question is: does a corresponding BIBD exist?

But let us now consider a practical example.

Example 6.38 – a practical block experiment

Wireworms, a popular name for the slender, hard-skinned grubs or larvae of the click beetles or *Elateridae*, are a family of the Coleoptera. These larvae pass a long life in the soil, feeding on the roots of plants, and the larvae often cause much damage to farm crops of all kinds, but especially to potatoes. In 2008 at the University of Natural Resources in Vienna an experiment was started to find the best plant variety which could be used as "catch crop" because the worms would prefer these plants when grown between the potato plants. Because the distribution of the worms in a field is not homogeneous and to avoid a bias in the judgment about the catching ability of plants (when unfortunately a good catch crop is in a part of a field where no worms are) a block experiment was recommended by the statistical department. Because for an effective experimentation plots on a field must be so large that it is possible to use machines for seeding and harvesting, a block could have only $k = 4$ plots to be still expected to be homogeneous in the distribution of the worms. A block consists of a rectangle of 2×2 plots.

The researcher wished to compare $v = 8$ varieties of catch crops. The problem is one of selecting the variety with a maximum (expected value) of worms caught. Using the R-program size.selection.bechhofer from Section 2.3 to determine the minimum number of experimental units needed from each of 8 populations and make sure that the probability of selecting the one with the largest expectation is at least 0.95, as long as the best population differs in expectation at least by 1 standard deviation from the second best, we find $n = 12$ for the size of each of the 8 samples. When we look at Table 6.1 we find design 4 which needs only 7 replications. Taking this design twice with 28 blocks gives us a sufficient ($n = 14$) sample size to solve our problem. In the practical layout of the experiment, the blocks of the R-output have to be randomly assigned to the real blocks on the field, and in any block the four treatments have to be randomly assigned to the 4 actual plots.

Table 6.1 gives an overview over the smallest (v, k)-designs for $v \leq 25$.

7

Constructing Fractional Factorial Designs

7.1 Introduction and basic notations

The classical method for studying the effect of several factors consists in varying one factor at a time while keeping the others constant. Apart from being very expensive, this method does not permit interaction effects between factors to be analysed. These drawbacks induced R. A. Fisher (1926, 1942) and F. Yates (1935) to develop the basics of the theory of factorial experiments. Experiments of this kind are used to study the effects of several factors simultaneously.

We call a factor a treatment factor if its effect on some observable variable y—modeled by a random variable $\boldsymbol{y}$—has to be investigated. An experiment is called a factorial experiment if at least two treatment factors are included in the experiment.

Example 7.1 The effects of different types of phosphate and nitrogenous fertilisers on the yield y of a crop are to be studied. In an experiment, factor P (phosphate) is applied at the levels (types) P_1, P_2 and P_3, and factor N (nitrogenous fertiliser) at the levels N_1 and N_2, each of the six plots in an experimental field being "treated" (fertilised) with one of the factor level combinations $P_1N_1, P_1N_2, P_2N_1, P_2N_2, P_3N_1$ and P_3N_2. To improve accuracy, this can be done r times, i.e., using a total of $r \cdot 6$ plots for each combination of levels of the two fertilisers. In this case, one would speak of r (complete) repetitions or an r-fold experiment. The choice of r will depend on the precision requirements and will be determined using the methods in Chapter two. All $r \cdot 6$ plots must have the same kind of soil to avoid the disturbing effects of soil differences. If no homogeneous field of the necessary size is available, then at least one set of treatment combinations should have the same kind of soil. The different soils will then act as levels of a disturbing or blocking factor, the effects of which must be eliminated during analysis. The combinations of factor levels tested on one kind of soil form a block, while all six possible combinations of levels correspond to the $v = 6$ treatments. The experiment will be done as a complete or incomplete (in cases where there are more factor level combinations as plots in a block) block experiment. The latter are described in Chapter six. A complete block (with all treatment combinations just once) is called a replication.

We can use Example 7.1 to illustrate the concept of interaction. An interaction is present if the effect on the characteristic being observed when we move from one level to another of a certain factor depends on the level(s) of one or more of the other factors. The number of other factors involved defines the order of the interaction. In Example 7.1 a first order interaction is present if the effect on the yield caused by changing from phosphate level P_1 to P_2 is affected by whether the nitrogen level is N_1 or N_2.

In general, the factors are denoted by $F_1, F_2, \ldots, F_p$ and their levels $F_{11}, \ldots, F_{1s_1}, \ldots, F_{p1}, \ldots, F_{ps_p}$. The factors being studied are treatment factors, and the combinations of levels of all treatment factors are the treatments (strictly treatment combinations) which for instance can be allocated into blocks as described in Chapter 6. Other factors that are of no interest, but whose possible effects mean that they have to be taken into account by blocking or other means, are called noise factors. The purpose of factorial experiments is to estimate the effects of the treatment factors and the interaction effects between factors, using a statistical model whose parameters can be tested or estimated. The number of factors and their levels are settled beforehand. A factor may be either qualitative or quantitative.

Factorial experiments are analysed either by analysis of variance (ANOVA) if all the factors are qualitative or by regression analysis if all factors are quantitative. If some of the factors are qualitative and others quantitative, sometimes covariance analysis is used. Most definitions can be given independently of the type of analysis.

The model equation of the observation of a factorial design can be written in the form

$$\boldsymbol{y} = X\beta + \boldsymbol{e} \tag{7.1}$$

In (7.1) $\boldsymbol{y}$ is an N-dimensional random vector as is $\boldsymbol{e}$. X is an $N * k$-matrix, and β is a vector with $k < N$ elements. X as well as β are fixed and non-random. If parts of X are random, we have a regression Model II or at least a mixed regression model with fixed and random cause variables. If parts of β are random, we have a Model II of ANOVA or at least a mixed model. In these latter cases the designs are of course also factorial designs, but the methods discussed in this book are mainly useful for the cases where both X and β are fixed. This will be assumed for the rest of this chapter.

If X in (7.1) is not of full (column) rank as in ANOVA models, so-called reparametrisation conditions have to be assumed for the parameter vector β. This is equivalent to make the generalized inverse of X^TX unique. By $(X^TX)^{-1}$ we mean either the normal inverse of X^TX or the unique generalized inverse obtained by the usual reparametrisation conditions in ANOVA models (see Chapter three). So we can write the least square estimators of the parameter vector β as will be discussed in Chapter eight as

$$\hat{\boldsymbol{\beta}} = (X^T X)^{-1} X^T \boldsymbol{y} \tag{7.2}$$

Optimal factorial designs are those which make some functional of $(X^T X)^{-1}$ to a minimum. We discuss this for regression models in detail later in Chapter eight.

We start with a small example of the two types of models for a factorial design with two factors with two levels each, i.e., a 2^2-design. If we have two observations at each factor level combination, we have $N = 8$ and $\boldsymbol{y}^T = (\boldsymbol{y}_1, \boldsymbol{y}_2, \ldots, \boldsymbol{y}_8), \boldsymbol{e}^T = (\boldsymbol{e}_1, \boldsymbol{e}_2, \ldots, \boldsymbol{e}_8)$.

ANOVA model: Let the effects of two levels of factor $F_1 = A$ be $\alpha_1 = -\alpha_2$ and of factor $F_2 = B$ be $\beta_1 = -\beta_2$ and the interaction effects be $(\alpha\beta)_{11} = -(\alpha\beta)_{12}; (\alpha\beta)_{21} = -(\alpha\beta)_{22}; (\alpha\beta)_{11} = -(\alpha\beta)_{21}$. The effects have been so defined that the reparametrisation conditions have already been included in the definitions. Thus $\beta^T = (\mu, \alpha_1, \alpha_2, \beta_1, \beta_2, (\alpha\beta)_{11}, (\alpha\beta)_{12}, (\alpha\beta)_{21}, (\alpha\beta)_{22})$ and the model equation is

$$\boldsymbol{y}_{ijk} = \mu + \alpha_i + \beta_j + (\alpha\beta)_{ij} + \boldsymbol{e}_{ijk};\ i = 1,2;\ j = 1,2, k = 1,2 \tag{7.3}$$

whereby $\boldsymbol{y}_1 = \boldsymbol{y}_{111}, \boldsymbol{y}_2 = \boldsymbol{y}_{112}, \ldots, \boldsymbol{y}_8 = \boldsymbol{y}_{222}$ and analogously the vector $\boldsymbol{e}$ has to be defined. The matrix X is

$$X = \begin{pmatrix} 1 & 1 & 0 & 1 & 0 & 1 & 0 & 0 & 0 \\ 1 & 1 & 0 & 1 & 0 & 1 & 0 & 0 & 0 \\ 1 & 1 & 0 & 0 & 1 & 0 & 1 & 0 & 0 \\ 1 & 1 & 0 & 0 & 1 & 0 & 1 & 0 & 0 \\ 1 & 0 & 1 & 1 & 0 & 0 & 0 & 1 & 0 \\ 1 & 0 & 1 & 1 & 0 & 0 & 0 & 1 & 0 \\ 1 & 0 & 1 & 0 & 1 & 0 & 0 & 0 & 1 \\ 1 & 0 & 1 & 0 & 1 & 0 & 0 & 0 & 1 \end{pmatrix}$$

The reparametrisation condition can be written as $K\beta = 0$ so that the generalised inverse of $X^T X$ is unique and equal to $(X^T X)^- = (X^T X + K^T K)^{-1}$.

Regression model: In a regression model we have for the eight observations the model equation

$$\boldsymbol{y}_{ijk} = \beta_0 + \beta_1 x_{1i} + \beta_2 x_{2j} + \beta_3 x_{1i} x_{2j} + \boldsymbol{e}_{ijk}; \tag{7.4}$$

$$i = 1,2;\ j = 1,2;\ k = 1,2;\ x_{11} \neq x_{12};\ x_{21} \neq x_{22}$$

and in (7.1) we have: $\beta^T = (\beta_0, \beta_1, \beta_2, \beta_3)$ and

$$X = \begin{pmatrix} 1 & x_{11} & x_{21} & x_{11}x_{21} \\ 1 & x_{11} & x_{21} & x_{11}x_{21} \\ 1 & x_{11} & x_{22} & x_{11}x_{22} \\ 1 & x_{11} & x_{22} & x_{11}x_{22} \\ 1 & x_{12} & x_{21} & x_{12}x_{21} \\ 1 & x_{12} & x_{21} & x_{12}x_{21} \\ 1 & x_{12} & x_{22} & x_{12}x_{22} \\ 1 & x_{12} & x_{22} & x_{12}x_{22} \end{pmatrix}$$

This matrix is under the conditions of (7.4) of full rank 4 and we obtain

$$X^T X = \begin{pmatrix} 8 & 4(x_{11}+x_{12}) & 4(x_{21}+x_{22}) & 2\sum_{ij} x_{1i}x_{2j} \\ 4(x_{11}+x_{12}) & 4\sum_j x_{1j}^2 & 2\sum_{ij} x_{1i}x_{2j} & 2\sum_{ij} x_{1i}^2 x_{2j} \\ 4(x_{21}+x_{22}) & 2\sum_{ij} x_{1i}x_{2j} & 4\sum_j x_{2j}^2 & 2\sum_{ij} x_{1i}x_{2j}^2 \\ 2\sum_{ij} x_{1i}x_{2j} & 2\sum_{ij} x_{1i}^2 x_{2j} & 2\sum_{ij} x_{1i}x_{2j}^2 & 2\sum_{ij} x_{1i}^2 x_{2j}^2 \end{pmatrix}$$

The factorial design in this example is complete. It allows us to estimate all the effects in the models. We can use half of this design deleting the 4 rows in X with an even number and the corresponding elements in (7.3) or (7.4). If we reduce the design in this way, all main effects can still be estimated.
It will be the aim of this chapter to reduce complete factorial designs in a proper way to make estimable as many effects as possible. The terms $\alpha_i, \beta_j;\ i = 1, 2;\ j = 1, 2$ and $\beta_1 x_{1i}, \beta_2 x_2;\ i = 1, 2;\ j = 1, 2$ in model (7.3) and (7.4), respectively, we will call main effects, and the terms $(\alpha\beta)_{ij;};\ i = 1, 2;\ j = 1, 2$ and $\beta_3 x_{1i} x_{2j};\ i = 1, 2;\ j = 1, 2$, respectively, we call interaction effects.
It will be easier to work in the next section with either the regression or the ANOVA model. We select the ANOVA model and give the definition in dependence on this.

7.2 Factorial designs—basic definitions

In this section we give a more detailed definition and some theorems on fractional factorial designs in general before considering special cases in the sections thereafter.

Definition 7.1 An experiment with $p > 1$ factors $F_1, F_2, \ldots, F_p$ and the corresponding numbers $s_1, s_2, \ldots, s_p$ of factor levels is called a **factorial experiment of type** $s_1 \cdot s_2 \cdot \ldots \cdot s_p$. If $s_1 = s_2 = s_p = s$, the experiment is a

symmetrical factorial experiment of type s^p, otherwise it is **asymmetric** or **mixed**. A factorial experiment is called **complete** if all combinations of the levels of all factors occur in the experiment at least once, otherwise it is called **incomplete**.

The *experimental design* of a factorial experiment with p factors specifies for each possible combination $F_{1i_1}, \ldots, F_{pi_p}$ of the levels of all factors the number $n_{i_1 \ldots i_p}$ of experimental units assigned to that combination of levels (i.e., the number of units "treated" with that combination). The total number $N = \sum\limits_{i_1 \ldots i_p} n_{i_1 \ldots i_p}$ of experimental units is the **size of the experiment**. This factorial experiment is thus complete if $n_{i_1 \ldots i_p} > 0$ for all combinations of the levels of all factors, otherwise it is incomplete. Complete factorial designs are seldom used if the number p of factors exceeds 4 or 5. Even if each factor has only 2 levels, we have already 16 and 32 treatment combinations, respectively. Each occurrence of a treatment combination is called a **run** for short. If each treatment combination appears exactly once in a design, this is called a single replicate design. Factorial designs can be laid out in blocks by treating the factor level combinations of the factorial design as the v treatments of the block design in Chapter six.

Notation:

We assume that the reader is familiar with the usual ANOVA notation and theory. Those who are not may read Chapter three, where a similar notation is used.

The levels of the i-th factor with s_i levels are written as $0, 1, \ldots, s_i - 1$. A factor level combination of p factors is a p-tupel $(a_1, a_2, \ldots, a_p)$ with elements $a_i \in \{0, 1, \ldots, s_i - 1\}$.

The triple $(2, 0, 1)$ thus means that factor 1 is on level 2, factor 2 on level 0 and factor 3 on level 1 in a design with $p = 3$ factors.

Let us consider a 2^p-design and denote the second level of the factor $F_i(i = 1, 2, \ldots, p)$ by f_i and the first one by unity. Further, let $n_{i_1 \ldots i_p} = n$. The main effects and interaction effects may be represented by

$$I\left(\prod_{j=1}^{p} k_j F_j\right) = \frac{1}{n \cdot 2^{p-1}} (f_1 \pm 1)(f_2 \pm 1) \ldots (f_p \pm 1) \tag{7.5}$$

In (7.5) $k_j = 0$ and the corresponding sign is positive, if the j-th factor is not contained in I and $k_j = 1$ and the corresponding sign is negative otherwise. The effects of the two levels of a factor $F_i(i = 1, 2, \ldots, p)$ are denoted by $\pm\phi_i$; the sign is negative for the lower level and positive for the upper level.

When we use Greek letters for the effects, the effect $I\left(\prod\limits_{j=1}^{p} k_j F_j\right)$ of the treatment combination $\prod\limits_{j=1}^{p} k_j F_j$ is then

$$\mu\left[\prod_{j=1}^{p} k_j F_{jl}, l = 1,2\right] =$$

$$\left\{\mu + \frac{1}{2}[\pm\varphi_1 \pm \ldots \pm \varphi_p], \pm\varphi_1\varphi_2 \pm \ldots \pm \varphi_{p-1}\varphi_p \pm \ldots \varphi_1\varphi_2 \ldots \varphi_p\right\} \quad (7.6)$$

In (7.6) we have $l = 1$ and the negative sign if the factor is on its lower level and $l = 2$ and the positive sign if it is on its upper level.

Example 7.2 In the case $p = 3$ and $n = 1$ (single replicate design) we have with $F_1 = A, F_2 = B$ and $F_3 = C$; $f_1 = a, f_2 = b$ and $f_3 = c$; $\phi_1 = \alpha,\ \phi_2 = \beta; \phi_3 = \gamma$.

$$I(A) = \frac{1}{4}(a-1)(b+1)(c+1) = \frac{1}{4}[a + ab + ac + abc - bc - b - c - 1]$$

and $\mu[A_1] = \left\{\mu - \frac{1}{2}\alpha\right\}$ and $\mu[A_2] = \left\{\mu + \frac{1}{2}\alpha\right\};\ \alpha = \varphi_1$.

A type of incomplete factorials, the ***fractional factorials***, is of special interest, and in this chapter we will mainly discuss construction methods for them. ***Fractional factorial designs*** are an important class of incomplete factorial designs, especially useful when the factors are qualitative. They consist of one k-th ($k = 2, 3, \ldots$) of all the possible treatment combinations. Fractional factorial designs are used when there are many factors to be considered in an experiment. Since the number of treatment combinations rises sharply with the number of factors, often there is an insufficient number of experimental units available for a complete factorial experiment, An incomplete factorial design has the consequence that not all the definable main and interaction effects in the ANOVA model can be estimated separately from each other, but in part only their sum. We call this confounding. Confounded effects are called *aliases.* One strives as far as possible to avoid confounding main effects and lower order interaction effects with each other. It is in general assumed that higher order interaction effects are negligible and so can be confounded with each other if necessary. We will come back to these problems in more detail later.

Important for constructing special fractional factorials with a given confounding structure is the concept of **resolution.**

Definition 7.2 Let us consider a fractional factorial design that ensures the estimability of the overall mean and complete sets of contrasts belonging to factorial effects involving at most u factors under the absence of factorial effects involving $w + 1$ or more factors. Such a fractional factorial design is said

to be of resolution $R = u + w + 1 (= II, III, IV, V, \ldots)$. This means that no u-factor interaction is aliased with another effect containing fewer than R-u factors. A slightly generalized definition of resolution is given by Dey and Mukerjee (1999).
In practice high order interaction effects (four-fold and more) are regarded as negligible; therefore, they may be confounded without important consequences.

Theorem 7.1 (Anderson and Thomas, 1979) Let w.l.o.g. $s_1 < s_2 < \ldots < s_p$. In a $s_1^{k_1} s_2^{k_2} \ldots s_p^{k_p}$ design of resolution IV, a lower bound on the number of runs needed is $s_p \{\sum (s_1 - 1)k_i - s_p + 2\}$.

Definition 7.3 A factorial experiment is called symmetric if all p factors have the same number s of levels; such a design is called an s^p-design. If at least two of the factors have different numbers of levels, the design is called asymmetric or mixed.
We let $R_s(p, k)$ denote the maximum possible resolution of an s^{p-k} design.

Definition 7.4 The non-random part of each row in equation (7.1) is called the effect of the corresponding treatment combination which was applied at the experimental unit of this row. We call this *treatment "effect"* for short; it is a function of the parameter vector β. If $\boldsymbol{\tau}$ is the vector of all treatment effects, then a linear function $c^T\boldsymbol{\tau}$ with a real vector c not all elements zero and summing up to 0 is called a (linear) treatment contrast or a contrast for short. The contrast is said to be normalised if $c^Tc = 1$. Two contrasts $c_1^T\boldsymbol{\tau}$ and $c_2^T\boldsymbol{\tau}$ are called orthogonal if $c_1^Tc_2 = 0$.

Example 7.3 Consider a single replicate 2^3-design with factors F_1, F_2, F_3 at levels 0 and 1 and corresponding effects $\varphi_i = -1$ and $+1$ each. Then the orthogonal contrasts are (up to multiplication of all entries by –1) uniquely determined by Table 7.1. The coefficients for the interaction can be obtained from those of the main effects of those factors defining the interaction by multiplying the corresponding entries in each column. For instance, the coefficients of $\varphi_1\varphi_2$ are the products of the coefficients for φ_1 and φ_2 in any column.

These contrasts are said to represent (or belong to) the main effects and interaction effects in the effect-column. If we rearrange the entries in Table 7.1 in such a way that the rows are defined by the coefficients of the contrasts (the first four rows by –1) and in the body of the table the corresponding factor level combinations are given, we obtain a partition as shown in Table 7.2.
We can generalise the results of Example 7.3 as follows.

Definition 7.5 In an s^p-design with factors $F_1, F_2, \ldots, F_p$ the main effect of

TABLE 7.1
Coefficients of the Orthogonal Contrasts for Example 7.3.

	Factor level combination							
Effect	(0,0,0)	(0,0,1)	(0,1,0)	(0,1,1)	(1,0,0)	(1,0,1)	(1,1,0)	(1,1,1)
φ_1	−1	−1	−1	−1	1	1	1	1
φ_2	−1	−1	1	1	−1	−1	1	1
φ_3	−1	1	−1	1	−1	1	−1	1
$\varphi_1\varphi_2$	1	1	−1	−1	−1	−1	1	1
$\varphi_1\varphi_3$	1	−1	1	−1	−1	1	−1	1
$\varphi_2\varphi_3$	1	−1	−1	1	1	−1	−1	1
$\varphi_1\varphi_2\varphi_3$	−1	1	1	−1	1	−1	−1	1

TABLE 7.2
A Partition of Factor Level Combinations.

	Part I				Part II			
Effect	−1	−1	−1	−1	1	1	1	1
φ_1	(0,0,0)	(0,0,1)	(0,1,0)	(0,1,1)	(1,0,0)	(1,0,1)	(1,1,0)	(1,1,1)
φ_2	(0,0,0)	(0,0,1)	(1,0,0)	(1,0,1)	(0,1,0)	(0,1,1)	(1,1,0)	(1,1,1)
φ_3	(0,0,0)	(0,1,0)	(1,0,0)	(1,1,0)	(0,0,1)	(0,1,1)	(1,0,1)	(1,1,1)
$\varphi_1\varphi_2$	(0,1,0)	(0,1,1)	(1,0,0)	(1,0,1)	(0,0,0)	(0,0,1)	(1,1,0)	(1,1,1)
$\varphi_1\varphi_3$	(0,0,1)	(0,1,1)	(1,0,0)	(1,1,0)	(0,0,0)	(0,1,0)	(1,0,1)	(1,1,1)
$\varphi_2\varphi_3$	(0,0,1)	(0,1,0)	(1,0,1)	(1,1,0)	(0,0,0)	(0,1,1)	(1,0,0)	(1,1,1)
$\varphi_1\varphi_2\varphi_3$	(0,0,0)	(0,1,1)	(1,0,1)	(1,1,0)	(0,0,1)	(0,1,0)	(1,0,0)	(1,1,1)

factor F_j; $j = 1, \dots, p$ is said to compare the responses (y-values in (7.1)) among the s sets $S_\kappa = \{(\nu_1, v_2, \dots, \nu_p) | \nu_j = \kappa\}, \kappa; \nu_i \in \{0, 1, \dots, s-1\}$. The s sets S_K partition the s^p treatment combinations into s sets of size s^{p-1} each. The j-th partition is denoted by $P(F_j;\ j = 1, \dots, p)$.

Definition 7.6 In an s^p-design with factors $F_1, F_2, \dots, F_p$ the interaction effect of the factors F_j; $j = 1, \dots, p$; F_κ; $j \neq \kappa = 1, \dots, p$ is said to compare the responses (y-values in (7.1)) among the s sets within the $s - 1$ partitions $S_{\kappa\lambda} = \{(\nu_1, \nu_2, \dots, \nu_p) | \nu_j = \kappa\nu_t = \lambda\}, \kappa; \lambda; \nu_j; \nu_t \in \{0, 1, \dots, s-1\}; \kappa \neq 0$. Analogously this definition can be generalised to higher order interaction effects.

Definition 7.7 A contrast belongs to the $(p-1)$-th order interaction of p factors $F_1, F_2, \dots, F_p$ ($p = 1$ means to the main effect of a factor), say, the i_1-th; i_2-th; ...; i_p-th if

- The coefficients in the linear function defining the contrast are independent of the factors other than the i_1-th; i_2-th; ...; i_p-th factor.
- The contrast is orthogonal to all contrasts belonging to all possible main effects and interaction effects of the i_1-th; i_2-th; ...; i_p-th factors.

TABLE 7.3
Coefficients of the Orthogonal Contrasts for Example 7.4.

	Factor level combination								
Effect	(0,0)	(0,1)	(0,2)	(1,0)	(1,1)	(1,2)	(2,0)	(2,1)	(2,2)
φ_1	−1	−1	−1	0	0	0	1	1	1
φ_2	−1	0	1	−1	0	1	−1	0	1
$\varphi_1\varphi_2$	−1	0	1	0	1	−1	1	−1	0
$\varphi_1\varphi_2\varphi_2$	−1	1	0	0	−1	1	1	0	−1

TABLE 7.4
A Partition of Factor Level Combinations for Example 7.4.

	Part I			Part II			Part III		
Effect	−1	−1	−1	0	0	0	1	1	1
φ_1	(0,0)	(0,1)	(0,2)	(1,0)	(1,1)	(1,2)	(2,0)	(2,1)	(2,2)
φ_2	(0,0)	(1,0)	(2,0)	(0,1)	(1,1)	(2,1)	(0,2)	(1,2)	(2,2)
$\varphi_1\varphi_2$	(0,0)	(1,2)	(2,1)	(0,1)	(1,0)	(2,2)	(0,2)	(1,1)	(2,0)
$\varphi_1\varphi_2\varphi_2$	(0,0)	(1,1)	(2,2)	(0,2)	(1,0)	(2,1)	(0,1)	(1,2)	(2,0)

There are $(s_{i1}-1)\cdot \ldots \cdot (s_{ip}-1)$ such contrasts. There is a connection between contrasts and partitions as we have already seen in Example 7.3.

Example 7.4 Let us consider a design with two factors at three levels each, a 3^2-design. Table 7.3 shows the orthogonal contrasts.

Remark: The regression model corresponding with an s^p-design will include p regressor variables, and the regression function is a polynomial up to order $s-1$. Thus for $s=2$ we have a p-fold linear regression, and for $s=3$ we have a p-fold quadratic regression. Therefore in Example 7.4 four further orthogonal contrasts which are orthogonal to those in Table 7.4 can be found (one belonging to each row in Table 7.3) which correspond with the non-linear terms in the regression function. The second contrast belonging to F_1 is

F_1	-1	-1	-1	2	2	2	-1	-1	-1

After defining when a contrast belongs to a lower order interaction, we can define by induction when it belongs to a higher order interaction. In the case of $p=2,3,\ldots r-1$ factors $F_1, F_2, \ldots, F_p$, if a contrast belongs to the $(p-1)$-th interaction we say that the contrast belongs to the p-th interaction of $r=p+1$ factors $F_1, F_2, \ldots, F_{p+1}$, say, i_1-th, i_2-th, $\ldots$, i_{p+1}-th if

- The coefficients of the linear function constituting the contrast are independent of the levels or the r factors other than the i_2-th, $\ldots$, i_{p+1}th.

- The contrast is orthogonal to all contrasts belonging to all possible main effects and interaction effects of the i_1-th, i_2-th, $\ldots, i_{p+1}$th factor.

There are of course $(s_{i_1} - 1), (s_{i_2} - 1), \ldots, (s_{i_{p+1}} - 1)$ contrasts representing the r-factor interaction.

The $s_1, s_2, \ldots, s_p$ treatment combinations are carrying $s_1 \cdot s_2 \cdots s_p - 1$ degrees of freedom between them. In ANOVA we can partition the treatment sum of squares into those for orthogonal contrasts possessing a single degree of freedom corresponding to the main effects and interaction effects of all factors in the experiment.

The intrablock subgroup is for prime s an additive group.

Each two effects are aliased (confounded) if, in the s^{p-k} fractional factorial design, the partitions corresponding to these two effects are undistinguishable.

Definition 7.8 In an s^p factorial experiment with a prime $s > 1$ we can identify the s^p treatment combinations with the points $(x_1, x_2, \ldots, x_p)$ of an $EG(p, s)$ (see Definition A.11). Any $(s-1)$-flat (with $s-1$ points) of the $EG(p, s)$ is defined by $(a_0 + a_1x_1 + a_2x_2 + \ldots + a_px_p) = 0$; $a_i \in GF(s)$ (see Definition A.7). By keeping $(a_0, a_1, a_2, \ldots, a_p)$ constant and varying a_0 over the elements of $GF(s)$, we generate s parallel (having no points in common) $(p-1)$-flats, called a pencil $P(a_1, a_2, \ldots, a_p)$. $P(a_1, a_2, \ldots, a_p)$ divides the s^p treatment combinations into $s-1$ degrees of freedom.

Theorem 7.2 The pencil $P(a_1, a_2, \ldots, a_p)$ represents the interaction of the i_1-th, i_2-th, $\ldots, i_{p+1}$-th factor if and only if $a_{i_1}, \ldots, a_{i_r}$ are non-zero and all the other coordinates of $P(a_1, a_2, \ldots, a_p)$ are zero.

We can consider a fractional factorial design; as being one of the partitions of all factor level combinations. In an s^p-factorial design we have just s^p possible factor level combinations. A fractional factorial design of the type s^{p-k} is one of the s^k blocks containing s^{p-k} factor level combinations; it is called a $\frac{1}{s^k}$-replicate of the s^p-factorial design. The problem is to find the partitions and then to select a proper partition (confounding just what we want to be confounded) as the fraction.

In this chapter we will not give general definitions of confounding and defining contrast but do this only for those cases for which R-programs are given. For the general approach see Wallis et al. (2008).

There are already programs for constructing fractional factorial designs. One is the module DESIGN of CADEMO[1] with 2^p-, 3^p- and 2^m3^{p-m}-designs up to 9 factors. R-programs for constructing such designs can be found in the library `conf.design`; for further illustrations see the next section. A computer program for determining defining contrasts (see 7.2) for 2^p-designs is given

[1] `www.Biomath.de`

in Turiel (1988) for $p < 13$, and a method for constructing s^{p-k}-designs is described in Franklin (1985).
In Section 7.3 we consider the 2^{p-k} and in Section 7.4 the 3^{p-m}-designs.
We shall try to find those fractional factorial designs in which main effects are not confounded and thus can be estimated independently from each others, or if they are confounded, then they are confounded with interaction effects of high order.

7.3 Fractional factorial designs with two levels of each factor (2^{p-k}-designs)

At first we give some special definitions for two-factor designs.
The number of effects in a complete 2^p-design is increasing rapidly with p. We have $\binom{p}{i+1}$ factorial effects of order $i = 0, \dots p-1$ (i-th order interaction, where main effects are to be understood as zero order interactions).

Definition 7.9 A $\frac{1}{2^k}$ replicate of a 2^p factorial design is called a 2^{p-k} fractional factorial design, having s^{p-k} treatment combinations appearing in it.

Remark. Do not calculate $p - k = x$ for special numbers p and k and write 2^x in place of 2^{p-k} because then you cannot distinguish a 2^{7-2}- from a 2^{9-4}-design!
There are 2^k such 2^{p-k} replicates which are also called blocks because each of the 2^k replicates could be arranged in an incomplete block of a block design containing in all its blocks all 2^p factor level combinations. The block containing the treatment combination $(0, 0, \dots, 0)$ is called the intrablock subgroup. Which runs to make and which to leave out is the subject of interest here. In general, we pick a fraction such as $\frac{1}{2}$ or $\frac{1}{4}$ of the runs called for by the full factorial. We will show how to choose an appropriate fraction of a full factorial design.
Tables of 2^{p-k}-design can be found in in Bisgaard (1994), Chen et al. (1993), Colbourn and Dinitz (2006), McLean and Anderson (1984) and Montgomery (2005). These designs can be constructed by using the R-functions from the libraries `faraway` and `alg.design`. We will use the library `conf.design` here.
Let us at first specialise the definition of resolution R designs for the case of 2^{p-k} fractional factorial designs.

Important are 2^{p-k} -design of resolution III, IV and V.
In a resolution III 2^{p-k}-design, no main effect ($u = 1$) is aliased with another main effect, but each main effect is aliased with a first order (two-factor) interaction.

In a resolution IV 2^{p-k} -design, no main effect is aliased with another main effect or any first order interaction, but some first order interaction effects are aliased with another first order interaction.
In a resolution V 2^{p-k}-design, no main effect is aliased with another main effect or any first order interaction. First order interaction effects are not confounded with each other but only with interaction effects of order 2 or higher.

For simplicity we will write for the factors $F_i(i = 1, 2, \dots, p)$ the p letters (in alphabetical order) $A, B, C, \dots$. We denote the lower level of any factor by 1 and the upper level by the same small letter as is the capital letter in the factor notation, i.e., $a, b, c, \dots$.

The resolution of a 2^{p-k}-design can easily be calculated by counting the number of letters in the smallest defining contrast.

Definition 7.10 A 2^{p-k}-design is of resolution R if the smallest defining contrast has exactly R letters.

To construct a 2^{p-k}-design ($p = 3, 4, \dots; k = 2, 3, \dots, p-1$) we first choose k independent interaction effects $I_1 \cdot I_2 \cdot, \dots, \cdot I_K$ such that none of them is a (formal) product of all others. We call the product $I_1 \cdot I_2 \cdot, \dots, \cdot I_k = I$ subject to the restriction that the square of any symbol is equal to the identity, the **defining contrast**. Then we select a set of factor level combinations (the **aliases**) each having an even number of symbols in common with each of the interaction effects $I_1, I_2, \dots, I_k$—the symbol 1 represents the factor level combination with all factors on the lower level. This set is a group under the binary operation of multiplication subject to the restriction that the square of any symbol is defined to be 1. We call this group the **initial block** or the **defining relation.**
The defining contrast in a fractional factorial design determines the aliasing structure as well as the effects about which no information is available. As already mentioned above we propose that defining contrasts correspond to interaction effects of order as high as possible.

Example 7.5 Consider the three factors A, B, C with $p = 3$. For $k = 1$ the defining contrast is $I = ABC$. The aliases are $C = AB, B = AC, A = BC$. Thus the initial block is
(1), ab, ac, bc. The 4 remaining factor level combinations can be obtained by formal multiplication of the elements of the initial block w.l.o.g. by a, which leads to the second 1/2-replicate a, b, c, abc (the same result is obtained when we multiply by b or c.)
The resolution is R = III.

Example 7.6 Consider the four factors A, B, C, D with $p = 4$.

a) For $k = 1$ the defining contrast is $I = ABCD$. The aliases are $D = ABC; C = ABD; B = ACD; A = BCD; AB = CD; AC = BD; AD = BC$. Thus the initial block is
(1), $ab, cd, abcd, ac, ad, bd, bc$. The remaining factor level combinations can be obtained by formal multiplication of the elements of the initial block by a, which leads to the second 1/2-replicate $a, b, acd, bcd, c, d, abd, abc$. The resolution is R = IV.

b) For $k = 2$ the defining contrasts are $I = AC = BCD = ABD$ and the resolution is R = II. The aliases are $A = C = BD = ABCD; B = ABC = AD = CD, D = ACD = AB = BC$. Thus the initial block is

(1), abc, bd, acd. The remaining 3 factor level combinations can be obtained by formal multiplication of the elements of the initial block by a, b and c which leads to the second, third and fourth 1/4-replicates:

- $a, bc, abd, cd,$
- $b, ac, d, abcd,$
- $c, ab, bcd, ad.$

Example 7.7 We consider the case $p = 7$ with the factors A, B, C, D, E, F, G. A complete 2^7 single replicated design needs 128 experimental units. We can split this design into 2 blocks, each containing a 2^{7-1} half fraction of the complete design. We can construct the fractions by using the defining relation $I = ABCDEFG$.
If we use any factor and multiply it by $X = XI = XABCDEFG$, we receive X in the symbolic multiplication (Again the product of a factor symbol with itself results in unity). If we use any t-factor-combination ($t = 1, \dots p - 1$) and multiply it with the defining contrast, we receive just this combination as an element of the first block aliased with a combination of all factors in the defining contrast not included in the corresponding combination. Let us take the combination ACF; it will be aliased with $BDEG$. Aliased are thus

a) main effects with 5th order interactions

b) first order interaction effects with 4th order interaction effects

c) 2nd order interaction effects with 3rd order interaction effects

One block (the first, say) thus includes besides (1) all pairs and quadruples and the sextuple of the 7 letters representing the factors and the other block contains besides the single letters all triples and quintuples as well as I. The resolution is R = VII.
Of course we can split the design into 4 parts by using the defining relation $I = ABCF = BCDEG = ADEFG$. One of the 4 blocks consists of all 32 treatment combinations having an equal number of letters in common with as well $BCDEG$ as also $ADEFG$ as a 2^{7-2}-design. The resolution is R =

IV. If we use a fractional factorial design, we can use the defining relation I = CDFG = BDEG = ADEF; the resolution is R = IV.
First we show how to construct a fractional factorial design by hand without using the R-package. The algorithm is as follows:

1. Specify the values of p and k.

2. Write a complete factorial design for $p-k$ factors (containing -1 and 1)

3. Add k further columns to be filled in the next steps.

4. Take the defining relations X containing only one letter (let us say W) other than in the already-filled columns. If no such relation can be found, multiply existing relations so that the result contains only one new letter.

5. Multiply this relation $I = X$ by the new letter W and use XW in the next step.

6. Calculate the entries of the next column by multiplying the entries of the original columns belonging to XW. Add the so-generated column to the set of filled-in columns.

7. Continue with Step 4 as long as all new columns are filled.

Example 7.8 Let us construct a 2^{8-3}-design with the 5 steps above:

- $p = 8; k = 3, p - k = 5$.
- The columns 2–6 of Table 7.5 show the complete factorial design with 5 factors.
- The defining relations are (see Table 7.6) $I = CDFG = BDEG = ADEFH$.
- We find $X = BDEG$ with $W = G$.
- $G\,BDEG = BDE$.
- Column $G = BDE$ now contains the entries for factor G.
- The next relation is $I = CDFG$.
- $FCDFG = CDG$.
- Column $F = CDG$ now contains the entries for factor F.
- The last relation is $I = ADEFH$.
- $HADEFH = ADEF$.
- Column $H = ADEF$ now contains the entries for factor H.

TABLE 7.5
The 2^5-Complete Factorial Design.

Run	Factors					Factor Combinations		
	A	B	C	D	E	$F = CDG$	$G = BDE$	$H = ADEF$
1	1	1	1	1	1	1	1	1
2	1	1	1	1	−1	−1	−1	1
3	1	1	1	−1	1	1	−1	−1
4	1	1	1	−1	−1	−1	1	−1
5	1	1	−1	1	1	−1	1	−1
6	1	1	−1	1	−1	1	−1	−1
7	1	1	−1	−1	1	−1	−1	1
8	1	1	−1	−1	−1	1	1	1
9	1	−1	1	1	1	−1	−1	−1
10	1	−1	1	1	−1	1	1	−1
11	1	−1	1	−1	1	−1	1	1
12	1	−1	1	−1	−1	1	−1	1
13	1	−1	−1	1	1	1	−1	1
14	1	−1	−1	1	−1	−1	1	1
15	1	−1	−1	−1	1	1	1	−1
16	1	−1	−1	−1	−1	−1	−1	−1
17	−1	1	1	1	1	1	1	−1
18	−1	1	1	1	−1	−1	−1	−1
19	−1	1	1	−1	1	1	−1	1
20	−1	1	1	−1	−1	−1	1	1
21	−1	1	−1	1	1	−1	1	1
22	−1	1	−1	1	−1	1	−1	1
23	−1	1	−1	−1	1	−1	−1	−1
24	−1	1	−1	−1	−1	1	1	−1
25	−1	−1	1	1	1	−1	−1	1
26	−1	−1	1	1	−1	1	1	1
27	−1	−1	1	−1	1	−1	1	−1
28	−1	−1	1	−1	−1	1	−1	−1
29	−1	−1	−1	1	1	1	−1	−1
30	−1	−1	−1	1	−1	−1	1	−1
31	−1	−1	−1	−1	1	1	1	1
32	−1	−1	−1	−1	−1	−1	−1	1

In the relations $I = AFGHJ = BEGHJ = CEFH = DEFG$ of the design 20 in Table 7.6 we find no relation with only one new letter. By multiplying $AFGHJ$ with $BEGHJ$ we obtain a derived defining relation ABEF containing only F as a new letter.

To construct 2^{p-k} designs with given defining relations we can use the R-package `conf.design`.

TABLE 7.6
Defining Relations and Resolutions for 2^{p-k}-Designs.

No of Design	p	k	R	Defining Relation
1	3	1	III	$I = ABC$
2	4	1	IV	$I = ABCD$
3		2	II	$I = AC = ABD = BCD$
4	5	1	V	$I = ABCDE$
5		2	III	$I = AC = ABD = BCD$
6		3	II	$I = CD = BDE = ADE = BCE = AB = ACE = ABCD$
7	6	1	VI	$I = ABCDEF$
8		2	IV	$I = ABCE = BCDF = ADEF$
9		3	III	$I = CDF = BDE = ADEF = BCEF = ABF = ACE = ABCD$
10		4	II	$I = AF = BE = CD = ABC$
11	7	1	VII	$I = ABCDEFG$
12		2	IV	$I = ABCF = BCDEG = ADEFG$
13		3	IV	$I = CDFG = BDEG = ADEF$
14		4	III	$I = ABD = ACE = BCF = CDG = BEG = AFG = DEF$
15	8	2	V	$I = ABCDG = CDEFH = ABEFGH$
16		3	IV	$I = CDFG = BDEG = ADEFH$
17		4	IV	$I = AFGH = BEGH = CEFH = DEFG$
18	9	2	VI	$I = ACDFGH = BCEFGJ = ABDEHJ$
19		3	IV	$I = CDFG = BDEGJ = ADEFH$
20		4	IV	$I = AFGHJ = BEGHJ = CEFH = DEFG$

Example 7.8. – continued As an example we construct by the R-program again the design number 16. The relation $I = CDFG = BDEG = ADEFH$ is first transformed into a character vector

```
> v = c("CDFG", "BDEG", "ADEFH")
```

and then into a corresponding matrix of zeroes and ones which is suitable as input to the R-built-in functions `conf.matrix` and `conf.set`:

```
> G = conf.matrix(v, p = 8)
> G

     A B C D E F G H
[1,] 0 0 1 1 0 1 1 0
[2,] 0 1 0 1 1 0 1 0
[3,] 1 0 0 1 1 1 0 1
```

```
> s = 2
> U = conf.set(G, s)
> U

     A B C D E F G H
[1,] 0 0 1 1 0 1 1 0
[2,] 0 1 0 1 1 0 1 0
[3,] 0 1 1 0 1 1 0 0
[4,] 1 0 0 1 1 1 0 1
[5,] 1 0 1 0 1 0 1 1
[6,] 1 1 0 0 0 1 1 1
[7,] 1 1 1 1 0 0 0 1
```

The function `conf.set` finds minimal complete sets of confounded effects from the above list v. It is useful for checking if a low order interaction will be unintentionally confounded with block. A more compact representation of the minimal sets is provided by our OPDOE-function `defining_relation. ...` to be used as follows:

```
> defining_relation.fractional_factorial.two_levels(U)

[1] "CDFG"  "BDEG"  "BCEF"  "ADEFH" "ACEGH" "ABFGH" "ABCDH"
```

Thus we have the complete list of $2^3 - 1 = 7$ defining relations
$I = CDFG = BDEG = BCEF = ADEFH = ACEGH = ABFGH = ABCDH$.

Example 7.9 We show how to derive the defining relations for design no 17 in Table 7.6. As described in the example before, the relation $I = AFGH = BEGH = CEFH = DEFG$ is processed as follows:

```
> v = c("AFGH", "BEGH", "CEFH", "DEFG")
> p = 8
> s = 2
> G = conf.matrix(v, p)
> U = conf.set(G, s)
```

With these inputs we get the complete list of $2^4 - 1 = 15$ defining relations:

```
> defining_relation.fractional_factorial.two_levels(U)

 [1] "AFGH"     "BEGH"     "ABEF"     "CEFH"     "ACEG"     "BCFG"
 [7] "ABCH"     "DEFG"     "ADEH"     "BDFH"     "ABDG"     "CDGH"
[13] "ACDF"     "BCDE"     "ABCDEFGH"
```

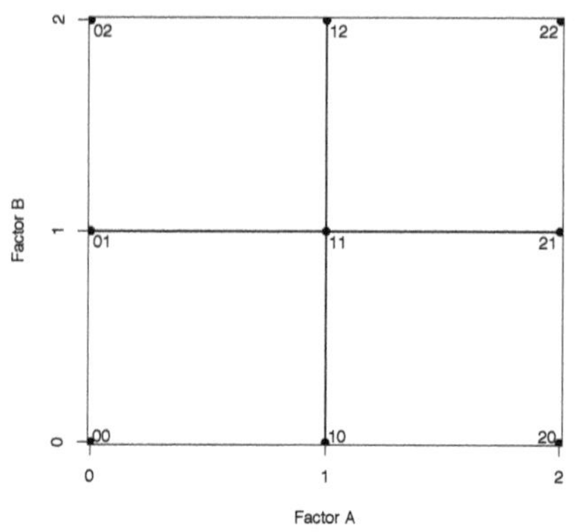

FIGURE 7.1
Treatment combinations for the 3^2-design

7.4 Fractional factorial designs with three levels of each factor (3^{p-m}-designs)

The two-level fractional factorial designs discussed before have wide applicability in medical, agricultural and industrial research and development. Sometimes, however, variations of these designs are of interest, such as designs where some or all of the factors are present at three levels. We start with the full 3^p-factorial design. Usually, two different notations are then used to represent the factor levels: one possibility is to use the digits 0 (low level), 1 (intermediate level) and 2 (high level), whereas in the case of *quantitative* factors one often denotes the low, intermediate and high levels by $-1, 0$ and $+1$, respectively. We illustrate the latter configuration for the simplest three-level full factorial design, the 3^2-design for two factors A and B having $3^2 = 9$ treatment combinations. See Figure 7.1.

Clearly, consideration of a third level of the factors allows us to include a quadratic term to model the relationship between the response and the factors. Note, however, that there are better alternatives to do this; e.g., with central composite designs we are able to model curvature, too (see Chapter nine).

Usually, the coding $0, 1, 2$ is used for the three levels, since these occur naturally in the modulus 3 calculus that is employed in the construction of three-level designs. For a better illustration let us consider $p = 3$ factors, A, B and C. Since the 3^3 design with 27 runs is a special case of a multi-way layout, the analysis of variance method introduced in Section 3.4 can be applied to

TABLE 7.7
Combinations of Factors $A(x_1\text{-levels})$ and $B(x_2\text{-levels})$.

x_1/x_2	0	1	2
0	αa	βc	γb
1	βb	γa	αc
2	γc	αb	βa

analyse this experiment. Thus, we can compute the sum of squares for the main effects A, B, C, two-factor interaction effects $A \times B, A \times C, B \times C$, and the residual sum of squares. Each main effect has $3 - 1 = 2$ degrees of freedom, each two-factor interaction has $(3-1)^2 = 4$ degrees of freedom and the residuals have $27 - 1 - 3 * 2 - 3 * 4 = 8$ degrees of freedom. Moreover, with one replication of each run we could additionally compute the three-factor interaction $A \times B \times C$, with $(3-1)^3 = 8$ degrees of freedom, and the residuals would have 27 degrees of freedom.

The sum of squares for a two-factor interaction can be further decomposed into two components, each with two degrees of freedom. For example, the $A \times B$ interaction is decomposed into two components denoted by AB and AB^2. Denoting, further, the levels of A and B by x_1 and x_2, respectively, AB represents the contrasts among the response values whose x_1 and x_2 satisfy

$$x_1 + x_2 = 0, 1, 2(\text{mod}3),$$

and AB^2 represents the contrasts among these values whose x_1 and x_2 sastisfy

$$x_1 + 2x_2 = 0, 1, 2(\text{mod}3).$$

The nine level combinations of A and B can be represented by the cells of the Graeco-Latin square in Table 7.7.

Thus, the nine combinations are arranged in two mutually orthogonal Latin squares, represented by (α, β, γ) and (a, b, c). Denoting by y_{ij} the observation (or average of replicates) in the cell $(i, j); i, j = 0, 1, 2$; then the letters $\alpha, \beta, \gamma, a, b, c$ represent groupings of the observations, e.g., α represents the group (y_{00}, y_{12}, y_{21}), β the group (y_{01}, y_{10}, y_{22}). Accordingly, let

$\overline{y}_\alpha = (y_{00} + y_{12} + y_{21})/3, \overline{y}_\beta = (y_{01} + y_{10} + y_{22})/3, \overline{y}_\gamma = (y_{02} + y_{11} + y_{20})/3,$
$\overline{y}_a = (y_{00} + y_{11} + y_{22})/3, \overline{y}_b = (y_{02} + y_{10} + y_{21})/3, \overline{y}_c = (y_{01} + y_{12} + y_{20})/3.$

Now, since $\overline{y}_\alpha, \overline{y}_\beta$ and $\overline{y}_\gamma$ represent the averages of the observations with $x_1 + 2x_2 = 0, 1, 2(\text{mod}3)$, respectively, the interaction component AB represents the contrasts among $\overline{y}_\alpha, \overline{y}_\beta$ and $\overline{y}_\gamma$.

The sum of squares for the AB interaction component can thus be interpreted as the treatment sum of squares for the Latin square design in the treatments α, β and γ. It follows that

$$SS_{AB} = 3r[(\overline{y}_\alpha - \overline{y}_G)^2 + (\overline{y}_\beta - \overline{y}_G)^2 + (\overline{y}_\gamma - \overline{y}_G)^2]$$

where $\overline{y}_G = (\overline{y}_\alpha + \overline{y}_\beta + \overline{y}_\gamma)/3$ is the "Greek average" and r is the number of replicates. Similarly, the AB^2 interaction compontent represents the contrasts among the three groups represented by the letters a, b and c; the corresponding values $\overline{y}_a, \overline{y}_b$ and $\overline{y}_c$ represent the averages of observations with $x_1 + x_2 = 0, 1, 2(\text{mod}3)$, respectively.

This implies

$$SS_{AB^2} = 3r[(\overline{y}_a - \overline{y}_L)^2 + (\overline{y}_b - \overline{y}_L)^2 + (\overline{y}_c - \overline{y}_L)^2]$$

where $\overline{y}_L = (\overline{y}_a + \overline{y}_b + \overline{y}_c)/3$ is the "Latin average". The decomposititon of the three-factor interaction $A \times B \times C$ has the four (orthogonal) components ABC, ABC^2, AB^2C and AB^2C^2, which represent the contrasts among the three groups of x_1, x_2, x_3 combinations satsifiying the systems of equations

$x_1 + x_2 + x_3 = 0, 1, 2(\text{mod}3)$,

$x_1 + x_2 + 2x_3 = 0, 1, 2(\text{mod}3)$,

$x_1 + 2x_2 + x_3 = 0, 1, 2(\text{mod}3)$,

$x_1 + 2x_2 + 2x_3 = 0, 1, 2(\text{mod}3)$.

Because of the fact that the decomposition of $A \times B$ and $A \times B \times C$ involves orthogonal components, the above system of parametrisation is also called the *orthogonal components system.*.

Note that in the above system we have used the convention that the coefficient for the first non-zero factor is 1. This is no restriction; e.g., A^2B^2C can be replaced by the equivalent component ABC^2, since the system $2x_1+2x_2+x_3 = 2(x_1+x_2+2x_3) = 0, 1, 2(\text{mod}3)$ is equivalent to $x_1+x_2+2x_3 = 0, 1, 2(\text{mod}3)$.

As for two-level fractional factorial designs, it is more efficient to use just a one-third or, more general, a 3^{p-m}-th fraction of a full factorial 3^p design. Note that for a 3^4 design with 81 runs, out of the 80 degrees of freedom, 48 are for estimating three-factor and four-factor interaction effects. Such high-order interaction effects are, however, difficult to interpret and usually insignificant. In the same way in which we have formed fractions of full factorial 2^p designs we can form fraction of 3^p designs by means of *defining relations*. For example, one-third fraction of a 3^4 design with factors A, B, C and D can be obtained by setting

$$D = ABC$$

Using $x_1, \ldots, x_4$ to represent the level combinations of these four factors, we have $x_4 = x_1 + x_2 + x_3(\text{mod}3)$, or equivalently,

$$x_1 + x_2 + x_3 + 2x_4 = 0(\text{mod}3) \tag{7.7}$$

which, in turn, can be written as

$$I = ABCD^2 \tag{7.8}$$

Here, again, I denotes the column of 0's and acts as (group-theoretic) identity, playing the same role as the column of +'s in two-level designs. The aliasing patterns can be deduced from the defining relation. For example, by adding $2x_2$ to both sides of (7.7) we get

$2x_2 = x_1 + 3x_2 + x_3 + 2x_4 = x_1 + x_3 + 2x_4(\text{mod}3)$,

which means that the three groups defined by $2x_2 = 0, 1, 2(\text{mod}3)$ are identical to the three groups defined by $x_1+x_3+2x_4 = 0, 1, 2(\text{mod}3)$. The corresponding contrasts among the three groups define the main effect B and the interaction ACD^2; thus, we can say that B and ACD^2 are *aliased.* Similarly, adding x_2 to both sides of (7.7) leads to the aliasing of B and AB^2CD^2. Continuing this way, we get a complete list of aliased effects:

$A = BCD^2 = AB^2C^2D, B = ACD^2 = AB^2CD^2,$

$C = ABD^2 = ABC^2D^2, D = ABC = ABCD,$

$AB = CD^2 = ABC^2D, AB^2 = AC^2D = BC^2D,$

$AC = BD^2 = AB^2CD, AC^2 = AB^2D = BC^2D^2,$

$AD = AB^2C^2 = BCD, AD^2 = BC = AB^2C^2D^2,$

$BC^2 = AB^2D^2 = AC^2D^2, BD = AB^2C = ACD,$

$CD = ABC^2 = ABD.$

If three-factor interaction effects are assumed negligible then, from the above aliasing relations, $A, B, C, D, AB^2, AC^2, AD, BC^2, BD$ and CD can be estimated. These main effects and two-factor interaction effects are then not alisased with any other main effects or two-factor interaction components.

Generally, a 3^{p-m} design is a fractional factorial design with p factors in 3^{p-m} runs. This fractional plan is the 3^{-m}-th fraction of the full 3^p design, and it is defined by m independent generators. As before, the resolution of the resulting designs can be obtained on the basis of the word lengths.

Example 7.10 Let $p = 5$, $m = 2$, i.e., we consider a 3^{5-2} design with five factors and 27 runs. Let the two generators be defined via the relations $D = AB$ and $E = AB^2C$, leading to $I = ABD^2 = AB^2CE^2$. From these generators, two additonal relations can be obtained:

$I = (ABD^2)(AB^2CE^2) = A^2CD^2E^2 = AC^2DE$

and

$I = (ABD^2)(AB^2CE^2)^2 = B^2C^2D^2E = BCDE^2$

Thus, the defining relation for this design reads

$I = ABD^2 = AB^2CE^2 = AC^2DE = BCDE^2$

Therefore, we have four words, whose word lengths are $3, 4, 4$ and 4. The word-length pattern is given by $(1, 3, 0)$, i.e., one word of length 3, three words of length 4, and no word of length 5.

We show how to generate the defining relations and the corresponding 3^{5-2} design with the R-built-in functions `conf.set` and `conf.design`. The relations $I = ABD^2 = AB^2CE^2$ are written as

```
> G = rbind(c(1, 1, 0, 2, 0), c(1, 2, 1, 0, 2))
```

Continuing, we get

```
> p = 5
> s = 3
> dimnames(G) = list(NULL, LETTERS[1:p])
> conf.set(G, s)

     A B C D E
[1,] 1 1 0 2 0
[2,] 1 2 1 0 2
[3,] 0 1 1 1 2
[4,] 1 0 2 1 1
```

From the third and fourth row of this matrix we deduce the remaining two relations $I = BCDE^2$ and $I = AC^2DE$, respectively. Finally, the corresponding 3^{5-2} design with $n = 27$ runs can be obtained as follows:

```
> m = 2
> n = s^(p - m)
> d = conf.design(G, s, treatment.names = LETTERS[1:p])
> d[1:n, ]

   Blocks A B C D E
1      00 0 0 0 0 0
2      00 1 2 1 0 0
3      00 2 1 2 0 0
4      00 2 2 0 1 0
5      00 0 1 1 1 0
6      00 1 0 2 1 0
7      00 1 1 0 2 0
8      00 2 0 1 2 0
9      00 0 2 2 2 0
10     00 2 1 0 0 1
11     00 0 0 1 0 1
12     00 1 2 2 0 1
```

13	00 1 0 0 1 1
14	00 2 2 1 1 1
15	00 0 1 2 1 1
16	00 0 2 0 2 1
17	00 1 1 1 2 1
18	00 2 0 2 2 1
19	00 1 2 0 0 2
20	00 2 1 1 0 2
21	00 0 0 2 0 2
22	00 0 1 0 1 2
23	00 1 0 1 1 2
24	00 2 2 2 1 2
25	00 2 0 0 2 2
26	00 0 2 1 2 2
27	00 1 1 2 2 2

In general, with a 3^{p-m} design we can study at most $(3^{p-m}-1)/2$ factors. For $p-m=3$, the $(3^3-1)/2=13$ factors are those we had considered in Example 7.10.

A general algebraic treatment of fractional factorial 3^{p-m} designs can be found in Kempthorne (1952). For tables and aliasing schemes for such designs we also refer to Wu and Hamada (2000).

8

Exact Optimal Designs and Sample Sizes in Model I of Regression Analysis

Whereas in Model II of regression analysis (see Chapter four) only the sample size determination is needed to design an experiment, we have in Model I of regression analysis the possibility to choose the x-values in such a way that for special precision requirements the number of observations is minimised.

8.1 Introduction

We consider the model $\boldsymbol{y}_i = f(x_i; \theta) + \boldsymbol{e}_i$; $i = 1, 2, \ldots n$; $\theta^T = (\theta_1 \ldots, \theta_p) \in \Omega \subset \mathbb{R}^p$; $p < n$ with i.i.d. random error terms and a continuously twice differentiable (with respect to x and θ) function f. The x_i-values have to be chosen by the experimenter before the experiment starts whereby at least p of them must be different from each other. We consider especially linear, quadratic and cubic polynomials and some intrinsically non-linear functions. Our aim is

- To find the exact Φ-optimal designs defined in Section 8.2
- To determine the minimum size of the experiment for a given structure and precision requirements.

We restrict ourselves in most cases of this chapter to one cause variable x. Because in Chapters 9 and 10 on second order and mixture designs, respectively, and Appendix A we assume at least two cause variables, a short introduction to more cause variables is given in Section 8.1.1. More about k cause variables can be found in Chapter seven about factorial designs.

Nevertheless we start in this introductory paragraph with the case of multiple regression (at least two cause variables) as a base of the polynomial regression and the mixture designs. We consider a special case of the *general regression model*

$$\boldsymbol{y} = f(x_1, \ldots, x_k; \theta) + \boldsymbol{e} \tag{8.1}$$

A character y (modelled by a random variable $\boldsymbol{y}$) may depend on one or more predictor or cause variables $x_1, x_2, \ldots, x_k$.

8.1.1 The multiple linear regression Model I

For the moment, we assume that there is a linear dependence of the form with the usual notation $\theta^T = \beta^T = (\beta_0, \beta_1, \ldots, \beta_k)$, $p = k + 1$, so Equation (8.1) becomes

$$\boldsymbol{y}_j = \beta_0 + \beta_1 x_{1j} + \beta_2 x_{2j} + \ldots + \beta_k x_{kj} + \boldsymbol{e}_j \quad (j = 1, \ldots, n) \tag{8.2}$$

where y_j is the j-th value of y which depends on the j-th values $x_{1j}, \ldots, x_{kj}$ of $x_1, \ldots, x_k$. The $\boldsymbol{e}_j$ are error terms with $E(\boldsymbol{e}_j) = 0, var(\boldsymbol{e}_j) = \sigma^2$ (for all j) and $cov(\boldsymbol{e}_{j'}, \boldsymbol{e}_j) = 0$ for $j' \neq j$. When we construct confidence intervals and test hypotheses, we also assume that the $\boldsymbol{e}_j$ are normally distributed.

As already said the $x_1, \ldots, x_k$ must all be fixed by the experimenter, i.e., (8.2) is a *Model I multiple linear regression.*

In the analysis we have to solve the following problems:

(1) estimate the regression coefficients $\beta_0, \beta_1, \ldots, \beta_k$

(2) estimate y by $\hat{y}$ for given values $x_1, \ldots, x_k$

(3) construct confidence intervals for each regression coefficient or for functions of the regression coefficients

(4) test hypotheses about the regression coefficients

The estimation of the regression coefficients is usually based on the Method of Least Squares (MLS). In this method we use, as estimates of $\beta_0, \beta_1, \ldots, \beta_k$, those values $b_0, b_1, \ldots, b_k$ which minimise the sum of squared deviations

$$S = \sum_{j=1}^{n} (y_j - \beta_0 - \beta_1 x_{1j} - \ldots - \beta_k x_{kj})^2 \tag{8.3}$$

Because S is a convex function of the β_i, a solution is reached if we solve the set of $k + 1$ equations

$$\frac{\partial S}{\partial \beta_l} = 0 (l = 0, \ldots, k) \tag{8.4}$$

We will now use matrix notation to write these solutions. We put $\boldsymbol{y}^T = (\boldsymbol{y}_1, \ldots, \boldsymbol{y}_n)$, $\beta^T = (\beta_0, \ldots, \beta_k)$,

$$X = \begin{pmatrix} 1 & x_{11} & \dots & x_{k1} \\ 1 & x_{12} & \dots & x_{k2} \\ \vdots & & & \\ 1 & x_{1n} & \dots & x_{kn} \end{pmatrix}$$

and $\boldsymbol{e}^T = (\boldsymbol{e}_1, \boldsymbol{e}_2, \dots, \boldsymbol{e}_n)$, then (8.2) becomes

$$\boldsymbol{y} = X\beta + \boldsymbol{e} \tag{8.5}$$

and this is the general linear regression equation with

$$E(\boldsymbol{e}) = \mathbf{0}_n, var(\boldsymbol{e}) = \sigma^2 I_n \tag{8.6}$$

as our side conditions. Here $\mathbf{0}_n$ is a vector with n zeros, and I_n is the idendity matrix of order n. The more general case with side conditions $E(\boldsymbol{e}) = 0_n, var(\boldsymbol{e}) = \sigma^2 V$ with a positive definite symmetric square matrix V of order n is not considered here. If V is known w.l.o.g. this case can be transformed into the one with side conditions (8.6). But in the practical more important case of unknown V, the generalised least squares method has to be used, and this is beyond the scope of this book.

The normal equations for obtaining the estimates b are then $X^T Xb = X^T y$. And after introducing the random variables into these equations the estimators become

$$X^T X\boldsymbol{b} = X^T \boldsymbol{y} \tag{8.7}$$

We can write the solution of our Least Squares problem as the *least squares estimate* (assuming that $(X^T X)^{-1}$ exists—what can always be achieved if we take at least $k+1$ different design points)

$$b = \hat{\beta} = (X^T X)^{-1} X^T y \tag{8.8}$$

The *least squares estimator (LSE)* is obtained by transferring to random variables, i.e., replacing y by $\boldsymbol{y}$:

$$\boldsymbol{b} = (X^T X)^{-1} X^T \boldsymbol{y} \tag{8.9}$$

From this we find that

$$E(\boldsymbol{y}) = E(X\beta) + E(\boldsymbol{e}) = X\beta + 0_n = X\beta$$

and

$$E(\boldsymbol{b}) = (X^TX)^{-1}X^TE\boldsymbol{y} = (X^TX)^{-1}X^TX\beta = \beta$$

and therefore $\boldsymbol{b}$ is an unbiased estimator of β.

The variance of $\boldsymbol{b}$ is given by the covariance (or variance-covariance) matrix

$$var(\boldsymbol{b}) = \sigma^2(X^TX)^{-1} = \begin{pmatrix} var(\boldsymbol{b}_0) & cov(\boldsymbol{b}_0,\boldsymbol{b}_1) & \dots & cov(\boldsymbol{b}_0,\boldsymbol{b}_k) \\ cov(\boldsymbol{b}_0,\boldsymbol{b}_1) & var(\boldsymbol{b}_1) & \dots & cov(\boldsymbol{b}_1,\boldsymbol{b}_k) \\ \vdots & & & \vdots \\ cov(\boldsymbol{b}_0,\boldsymbol{b}_k) & cov(\boldsymbol{b}_1,\boldsymbol{b}_k) & \dots & var(\boldsymbol{b}_k) \end{pmatrix} \quad (8.10)$$

On this matrix or its inverse, the so-called information matrix, the definitions of optimal designs are based.

The error variance σ^2 of the $\boldsymbol{e}_j$ in the model (8.5) is unbiasedly estimated by

$$s^2 = \frac{1}{n-k-1}\sum_{j=1}^{n}(y_j - b_0 - b_1x_{1j} - \ldots - b_kx_{kj})^2$$

$$= \frac{1}{n-k-1}y^T(I_n - X(X^TX)^{-1}X^T)y \quad (8.11)$$

and is used to evaluate $s^2(X^TX)^{-1}$, the estimate of $\sigma^2(X^TX)^{-1}$.

We have the well-known Gauß-Markov Theorem.

Theorem 8.1 (Gauß-Markov Theorem) For any vector $c^T = (c_1, c_2, \ldots, c_{k+1})$ the least squares estimator $c^T\boldsymbol{b}$ in (8.9) of the linear function $c^T\beta$ of β in (8.5) is under the side conditions in (8.6) the unique best (i.e., with minimal variance) linear unbiased estimator (BLUE).

This means that all the sample sizes calculated below are the smallest possible ones for an unbiased estimation of a scalar function of β which has a variance below a given bound.

8.1.2 Simple polynomial regression

We consider the case $k = 1$ in the model (8.1) and therefore denote x_1 by x, but we now take f to be a *polynomial* in x.

$$f(x) = \beta_0 + \beta_1x + \beta_2x^2 + \ldots + \beta_kx^k. \quad (8.12)$$

Here f is a k-th order polynomial.

In practice quadratic ($k = 2$) or cubic ($k = 3$) functions are usually sufficient. Equation (8.12) gives us the following matrix expressions:

$$\beta^T = (\beta_0, \beta_1, \ldots, \beta_k),$$

$$X = \begin{pmatrix} 1 & x_1 & \ldots & x_l^k \\ 1 & x_2 & \ldots & x_2^k \\ \vdots & \vdots & \vdots & \\ 1 & x_n & \ldots & x_n^k \end{pmatrix}$$

with $\boldsymbol{e}$ and $\boldsymbol{y}$ as in (8.5), and we obtain Equations (8.7) to (8.9); however we have the following slightly changed notations:

$$X^T X = \begin{pmatrix} n & \sum x_j & \ldots & \sum x_j^k \\ \sum x_j & \sum x_j^2 & \ldots & \sum x_j^{k+1} \\ \ldots & \ldots & \ldots & \ldots \\ \sum x_j^k & \sum x_j^{k+1} & \ldots & \sum x_j^{2k} \end{pmatrix}$$

and

$$X^T y = \begin{pmatrix} \sum y_j \\ \sum x_j y_j \\ \ldots \\ \sum x_j^k y_j \end{pmatrix}$$

The side conditions for the regression model (8.5) with the X-matrix above are again given by (8.6).

8.1.3 Intrinsically non-linear regression

In addition to polynomials, there are also non-linear functions which are non-linear not only in the regressor variables but also in the parameters. We shall call such functions *intrinsically non-linear*, because they cannot be handled by the methods of multiple linear regression.

Definition 8.1 (Rasch, 1995) A differentiable regression function $f(x;\theta); \theta^T = (\theta_1, \ldots, \theta_p) \in \Omega \subset \mathbb{R}^p$ is called *partially intrinsically non-linear* if

$$\frac{\partial f(x;\theta)}{\partial \theta} = C(\theta) \cdot g(x, \varphi); \varphi^T = (\theta_{i_1}, \ldots, \theta_{i_r}) \tag{8.13}$$

and $0 < r < p$ with a matrix $C(\theta)$ not depending on x, a function g and a vector φ which is chosen in such a way that r is minimum ($r = 0$, not included here leads to a linear or quasi-linear regression function). $\varphi^T = (\theta_{i_1}, \ldots \theta_{i_r})$ is called the *vector of non-linearity parameters*; the remaining elements of θ are called *linearity parameters*. If $r = p$, the function is called *totally intrinsically non-linear*.

The model equation is now

$$\boldsymbol{y}_i = f(x_i;\theta) + \boldsymbol{e}_i; i = 1, \ldots, n \tag{8.14}$$

Again we assume i.i.d. error terms $\boldsymbol{e}_j$ with $E(\boldsymbol{e}_j) = 0, var(\boldsymbol{e}_j) = \sigma^2$ (for all j). The following functions play a special role in the theory of growth curves as well as in many other applications:

the exponential function $f_E(x;\theta) = \theta_1 + \theta_2 e^{\theta_3 x}$

the logistic function $f_L(x,\theta) = \frac{\theta_1}{1+\theta_2 e^{\theta_3 x}}$

the Gompertz function $f_G(x,\theta) = \theta_1 e^{\theta_2 e^{\theta_3 x}}$

the Bertalanffy function $f_B(x,\theta) = [\theta_1 + \theta_2 e^{\theta_3 x}]^3$

In biochemistry but also in other applications the Michaelis-Menten function $f_M(x,\theta) = \frac{\theta_1 x}{1+\theta_2 x}$ is often used.

There are many other intrinsically non-linear functions which in special applications are more appropriate than polynomials, especially when extrapolation over the experimental region is needed. We will demonstrate this by an example before we discuss the numerical and statistical problems in intrinsically non-linear regression.

First we must solve the problem of selecting an adequate model for the given data. By selecting a model we mean a regression function out of a well-defined set of possible functions. For this purpose, several model selection criteria have been developed.

The simplest criterion is the estimate s^2 of the error variance σ^2. But due to the well-known Weierstrass theorem, a polynomial of degree $n-1$ always passes exactly through n given points with $s^2 = 0$.

Therefore model selection criteria have been developed including a penalty function for the number of parameters in the regression function. We describe here a modified Akaike criterion and the Schwartz criterion. All the criteria are based on the residual sum of squares $SS_j = \sum_{i=1}^{n} (y_i - f_j[x_i;\theta])^2$ between the observed y_i and the values of the candidate function f_j at x_i and on the numbers p_j of unknown parameters in the function f_j.

The Akaike criterion in the modification of Hurvich and Tsai (1989) is defined as $AIC_j = n \cdot \ln(SS_j) + \frac{n(n+p_j)}{n-p_j-2}$.

The Schwartz criterion is defined as $SCC_j = n \cdot \ln(SS_j) + p_j \cdot \ln(n)$.

Example 8.1 We now determine the linear and non-linear parameters of the

logistic function $f_L(x,\theta) = \frac{\theta_1}{1+\theta_2 e^{\theta_3 x}}$.
At first we derive the function

$$\frac{\partial}{\partial\theta} f_L(x,\theta) = \begin{pmatrix} \frac{1}{1+\theta_2 e^{\theta_3 x}} \\ \frac{-\theta_1 e^{\theta_3 x}}{(1+\theta_2 e^{\theta_3 x})^2} \\ \frac{-\theta_1\theta_2 x e^{\theta_3 x}}{(1+\theta_2 e^{\theta_3 x})^2} \end{pmatrix}$$

With $\varphi^T = (\theta_2, \theta_3)$ and thus $r = 2$ and $C(\theta) = \begin{pmatrix} 1 & 0 & 0 \\ 0 & -\theta_1 & 0 \\ 0 & 0 & -\theta_1 \end{pmatrix}$, the vector g in (8.13) is

$$g(x,\varphi) = \begin{pmatrix} \frac{1}{1+\theta_2 e^{\theta_3 x}} \\ \frac{e^{\theta_3 x}}{(1+\theta_2 e^{\theta_3 x})^2} \\ \frac{\theta_2 x e^{\theta_3 x}}{(1+\theta_2 e^{\theta_3 x})^2} \end{pmatrix}$$

Therefore the logistic function is partially intrinsically non-linear with linear parameter θ_1 and non-linear parameters θ_2, θ_3.

Definition 8.2 In a regression Model I with one cause variable, the interval $[a, b]$ in which the values of the cause variable x can or should be chosen as part of designing the experiment is called the *design region.*

Example 8.2 (Rasch et al. 2007b) We consider data of hemp growth in the design region $[a, b] = [1, 14]$ (weeks) (see Table 8.1).

TABLE 8.1
The Height of Hemp Plants (y in cm) during Growth (x Age in Weeks).

x_i	y_i	x_i	y_i
1	8.30	8	84.40
2	15.20	9	98.10
3	24.70	10	107.70
4	32.00	11	112.00
5	39.30	12	116.90
6	55.40	13	119.90
7	69.00	14	121.10

From the scatter plot in Figure 8.1 we learn that the relation is non-linear. When we fit a linear, quadratic and a cubic polynomial we obtain the following estimated regression equations and estimated residual variances:

Linear: $\hat{y}_{lin}(x) = -2.3527 + 9.8756x$; $s^2_{lin} = 59.76$
Quadratic $\hat{y}_{quad}(x) = -13.7154 + 14.1366x - 0.2841x^2$; $s^2_{quad} = 43.83$
Cubic: $\hat{y}_{cub}(x) = 8.563 - 1.1197x + 2.173x^2 - 0.1092x^3$; $s^2_{cub} = 6.46$

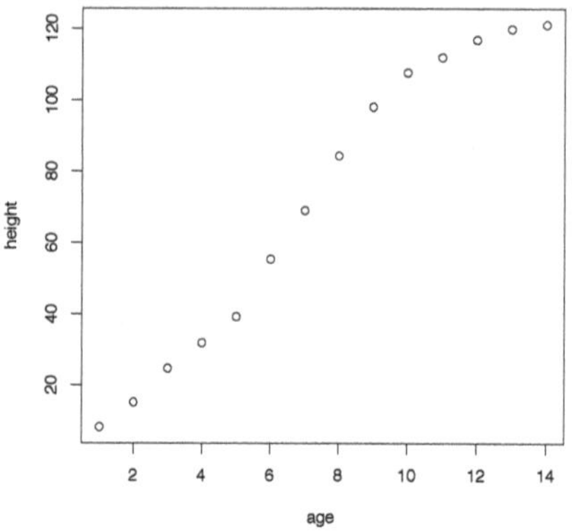

FIGURE 8.1
Scatter-plot for the association between age and height of hemp plants

The cubic polynomial gives us the best fit if we use the estimated residual variance as selection criterion. But also with the Akaike and the Schwartz criterion which include a penalty function for the number of parameters in the model, the cubic polynomial gives the best fit. Figure 8.2 shows the fitted cubic polynomial.

When we try to predict the height of the hemp plants after 16 weeks of growth by extrapolating the cubic polynomial, we obtain the graph in Figure 8.3.

It is quite clear that we should not do any extrapolation of polynomials outside the experimental region. When we fit several growth function to the data (including the exponential, logistic, Gompertz and Bertalanffy functions mentioned above and some four parametric functions) and use the Akaike criterion to select that one fitting best, we have to take the logistic function with the following estimated regression equation and estimated residual variance:

$$f_{log}(x,\theta) = \frac{126.19}{1 + 19.7289e^{-0.4607x}};\, s_{log}^2 = 3.704$$

This gives the best fit of all functions used in this example. When we now extrapolate to 16 weeks, we get the graph in Figure 8.4; this shows us that intrinsically non-linear functions can be better adapted to specific problems than polynomials.

Let us now demonstrate Definition 8.1 by other examples of intrinsically non-linear functions given above.

Example 8.3 The exponential function $f_E(x,\theta) = \theta_1 + \theta_2 e^{\theta_3 x}$ has the following derivative with respect of θ:

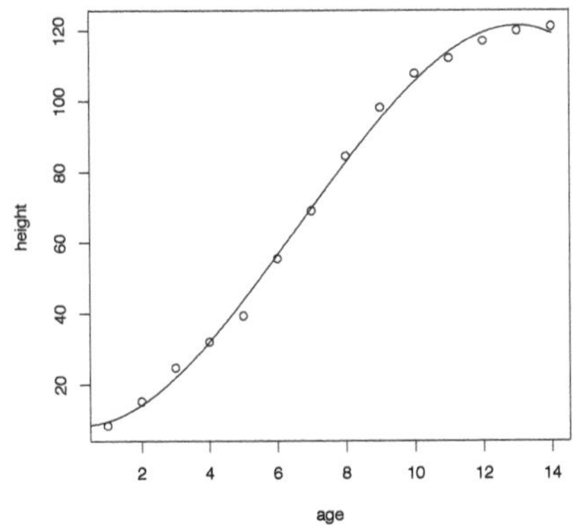

FIGURE 8.2
The graph of the fitted cubic polynomial

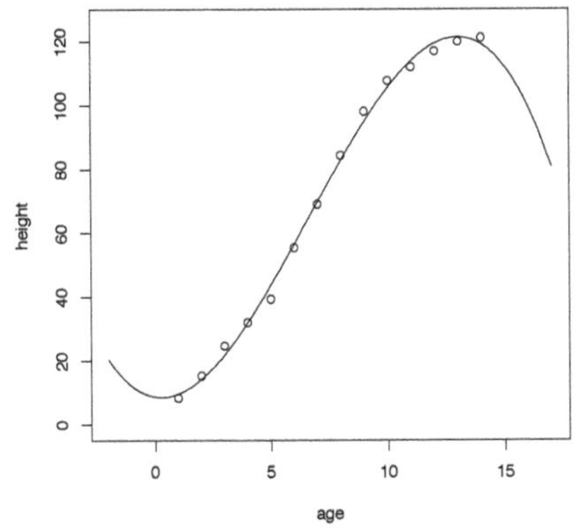

FIGURE 8.3
The graph of the fitted cubic polynomial over 16 weeks

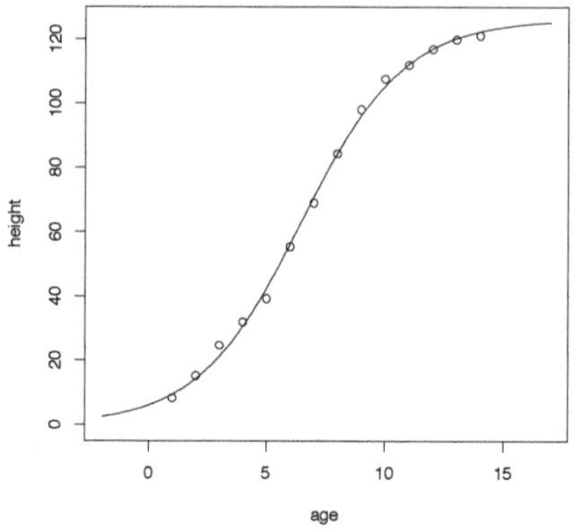

FIGURE 8.4
The graph of the fitted logistic function over 16 weeks

$$\frac{\partial f_E(x\theta)}{\partial \theta} = \begin{pmatrix} 1 \\ e^{\theta_3 x} \\ x\theta_2 e^{\theta_3 x} \end{pmatrix} = \begin{pmatrix} 1 & 0 & 0 \\ 0 & 1 & 0 \\ 0 & 0 & \theta_2 \end{pmatrix} \cdot \begin{pmatrix} 1 \\ e^{\theta_3 x} \\ x e^{\theta_3 x} \end{pmatrix}$$

We see from this decomposition in the C-matrix and the g-vector that only θ_3 is a non-linear parameter.

Example 8.4 The Bertalanffy function $f_B(x,\theta) = [\theta_1 + \theta_2 e^{\theta_3 x}]^3$ has partial derivatives with respect to θ such that all three parameters occur in the φ-vector, and therefore all parameters are non-linear.

Using the least squares method to estimate the parameters in the intrinsically non-linear case, we have to replace Equations (8.3) and (8.4) by

$$S = \sum_{j=1}^{n} [y_j - f(x_i;\theta)]^2 \tag{8.15}$$

and

$$\frac{\partial S}{\partial \theta_i} = 0, (i = 1, \ldots, p) \tag{8.16}$$

Because at least one of the parameters in an intrinsically non-linear regression function occurs in the vector of derivatives of f with respect to θ, we obtain normal equations analogous to (8.7) which include all non-linear parameters. If we replace them by their estimates, a non-linear system of equations is the result, which can only be solved iteratively, a procedure which sometimes causes numerical problems (no program exists which can guarantee that in every case a solution will be found which really is the overall minimum). The consequence is that neither the estimates nor the estimators can be written in closed form. The consequence of this is that the distributions of the (iteratively obtained) estimators are unknown. Details can be found in Rasch (1990, 1995) and Melas (2006).

We denote the least squares estimator (even if it is not known in closed form) of θ based on n observations by $\hat{\boldsymbol{\theta}}_n$. For this least squares estimator Theorem 8.1 is no longer valid; the estimator is no longer unbiased.

What we know from a theorem of Jennrich (1969) is that under mild conditions about f (which are fulfilled for all functions considered in this book), the sequence $\sqrt{n}(\hat{\boldsymbol{y}}_n - \theta)$ inside any compact Ω is asymptotically $N(0_p; \sigma^2 I^{-1}(\theta))$ where

$$I^{-1}(\theta) = \lim_{n\to\infty} \frac{1}{n} F_n^T(\theta) F_n(\theta) \text{ with}$$

$$F_n(\theta) = \begin{pmatrix} \frac{\partial f(x_1;\theta)}{\partial \theta_1} & \frac{\partial f(x_1;\theta)}{\partial \theta_2} & \cdots & \frac{\partial f(x_1;\theta)}{\partial \theta_p} \\ \frac{\partial f(x_2;\theta)}{\partial \theta_1} & \frac{\partial f(x_2;\theta)}{\partial \theta_2} & \cdots & \frac{\partial f(x_2;\theta)}{\partial \theta_p} \\ \cdots & \cdots & \cdots & \cdots \\ \frac{\partial f(x_n;\theta)}{\partial \theta_1} & \frac{\partial f(x_n;\theta)}{\partial \theta_2} & \cdots & \frac{\partial f(x_n;\theta)}{\partial \theta_p} \end{pmatrix} \tag{8.17}$$

and σ^2 is the limiting value of the estimated residual variance based on n observations.

We will see in the following section that optimal design criteria for intrinsically non-linear functions are based on $\frac{1}{n} F_n^T(\theta) F_n(\theta)$ for any selected n. The consequence of this is that the optimal designs depend on the non-linearity parameters of these functions.

8.2 Exact Φ-optimal designs

Most optimal experimental designs are concerned with estimation criteria. Thus the x_i-values will often be chosen in such a way that the variance of an estimator is minimised (for a given number of measurements) amongst all possible allocations of x_i-values.

At first we remember the definition of the experimental region $[x_l, x_u] = [a, b]$ in which measurements can or will be taken. Here $x_l = a$ is the lower bound and $x_u = b$ the upper bound of an interval on the x-axis.

A disadvantage of optimal designs is their dependence on the accuracy of the

chosen model; they are only certain to be optimal if the assumed model is correct. For example, in the cases discussed in this section, we must know that the chosen model is adequate, i.e., that the regression function actually is linear, quadratic, cubic or of a special intrinsically non-linear type. Optimal designs are often selected in a way that does not allow the model assumptions to be tested; for example, in the designs that we shall present below in the linear case, we can only test whether a linear or quadratic function describes the relationship better for the G-optimal design if n is odd.
Often the criteria of the following definition are used in multiple linear regression of the model (8.2) where the $\boldsymbol{b}_i$ are the least squares estimators of the β_i.

We use the term *support points* for the different places where measurements are made.

Definition 8.3 For any $n > k + 2$ and $r > k$ we write an exact design in the form $\begin{pmatrix} x_1 & x_2 & \dots & x_r \\ n_1 & n_2 & \dots & n_r \end{pmatrix}$ with the r *support points* all within $[a, b]$ and the (integer) *frequencies* $n_1, n_2, \dots, n_r$; $\sum\limits_{i=1}^{r} n_i = n$.

Definition 8.4 An exact design is called

(C_i)-*optimal* if it minimises $var(\boldsymbol{b}_i)$

D-optimal if it minimises the determinant D of the covariance matrix

A-optimal if it minimises the trace of the covariance matrix

G-optimal if it minimises the maximum [over (x_l, x_u)] of the expected width of the confidence interval for $E(\boldsymbol{y})$.

8.2.1 Simple linear regression

In the simple linear regression we have $k = 1$ and (8.2) becomes $\boldsymbol{y}_j = \beta_0 + \beta_1 x_j + \boldsymbol{e}_j$ $(j = 1, \dots, n)$. Further we obtain $b_0 = \overline{y} - b_1\overline{x}$ and $b_1 = \frac{SP_{xy}}{SS_x}$. The special case of $var(\boldsymbol{b})$ in (8.10) for $k = 1$ gives

$$var\begin{pmatrix} \boldsymbol{b}_0 \\ \boldsymbol{b}_1 \end{pmatrix} = \frac{\sigma^2}{SS_x}\begin{pmatrix} \frac{1}{n}\sum\limits_{j=1}^{n} x_j^2 & -\overline{x} \\ -\overline{x} & 1 \end{pmatrix} \tag{8.18}$$

and the determinant of this covariance matrix is

$$D = \frac{\sigma^4}{SS_x^2}\left(\frac{1}{n}\sum_{j=1}^{n} x_j^2 - \overline{x}^2\right) = \frac{\sigma^4}{nSS_x^2} \tag{8.19}$$

The $(1-\alpha)$-confidence interval for $E(\hat{\boldsymbol{y}}) = \beta_0 + \beta_1 x$ with $\hat{\boldsymbol{y}} = \boldsymbol{b}_0 + \boldsymbol{b}_1 x = \overline{\boldsymbol{y}} + \boldsymbol{b}_1(x-\overline{x})$ is for any $x \in [a,b]$ given by

$$\left[\boldsymbol{b}_0 + \boldsymbol{b}_1 x - t\left(n-2;\ 1-\frac{\alpha}{2}\right)\boldsymbol{s}\sqrt{\frac{1}{n}+\frac{(x-\overline{x})^2}{SS_x}}\ ;\right.$$

$$\left.\boldsymbol{b}_0 + \boldsymbol{b}_1 x + t\left(n-2;\ 1-\frac{\alpha}{2}\right)\boldsymbol{s}\sqrt{\frac{1}{n}+\frac{(x-\overline{x})^2}{SS_x}}\right] \tag{8.20}$$

with $\boldsymbol{s}^2 = \frac{1}{n-2}\sum_{i=1}^{n}(\boldsymbol{y}_i - \boldsymbol{b}_0 - \boldsymbol{b}_1 x_i)^2$ analogous to (8.11). Thus to obtain the C_1- and the D-optimum design for any $n > 2$ we have to maximise SS_x in the denominator of (8.18) or (8.19). The two criteria give the same exact optimal design solution, and the resulting design also minimises the *expected width of the confidence interval* for β_1. On the other hand, the power of a test of $H_0: \beta_1 = \beta_{10}$ is for any first kind risk $0 < \alpha < 1$ and given $n > 2$ is maximum for the corresponding designs.

Theorem 8.2 If n is even, the C_1- and the D-optimum design is given by $\begin{pmatrix} a & b \\ \frac{n}{2} & \frac{n}{2} \end{pmatrix}$.
If $n = 2t+1$ with an integer $t > 0$ is odd, the C_1- and the D-optimum design is not unique and is given by either $\begin{pmatrix} a & b \\ t+1 & t \end{pmatrix}$ or $\begin{pmatrix} a & b \\ t & t+1 \end{pmatrix}$.

We propose to use the first version because it is often cheaper than the second one.

In the case of normally distributed errors, the D-optimality criterion can be interpreted as follows: If we construct a confidence region for the vector of coefficients in the simple linear regression model, we get an ellipse. The D-optimal experimental design minimises the area of this ellipse, among all such ellipses arising from any design having the same number n of observations.

It is easy to see that the interval (8.20) has its maximum expected width in $[a,b]$ as well at a as also at b what means that we obtain the G-optimal design by minimising the expected width at a or b. The G-optimum design is for even n identical with the C_1- and the D-optimum design (due to the equivalence theorem of Kiefer and Wolfowitz; see, for instance, Melas, 2006). For odd $n = 2t+1$ the G-optimum design is given by $\begin{pmatrix} a & \frac{a+b}{2} & b \\ t & 1 & t \end{pmatrix}$.

8.2.2 Polynomial regression

Let us now construct exact D-optimal experimental designs for polynomial regression of degree k in the interval $[a, b]$. At first we determine exact D-optimal experimental designs for polynomial regression of degree k in the standard interval $[-1, 1]$.
Following, for instance, Pukelsheim (1993) the exact D-optimal designs have $k+1$ support points. Their values in $[-1, 1]$ are obtained by $(1-x^2)\frac{dP_k(x)}{dx}$, where $\frac{dP_k(x)}{dx}$ is the derivative of the Legendre polynomial of degree k (see Appendix A).
The frequencies of the exact D-optimal designs are for $n = (k+1)t$ equal to t. Otherwise the frequencies are chosen as nearly equal as possible (thus two differences of adjacent support frequencies differ by not more than 1). For $n = (k+1)t$ the exact D-optimal designs are also exact G-optimal.

Example 8.5 Let $k = 2$. The quadratic Legendre polynomial equals following the notation given at the start of Appendix A is $P_2(x) = \frac{1}{8}\frac{d^2}{dx^2}\left[(x^2-1)^2\right] = \frac{3x^2-1}{2}$ $\frac{1}{2}(3x^3-1)$ $\frac{d\frac{1}{2}(3x^3-1)}{dx} = \frac{9}{2}x^2$ and because $(1-x^2)\frac{dP_2(x)}{dx} = 3x(1-x^2)$, we have the roots $0, -1$ and 1 in $[-1, 1]$ of this expression $(x^4 - x^2) = 0$. These are the values in the upper row of the design matrix.
If $n = 3t$, the second row gives the corresponding frequencies. If $n = 3t+1$, any of the following designs is D-optimal:

$$\begin{pmatrix} -1 & 0 & 1 \\ t+1 & t & t \end{pmatrix}; \begin{pmatrix} -1 & 0 & 1 \\ t & t+1 & t \end{pmatrix}; \begin{pmatrix} -1 & 0 & 1 \\ t & t & t+1 \end{pmatrix}$$

If $n = 3t+2$, any of the following designs is D-optimal:

$$\begin{pmatrix} -1 & 0 & 1 \\ t+1 & t+1 & t \end{pmatrix}; \begin{pmatrix} -1 & 0 & 1 \\ t & t+1 & t+1 \end{pmatrix}; \begin{pmatrix} -1 & 0 & 1 \\ t+1 & t & t+1 \end{pmatrix}$$

To obtain exact D-optimal designs in our interesting interval $[a, b]$, we transform any value $\nu \in [-1; 1]$ by the transformation $\frac{a+b}{2} + \frac{b-a}{2}\nu$ into a corresponding value in $[a, b]$. In our example, this transforms the design $\begin{pmatrix} -1 & 0 & 1 \\ t & t & t \end{pmatrix}$ into $\begin{pmatrix} a & \frac{a+b}{2} & b \\ t & t & t \end{pmatrix}$. Table 8.2 (Rasch et al. 2007b) gives D- and exact G-optimal designs for polynomial regression for $x \in [a, b]$ and $n = t(k+1)$.

The R-program

The R-program needs as input the values a,b,k and n. The output gives as well the kind of criterion (D- or D-/G-), the design matrix and the criterion value. In case of $n \neq t(k+1)$ at first the higher frequencies and then the lower are given.

TABLE 8.2
Exact D- and G-optimal Designs for Polynomial Regression for $x \in [a, b]$ and $n = t(k+1)$.

k	Exact D- and G-optimal design
1	$\begin{pmatrix} a & b \\ t & t \end{pmatrix}$
2	$\begin{pmatrix} a & \frac{a+b}{2} & b \\ t & t & t \end{pmatrix}$
3	$\begin{pmatrix} a & 0.7236a+0.2764b & 0.2764a+0.7236b & b \\ t & t & t & t \end{pmatrix}$
4	$\begin{pmatrix} a & 0.82735a+0.17265b & \frac{a+b}{2} & 0.17265a+0.82735b & b \\ t & t & t & t & t \end{pmatrix}$
5	$\begin{pmatrix} a & 0.88255a+0.117745b & 0.6426a+0.3574b & 0.3574a+0.6426b \\ t & t & t & t \end{pmatrix}$ $\begin{pmatrix} 0.11745a+0.88255b & b \\ t & t \end{pmatrix}$

Example 8.6 $a = 10, b = 100, k = 3$ and $n = 22$. Design

$$\begin{pmatrix} 10 & 34.88 & 75.12 & 100 \\ 6 & 6 & 5 & 5 \end{pmatrix}$$

```
> design.reg.polynom(10, 100, 3, 22)

     [,1]     [,2]     [,3] [,4]
[1,]   10 34.87539 75.12461  100
[2,]    6  6.00000  5.00000    5
```

8.2.3 Intrinsically non-linear regression

In the intrinsically non-linear regression we consider only the criterion of D- and C-optimality. Because we know only the asymptotic distributions of the estimator of θ we use the asymptotic covariance matrices $var_A(\theta)$ (the A in the subscript means asymptotic, the estimator not occurring in the argument (θ) means that the matrix depends on the unknown θ).

Definition 8.5 Let f in (8.14) be an intrinsically non-linear function and

$$var_A(\theta) = \lim_{n\to\infty}\left[\frac{1}{n}F_n^T(\theta)F_n(\theta)\right]^{-1} = \lim_{n\to\infty} var_{A,n}(\theta) \qquad (8.21)$$

with $F_n(\theta)$ from (8.17).

An experimental design minimising $var_{A,n}(\theta)$ for a given $n > p$ is called exact locally D-optimal for that n. *Locally* means for a specific value of $\theta \in \Omega \subset \mathbb{R}^p$. In this section we always assume that $F_n^T(\theta)F_n(\theta)$ is non-singular.

Definition 8.6 Let f in (8.14) be an intrinsically non-linear function and

$$var_{A,n}(\theta) = \left[\frac{1}{n}F_n^T(\theta)F_n(\theta)\right]^{-1}$$

with $F_n(\theta)$ from (8.17). An exact experimental design minimising the i-th diagonal element of $var_{A,n}(\theta)$ defined above for a given $n > p$ is called exact locally C_i-optimal for that n.

The proofs of the theorems below can be found in Rasch (1995).

Theorem 8.3 The support of a locally D-optimum designs for any continuous differentiable intrinsically non-linear function does not depend on n, and reparametrisation of the function does not change the locally D-optimum design.

What we mean by reparametrisation is shown by an example.

Example 8.7 We consider the logistic function

$$f_L(x,\theta) = \frac{\theta_1}{1 + \theta_2 e^{\theta_3 x}}$$

and let $\theta_1 \cdot \theta_2 \cdot \theta_3 \neq 0$, then the parameter transformation $\beta^T = \left(\beta_1 = \frac{\theta_1}{2};\ \beta_2 = \frac{-\theta_3}{2};\ \beta_3 = -\frac{1}{\theta_3}\ln\theta_2\right)$ transforms the logistic function into $f_{\tanh}(x,\beta) = \beta_1\{1 + \tanh[\beta_2(x - \beta_3)]\}$ (Rasch, 1995, p. 602).

For both forms the D-optimum design is the same.

Theorem 8.4 For any continuous differentiable partially or totally intrinsically non-linear function, the exact locally D-optimal design depends on the non-linear parameters only.

Theorem 8.5 Let $f(x,\theta)$ be an intrinsically non-linear function with non-linear parameter vector φ. The exact locally D-optimal design of size $n > p$ depends on φ only.

Theorem 8.6 Let $f(x,\theta)$ be an intrinsically non-linear function with parameter vector $\theta^T = (\theta_1, \dots, \theta_p) \in \Omega \subset \mathbb{R}^p$. Let $\mathbb{R}_{n,p}$ be the set of all exact designs with support size p (all p support points are assumed to be different) and size $n = \sum_{i=1}^{p} n_i;\ n_i > 0$ of the form $V_{n,p} = \begin{pmatrix} x_1 & ,\dots, & x_p \\ n_1 & ,\dots, & n_p \end{pmatrix}$. Then the support of the $\mathbb{R}_{n,p}$ in D-optimal design $V_{n,p}^* = \begin{pmatrix} x_1^* & ,\dots, & x_p^* \\ n_1^* & ,\dots, & n_p^* \end{pmatrix}$ is independent of n and invariant against permutations of the n_i^*, and the n_i^* are as equal as possible.

Theorem 8.6 allows us to find the support of an exact D-optimum design for all possible n independent of n.

There are only a few intrinsically non-linear functions for which the exact locally D- and C-optimal design can be given in closed form. We give four examples of this case.

Example 8.8 The Michaelis-Menten function $f_M(x,\theta) = \frac{\theta_1 x}{1+\theta_2 x}$ can by reparametrisation also be written as $f_M(x,\beta) = \frac{\beta_1 x}{\beta_2 + x}$; $\beta_1 = \frac{\theta_1}{\theta_2}$; $\beta_2 = \frac{1}{\theta_2}$ or as $f_M(x,\delta) = \frac{x}{\delta_1 + \delta_2 x}$; $\delta_1 = \frac{\beta_2}{\beta_1}$; $\delta_2 = \frac{1}{\beta_1}$.

The locally D-optimum design for any parametrisation of this function in $[a,b]$ is for even $n = 2t$ given by (Ermakov and Zhiglyavskii, 1987): $\begin{pmatrix} a + \frac{b-a}{2+b\theta_2} & b \\ t & t \end{pmatrix}$. For odd $n = 2t+1$, we have two locally D-optimum designs, namely $\begin{pmatrix} a + \frac{b-a}{2+b\theta_2} & b \\ t+1 & t \end{pmatrix}$ and $\begin{pmatrix} a + \frac{b-a}{2+b\theta_2} & b \\ t & t+1 \end{pmatrix}$

The proofs of the results of the three following examples are given in Han and Chaloner (2003).

Example 8.9 Consider the two-parameter exponential function $\theta_1 e^{\theta_2 x}$.

The locally D-optimum design for the function $\theta_1 e^{\theta_2 x}(\theta_2 < 0)$ in $[a,b]$ is for even $n = 2t$ given by $\begin{pmatrix} a & \min\left\{b, a - \frac{1}{\theta_2}\right\} \\ t & t \end{pmatrix}$. For odd $n = 2t+1$, we have two locally D-optimum designs, namely $\begin{pmatrix} a & \min\left\{b, a - \frac{1}{\theta_2}\right\} \\ t+1 & t \end{pmatrix}$ and $\begin{pmatrix} a & \min\left\{b, a - \frac{1}{\theta_2}\right\} \\ t & t+1 \end{pmatrix}$

The locally C-optimum design for estimating $\theta_2 < 0$ is given by

$$\begin{pmatrix} a & \min\left\{b, a - \frac{1.278465}{\theta_2}\right\} \\ \left\lceil \frac{e^{\theta_2 b}}{e^{\theta_2 b} + e^{\theta_2 a}} \right\rceil & n - \left\lceil \frac{e^{\theta_2 b}}{e^{\theta_2 b} + e^{\theta_2 a}} \right\rceil \end{pmatrix}$$

Example 8.10 Consider the three-parameter exponential function $f_E(x,\theta) = \theta_1 + \theta_2 e^{\theta_3 x}$.

We have already discussed the function $f_E(x,\theta) = \theta_1 + \theta_2 e^{\theta_3 x}$ $(\theta_3 < 0)$. The locally D-optimum design for this function in $[a,b]$ and for $n = 3t$ is given by $\begin{pmatrix} a & \frac{-1}{\theta_3} + \frac{a \cdot e^{a\theta_3} - b \cdot e^{b\theta_3}}{e^{a\theta_3} - e^{b\theta_3}} & b \\ t & t & t \end{pmatrix}$; otherwise the weights of the support points

have to be chosen as nearly equal as possible, which leads to one of three possible locally D-optimum designs.
The locally C-optimum design for estimating in $[a, b]$ and for $n = 4t$ is given by

$$\begin{pmatrix} a & \frac{-1}{\theta_3} + \frac{a\cdot e^{a\theta_3} - b\cdot e^{b\theta_3}}{e^{a\theta_3} - e^{b\theta_3}} & b \\ \left\lceil \frac{e^{-1+\theta_3\cdot\frac{a\cdot e^{a\theta_3} - b\cdot e^{b\theta_3}}{e^{a\theta_3} - e^{b\theta_3}}} - e^{b\theta_3}}{2\cdot(e^{a\theta_3} - e^{b\theta_3})} \right\rceil & 2t & \left\lceil \frac{e^{a\theta_3} - e^{-1+\theta_3\cdot\frac{a\cdot e^{a\theta_3} - b\cdot e^{b\theta_3}}{e^{a\theta_3} - e^{b\theta_3}}} - e^{b\theta_3}}{2\cdot(e^{a\theta_3} - e^{b\theta_3})} \right\rceil \end{pmatrix}.$$

Thus it can happen that

$$\left\lceil \frac{e^{-1+\theta_3\cdot\frac{a\cdot e^{a\theta_3} - b\cdot e^{b\theta_3}}{e^{a\theta_3} - e^{b\theta_3}}} - e^{b\theta_3}}{2\cdot(e^{a\theta_3} - e^{b\theta_3})} \right\rceil + \left\lceil \frac{e^{a\theta_3} - e^{-1+\theta_3\cdot\frac{a\cdot e^{a\theta_3} - b\cdot e^{b\theta_3}}{e^{a\theta_3} - e^{b\theta_3}}} - e^{b\theta_3}}{2\cdot(e^{a\theta_3} - e^{b\theta_3})} \right\rceil \neq 2t \text{ due to}$$

rounding up.

For other functions, a search procedure has to be applied which gives the exact locally D- and C-optimal design (the latter concerning the non-linear parameter) for a given θ.

The R- program

We have to put in the function which must be continuously differentiable with respect to θ, n and the value of θ for which the exact optimal design should be calculated. If possible the program gives an anlytical solution. Otherwise a search algorithm is applied to find the exact locally optimum design.

Example 8.11 Table 8.3 gives the average wither height in cm of 112 heifers at 11 support points (age in months).

TABLE 8.3
Average Withers Heights (in cm) of 112 Cows in the First 60 Months of Life.

Age (Months)	Height (cm)
0	77.20
6	94.50
12	107.20
18	116.00
24	122.40
30	126.70
36	129.20
42	129.90
48	130.40
54	130.80
60	131.20

Let the design region for an analogue experiment with another population be $[0; 60]$ months. We fit several specific functions and calculate exact (locally) D-optimal designs for $n = 12$.

The quadratic and cubic polynomials fitted are $\hat{y}_{quad} = 80.76 + 2.28x - 0.2248x^2$ and $\hat{y}_{cub} = 77.427 + 3.159x - 0.06342x^2 + 0.00043x^3$, respectively, and the fitted intrinsically non-linear exponential, logistic and Gompertz functions are $\hat{y}_E = 132.96 - 56.42e^{-0.677x}$, $\hat{y}_L = \frac{131.62}{1+0.7012e^{-0.0939x}}$ and $\hat{y}_G = 132.19 \cdot e^{-0.541e^{-0.00808x}}$, respectively.

The corresponding exact D-optimal designs are (in case of intrinsically non-linear functions, the fitted parameters have been used for the exact locally optimal designs)

Type of Function	Exact D-Optimal Design
Quadratic	$\begin{pmatrix} 0 & 30 & 60 \\ 4 & 4 & 4 \end{pmatrix}$
Cubic	$\begin{pmatrix} 0 & 16.6 & 43.4 & 60 \\ 3 & 3 & 3 & 3 \end{pmatrix}$
Exponential	$\begin{pmatrix} 0 & 13.7 & 60 \\ 4 & 4 & 4 \end{pmatrix}$
Gompertz	$\begin{pmatrix} 0 & 14.2 & 60 \\ 4 & 4 & 4 \end{pmatrix}$
Logistic	$\begin{pmatrix} 0 & 14.7 & 60 \\ 4 & 4 & 4 \end{pmatrix}$

Of course it is a disadvantage that after selecting a proper regression model, we have to know the values of the non-linear parameters. The following examples show how we could proceed in some special cases.

Example 8.12 In a lecture on non-linear regression a PhD student had to solve the following problem for a soil experiment on the next day.

Heavy metal extracts have to be observed for 24 hours. The student was looking for an optimal experimental design which would allow estimation of the speed of the increase of extracted metal over time as accurately as possible with 10 measurements. He had the following prior information: In the beginning an extract of 0.1 units was expected, and after 24 hours a maximum of 125 units was expected. The increase of the extract was guessed to be a sigmoid curve.

We selected a logistic regression function as our model. If we put $x = 0$ in $f_L(x) = \frac{\theta_1}{1+\theta_2 e^{\theta_3 x}}$, we receive with the data given $f_L(0) = \frac{\theta_1}{1+\theta_2} = 0.1$ and for $x \to \infty$, we obtain for negative θ_3 $f_L(\infty) = \theta_1 = 125$ and from both, we further have $\theta_2 = 1249$. With a rough guess of $\theta_3 = -0.5$, the curve was as shown in Figure 8.5 but for our student the curve in the beginning was not

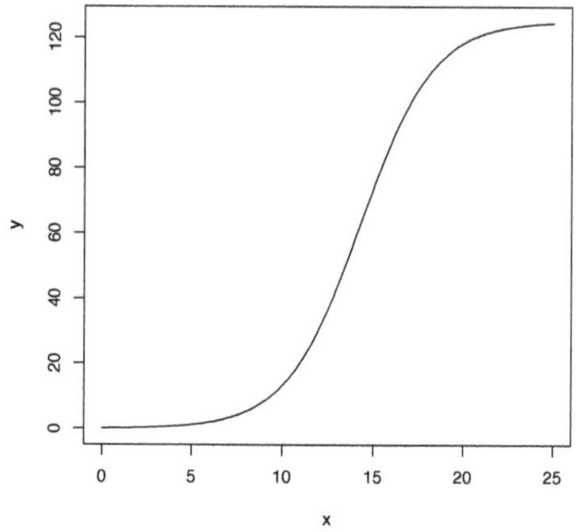

FIGURE 8.5
The graph of the sigmoid curve modelling extracted metal

steep enough. Therefore we decreased θ_3 to $\theta_3 = -0.8$ and received the curve shown in Figure 8.6.

Now we had all the information to find the $var(\hat{\boldsymbol{\theta}})$-optimal design.

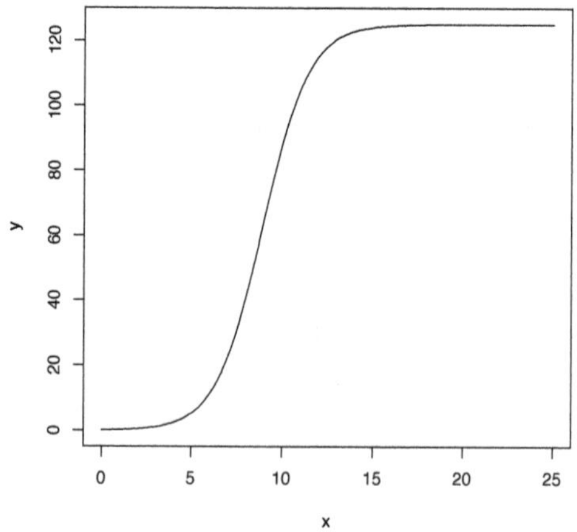

FIGURE 8.6
The graph of the improved sigmoid curve modelling extracted metal

8.2.4 Replication–free designs

Exact optimal designs—as we have seen—mostly use more than one measurement at the support points. There are, however, many applications where only one measurement per support point is allowed.

Definition 8.7 If at each support point of a design in regression experiments we can measure only once, we have to use replication–free designs (Rasch et al. 1996); these are designs with frequency for each support point.

It is clear that replication–free designs have exactly n support points. Because the design region is a finite interval for a given respective distance between two consecutive measurements in time or neighbouring measurements in two-dimensional intervals in a plane (as it may occur in spatial statistics), the number of possible support points N is finite, and principally the exact optimal design can be found by full enumeration looking at all $\binom{N}{n}$ possible designs, calculating for each the design criterion and selecting the design with the smallest one. For large N even high-speed computers cannot solve this problem; therefore, search procedures are in use (Rasch et al. 1997a).

The R-program offers both possibilities, if the time for the full enumeration is estimated by the program as too large, a search program is used.

8.3 Determining the size of an experiment

In regression analysis Model I the precision requirements partly depend on the experimental design $\begin{pmatrix} x_1 & x_2 & \dots & x_r \\ n_1 & n_2 & \dots & n_r \end{pmatrix}$. Therefore we have to first select the design and after that we can minimise the size of the experiment for this design. Furthermore, in intrinsically non-linear regression the design is locally which means it depends on the parameter vector φ in Definition 8.1 and by this the size of the experiment is locally minimal.

8.3.1 Simple linear regression

In the simple linear regression we have $\boldsymbol{y}_j = \beta_0 + \beta_1 x_j + \boldsymbol{e}_j (j = 1, \dots, n)$ and

$$var \begin{pmatrix} \boldsymbol{b}_0 \\ \boldsymbol{b}_1 \end{pmatrix} = \frac{\sigma^2}{SS_x} \begin{pmatrix} \frac{1}{n} \sum_{j=1}^{n} x_j^2 & -\overline{x} \\ -\overline{x} & 1 \end{pmatrix} \tag{8.22}$$

and the determinant of this covariance matrix is

$$D = \frac{\sigma^4}{SS_x^2}\left(\frac{1}{n}\sum_{j=1}^{n} x_j^2 - \overline{x}^2\right) = \frac{\sigma^4}{nSS_x} \tag{8.23}$$

If $n = 2t$ is even, the C_1- and the D-optimum design is given by $\begin{pmatrix} a & b \\ t & t \end{pmatrix}$.

When our aim is to estimate the slope so that either the variance $var(\boldsymbol{b}_1) < c$ or that the $(1-\alpha)$-confidence interval for β has an expected length $< \delta$, the smallest minimum size is needed if we choose the D-optimum design. This also gives the maximum power of a test about β_1.

Generally we proceed as follows:

1. Select the experimental region $[a, b]$.
2. Determine the type of analysis planned, i.e., select among
   - point estimation for β_1,
   - interval estimation for β_1,
   - testing $H_0. : \beta_1 = \beta_{10}$ and a one- or two-sided alternative,
   - interval estimation for $\hat{\boldsymbol{y}}$ and
   - minimising the area of the confidence ellipse for (β_0, β_1) (in the normal case).
3. Specify the precision requirement
   - by an upper bound c for $var(\boldsymbol{b}_1)$,
   - by the confidence coefficient $(1-\alpha)$ and an upper bound for the half expected width δ of the interval,
   - by the risk of the first kind and the risk of the second kind at a given distance δ from β_{10},
   - by the confidence coefficient $(1-\alpha)$ and an upper bound for the half expected width δ of the interval or
   - by an upper bound c for the determinant.
4. Select the D-optimum design and calculate the minimum size of the experiment.

The R-program then gives the minimal size of the experiment needed. We demonstrate the underlying algorithm by some examples.

In case of point estimation of the slope in $[5, 9]$ let the precision requirement be given by $var(\boldsymbol{b}_1) < c = 0.01\sigma^2$. For the exact D-optimal design we have for $n = 2t$ $var(\boldsymbol{b}_1) < c = 0.01\sigma^2$.

From $var(\boldsymbol{b}_1) \leq c = 0.01\sigma^2$ it follows that $\frac{\sigma^2}{8t} \leq 0.01\sigma^2$ or $t = 13$. The size of the experiment is therefore 26 and the exact optimal design is given by $\begin{pmatrix} 5 & 9 \\ 13 & 13 \end{pmatrix}$. The R-program tries to find a design whose size is smaller by 1 and still fulfills the requirement. In our case we could try whether $\begin{pmatrix} 5 & 9 \\ 13 & 12 \end{pmatrix}$ is also sufficient, which is not the case.

In the next example the maximal half expected width of a 0.95-confidence interval for $E(\hat{\boldsymbol{y}})$ in $x \in [5, 9]$ should be not larger than 0.5σ. We receive the smallest sample size if we use the G-optimal design. The 0.95-confidence interval for $E(\hat{\boldsymbol{y}})$ in (8.20) is given by

$$\left[\boldsymbol{b}_0 + \boldsymbol{b}_1 x - t(n-2;\ 0.975)\boldsymbol{s}\sqrt{\frac{1}{n} + \frac{(x-\overline{x})^2}{SS_x}}\right.$$

$$\left.\boldsymbol{b}_0 + \boldsymbol{b}_1 x + t(n-2;\ 0.975)\boldsymbol{s}\sqrt{\frac{1}{n} + \frac{(x-\overline{x})^2}{SS_x}}\right] \tag{8.24}$$

For $n = 2z$ and the exact G-optimal design, the maximal expected width is at either 5 or 9. Let us choose 9, which gives us $(x - \overline{x})^2 = 4$; then the expected width of $\left[\boldsymbol{b}_0 + \boldsymbol{b}_1 x - t(2z-2;\ 0.975)\boldsymbol{s}\sqrt{\frac{1}{2z} + \frac{4}{8z}};\ \boldsymbol{b}_0 + \boldsymbol{b}_1 x\right.$ $\left. + t(2z-2;\ 0.975)\boldsymbol{s}\sqrt{\frac{1}{2z} + \frac{1}{2z}}\right]$ is $2t(2z-2;\ 0.975)E(\boldsymbol{s})\sqrt{\frac{1}{z}}$. In the R-program we use the exact value

$$E(\boldsymbol{s}) = \frac{\Gamma\left(\frac{n-1}{2}\right)\sqrt{2}}{\Gamma\left(\frac{n-2}{2}\right)\sqrt{n-2}}\sigma \tag{8.25}$$

but for hand calculation here we use the approximation $E(\boldsymbol{s}) = \sigma$. This gives the equation for t (limiting the half expected width) as $t(2z-2;\ 0.975)\sqrt{\frac{1}{z}} \leq 0.5$ or $z = \lceil 4t^2(2z-2;\ 0.975) \rceil$ and this is $z = 17$.

Part III

Special Designs

In this part designs are described which aid in planning experiments for special applications. Chapter nine deals with methods of construction of special designs in quadratic regression models, the central composite second order designs. In Chapter 10 again factorial designs are discussed with the special situation that the factors are quantitative and their values are components of a mixture which must add up to 100%. These mixture designs have many applications in industry and in both animal and human food production industries.

9

Second Order Designs

In Chapter three the analysis of variance is shown to be a method by which *factorial experiments*, as introduced in Section 7.1, can be analysed, provided all the factors are *qualitative*. If all the factors are *quantitative*, we can regard them as regressors in a *Model I multiple regression*. This topic is considered in Chapter eight.

In applications, factorial experiments with quantitative factors are mostly used to search for the optimum on a *response surface*. The description of such non-linear relationships can be achieved most simply by a quadratic function of the regressor variables. In this chapter we shall describe the design and analysis of such experiments. See also Myers (1976) and Khuri and Cornell (1987).

This chapter combines the ideas from Sections 8.1.1 and 8.2.2. We assume that we have more than one regressor variable and a quadratic regression function. In the case of $p = 2$ regressors (x_1, x_2), the complete regression model takes the form (with i.i.d. error terms $\boldsymbol{e}_j$ with $E\boldsymbol{e}_j$, $Var(\boldsymbol{e}_j) = \sigma^2$ for all j):

$$\boldsymbol{y}_j = \beta_0 + \beta_1 x_{1j} + \beta_2 x_{2j} + \beta_{11} x_{1j}^2 + \beta_{22} x_{2j}^2 + \beta_{12} x_{1j} x_{2j} + \boldsymbol{e}_j \tag{9.1}$$

or for p factors the form

$$\boldsymbol{y}_j = \beta_0 + \sum_{i=1}^{p} \beta_i x_{ij} + \sum_{i=1}^{p} \beta_{ii} x_{ij}^2 + \sum_{i<l}^{p} \beta_{il} x_{ij} x_{lj} + \boldsymbol{e}_j; j = 1, \ldots, n \tag{9.2}$$

or

$$\boldsymbol{y} = X\beta + \boldsymbol{e} \tag{9.3}$$

In (9.3) we have $j = 1, \ldots, n > 1 + p(p+3)/2$ and

$$\boldsymbol{X} = \begin{pmatrix} 1 & x_{11} & x_{21} \ldots & x_{p1} & x_{11}^2 & x_{21}^2 \ldots & x_{p1}^2 & x_{11}x_{21} \ldots & x_{11}x_{p1} \ldots & x_{p-1}x_{p1} \\ \vdots & \vdots & \vdots & \vdots & \vdots & \vdots & \vdots & \vdots & & \\ 1 & x_{1n} & x_{2n} \ldots & x_{pn} & x_{1n}^2 & x_{2n}^2 \ldots & x_{pn}^2 & x_{1n}x_{2n} \ldots & x_{1n}x_{pn} \ldots & x_{p-1n}x_{pn} \end{pmatrix};$$

$\beta^T = (\beta_a, \beta_1, \ldots, \beta_p, \beta_{11}, \ldots, \beta_{pp}, \beta_{12}, \ldots, \beta_{p-1p})$ and the other vectors as defined in Chapter eight.

If any of the β's on the right hand side of (9.2) are negligibly small, a reduced model can be employed having the corresponding terms omitted. Model (9.1) without the term $\beta_{12}x_{1j}x_{2j}$ is called a *pure quadratic model* in x_1 and x_2 (without an *interaction term* x_1x_2). The analysis proceeds as described in Section 8.1.1; we merely adapt X and the vector β accordingly.

The expectation $E(\boldsymbol{y}_j)$ of $\boldsymbol{y}_j$ in (9.2) is called the expected quadratic *response surface.* Equation (9.2) may be written in an alternative matrix version as

$$E(\boldsymbol{y}_j) = \beta_0 + x_j^T A x_j + B^T x_j \; (j = 1, 2, \ldots, n) \tag{9.4}$$

where $x_j = (x_{1j}, x_{2j}, \ldots, x_{kj})^T$ is a vector of the k factor-variables for the j-th observation; A is the symmetric $k \times k$ matrix of non-linear regression-parameters

$$A = \begin{pmatrix} \beta_{11} & \cdots & \frac{1}{2}\beta_{1p} \\ \vdots & & \vdots \\ \frac{1}{2}\beta_{p1} & \cdots & \beta_{pp} \end{pmatrix} \text{ and } B = (\beta_1, \beta_2, \ldots, \beta_p)^T$$

The parameters in (9.2) or (9.3) are estimated using the method of least squares. The estimates of the parameters are $\hat{A}, \hat{B}$, and $\hat{E}(\boldsymbol{y})$. We can use the expression for $\hat{E}(\boldsymbol{y})$ to find the optimum as a function of the independent variables $x_1, x_2 \ldots, x_p$.

When the number of regressor variables increases, it is often not possible to include all the combinations of factor levels in the experiment. One must then use an incomplete factorial design, as mentioned in Chapter seven. Having said in Section 8.2.2 that we must ensure that in a quadratic model three different levels of each of the p factors are included, an especially efficient way of accomplishing this without including all 3^p possible factor level combinations is to use a *second order central composite design* and *Doehlert designs*.

9.1 Central composite designs

The phrase *second order* means that the analysis is based on a quadratic model (which can also be called a model of the second degree or second order). This second order central composite design consists of three components. The *kernel* is either a complete factorial design with two levels for each factor or, when there is a large number of factors, a fractional two-level design. We therefore

describe the kernel as a 2^{p-k} design, in which we allow k to take the value 0 (when the design is complete). We can thus summarise the components of the full design as follows:
Basis design 1 (cube or kernel):
2^{p-k}-design with levels $+1$ and -1 with n_K replicates, often we have $n_K = 1$.
Basis design 2 (star):
This design contains for each factor two "star points" with the levels $+\alpha$ and $-\alpha$ for the corresponding factor and zero for the remaining factors (number of replicates $= n_S$; star point distance $= \alpha$).
Basis design 3 (centre):
A design with the factor level combination $(0, 0, \ldots, 0)$ replicated n_C times.

The total number of experimental units is therefore $n = 2^{p-k} \cdot n_K + 2 \cdot p \cdot n_S + n_C$. The levels above are given in the standardised notation for the treatment factors, analogous to the notation for two level designs. In this notation the factors are transformed in such a way that the levels of the kernel correspond with the numbers $+1$ and -1. If $\alpha \leq 1$, the standardised experimental region for each factor is $(-1; 1)$; otherwise it is $(-\alpha; \alpha)$. All basis designs have the origin as their centre of symmetry of the coordinates, hence the phrase *central composite*.
In practical applications, of course, the design is given in the non-standardised way when each factor can take values in its experimental region $a_i; b_i; i = 1, \ldots, p$.
If $\alpha \neq 1$, each factor in a composite design occurs at five levels. The matrix of the coefficients of such a design with three factors and a complete kernel is given in Table 9.1.

If interaction effects of three or more factors are negligible, all effects besides the pure quadratic ones can be estimated and unconfounded if the kernel has a resolution $R > IV$ (for confounding and resolution see Chapter seven).
The construction of optimal second order designs is difficult and such designs need a relative large number of experimental units. Therefore in place of the D- or G-optimality, designs with other useful properties like *rotatability* and/or *orthogonality* have been used.
The value of α can be chosen such that the variance of the estimator of $E(\boldsymbol{y})$ at all points equidistant from the centre point is constant. Such a design is called *rotatable*. One gets a rotatable design if the kernel has a resolution $R > IV$ and

$$\alpha = \sqrt[4]{\frac{n_K}{n_S} 2^{p-k}} = \alpha_R \tag{9.5}$$

We get an orthogonal design (hence the columns of the matrix X in $E(\boldsymbol{y}) = X\beta$ are orthogonal), if the kernel has a resolution $R > IV$ and

TABLE 9.1
Matrix of the Coefficients of a Standard Composite Design with Three Factors.

I	x_1	x_2	x_3	x_1x_2	x_1x_3	x_2x_3	x_1^2	x_2^2	x_3^2	Design Component	Number of Replicates
1	1	1	1	1	1	1	1	1	1		
1	1	1	−1	1	−1	−1	1	1	1		
1	1	−1	1	−1	1	−1	1	1	1		
1	1	−1	−1	−1	−1	1	1	1	1	kernel	n_K
1	−1	1	1	−1	−1	1	1	1	1		
1	−1	1	−1	−1	1	−1	1	1	1		
1	−1	−1	1	1	−1	−1	1	1	1		
1	−1	−1	−1	1	1	1	1	1	1		
1	$-\alpha$	0	0	0	0	0	α^2	0	0		
1	α	0	0	0	0	0	α^2	0	0		
1	0	$-\alpha$	0	0	0	0	0	α^2	0	star	n_S
1	0	α	0	0	0	0	0	α^2	0		
1	0	0	$-\alpha$	0	0	0	0	0	α^2		
1	0	0	α	0	0	0	0	0	α^2		
1	0	0	0	0	0	0	0	0	0	centre	n_C

$$\alpha = \frac{1}{\sqrt{2n_S}}\sqrt{\sqrt{n \cdot n_K \cdot 2^{p-k}} - n_K \cdot 2^{k-1}} = \alpha_0 \tag{9.6}$$

and we get a rotatable and orthogonal design by choosing $\alpha = \alpha_R$ and $n_C = 4\sqrt{n_K \cdot n_S \cdot 2^{p-k}} + 2n_S(2-p)$.

If $p-k$ is equal to $2r$ and $n_K = n_S = 1$, the smallest rotatable and orthogonal design is obtained for

$$\alpha = \sqrt{2^r}, r = (p-k)/2; \;\; n_C = 4 \cdot 2^r + 2(2-p) \tag{9.7}$$

Smaller designs which are rotatable but not orthogonal can be constructed for $n_K = n_S = 1, \alpha$ und n_C from (9.7) (n_C must be rounded). In Table 9.2 the parameters of the smallest *rotatable* and orthogonal designs are shown.

By the OPDOE-program `> design.centralcomposite(factors = a, alpha = x)`, we can construct central composite designs with a factors if the experimental region is a hypercube in the p-dimensional Euclidian space.

If the experimenter is not able to take as many measurements as needed in the n-column of Table 9.3, we can use the Hartley design(s) (see Hartley, 1959).

Example 9.1 In a laboratory experiment to investigate the time t (in minutes) and temperature T (in °C) of a chemical process, we wish to get the

TABLE 9.2
Parameters of the Smallest Rotatable and Orthogonal Second Order Designs.

p	2^p	$p-k$	n_k	n_S	n_C	n	α
2	4	2	1	1	8	16	$\sqrt{2}$
3	8	3	1	2	12	32	$\sqrt{2}$
4	16	4	1	1	12	36	2
5	32	4	1	1	10	36	2
6	64	5	1	2	16	72	$\sqrt{8}$
7	128	6	1	1	22	100	$\sqrt{8}$
8	256	6	1	1	20	100	$\sqrt{8}$
9	512	7	1	2	36	200	$\sqrt{8}$

TABLE 9.3
Hartley Designs for 3 and 4 Factors.

p	n	n_K	n_S	n_Z	α, if the cube is $[-1;+1]^p$
3	11	1	1	1	1.1474
4	17	1	1	1	1.3531
	20	1	1	4	1.5246

maximum output. The response y is the chemical output in grams. We shall use the standardised notation, and the transformed quantitative variables are $x_1 = (t - 90 \text{ min.})/(10 \text{ min.})$ and $x_2 = (T - 145°C)/(5°C)$.
First a 2^2 experiment with factors x_1 and x_2 at the levels -1 and $+1$ is done. Then $n_c = 4$ observations are done at the centre point $(x_1, x_2) = (0, 0)$. Further observations are done once at each of the 4 star points (x_1, x_2) : $(-\sqrt{2}, 0), (+\sqrt{2}, 0), (0, -\sqrt{2})$ and $(0, +\sqrt{2})$. The experimental region is therefore

$(-\sqrt{2} \cdot 10 + 90; \sqrt{2} \cdot 10 + 90) = (75.86; 104.14)$ and
$(-\sqrt{2} \cdot 5 + 145; \sqrt{2} \cdot 5 + 145) = (137.93; 152.07)$.

The experiment is conducted in a randomised order. Note that the design is rotatable. The total number of experimental units is $2^2 + 2 \cdot 2 + 4 = n = 12$. Because $n_C = 4$, the design is not orthogonal; for orthogonality we need $n_C^* = 4 \cdot \sqrt{1 \cdot 1 \cdot 2^2} + 2 \cdot 1 \cdot 0 = 8$ replications is the centre.

The observations are

j	x_{1j}	x_{2j}	y_j
1	−1	−1	78.8
2	−1	1	91.2
3	1	−1	84.5
4	1	1	77.4
5	0	0	89.7
6	0	0	86.8
7	−1.4142	0	83.3
8	0	−1.4142	81.2
9	0	1.4142	79.5
10	1.4142	0	81.2
11	0	0	87.0
12	0	0	86.0

From this we obtain the estimated regression coefficients in (9.1) as $b_0 = 87.37$, $b_1 = -1.38$, $b_2 = 0.36$, $b_{11} = -2.14$, $b_{22} = -3.09$, $b_{12} = -4.88$ and the estimated regression function

$$\hat{E}(\boldsymbol{y}) = 87.37 - 1.38x_1 + 0.36x_2 - 2.14x_1^2 - 3.09x_2^2 - 4.88x_1x_2$$

and SS(Error) $= 24.0884$ with $12 - 6 = 6$ degrees of freedom. The estimate of σ^2 is $s^2 = \frac{24.0884}{6} = 4.0144$ (see (8.11)).

From the estimated regression coefficients we find $\hat{A} = \begin{pmatrix} -2.14 & -\frac{1}{2} \cdot 4.88 \\ -\frac{1}{2} \cdot 4.88 & -3.09 \end{pmatrix}$ and $\hat{B}^T = (-1.38;\ 0.36)$.

Since the two eigenvalues of (-0.135 and -5.102) are negative, we have a maximum. If both the eigenvalues are positive we have a minimum, and if they have opposite signs, there is a saddle point.

The maximum of the 2nd degree polynomial is found as $(x_1; x_2) = (-3.74; 3.00)$. Because this point lies outside the experimental range, a new experiment is recommended with $(x_1; x_2) = (-3.74; 3.00)$ as new centre point or in the experimental region at the point $(-3.74 \cdot 10 + 90; 3.00 \cdot 5 + 145) = (52.6; 160.0)$.

9.2 Doehlert designs

We introduce here the Doehlert designs at first for $p = 2$ factors.

Doehlert (1970) proposed, starting from an equilateral triangle with sides of length 1, to construct a regular hexagon with a centre point at (0,0). The $n = 7$ experimental points in the experimental region $[-1; +1] \times [-1; +1]$ of $(x_1; x_2)$ are (1, 0), (0.5, 0.866), (0, 0), (−0.5, 0.866), (−1, 0), (−0.5, −0.866) and (0.5, −0.866). The 6 outer points lie on a circle with a radius 1 and centre (0.0). The factor x_1 has 5 levels (−1, −0.5, 0, 0.5, 1) and x_2 has 3 levels (−0.866, 0, 0.866).

This Doehlert design has an equally spaced distribution of points over the experimental region, a so-called uniform space filler, where the distances between neighboring experiments are equal. The $Var(\boldsymbol{b}) = \sigma^2(X'X)^{-1}$ with $(X'X)^{-1}$:

$$\begin{pmatrix} 1 & 0 & 0 & -1 & -1.00006 & 0 \\ 0 & 0.333333 & 0 & 0 & 0 & 0 \\ 0 & 0 & 0.333353 & 0 & 0 & 0 \\ -1 & 0 & 0 & 1.5 & 0.833382 & 0 \\ -1.00006 & 0 & 0 & 0.833382 & 1.500176 & 0 \\ 0 & 0 & 0 & 0 & 0 & 1.333412 \end{pmatrix}$$

To increase the degrees of freedom for the estimator $\boldsymbol{s}^2$ of σ^2 we can add two centre points (0,0) to the Doehlert design. The experimental points of $(x_1; x_2)$ are then (1, 0), (0.5, 0.866), (0, 0) three times, (−0.5, 0.866), (−1, 0), (−0.5, −0.866) and (0.5, -0.866). This design remains a regular hexagon with a centre point at (0,0).
For this Doehlert design $Var(\boldsymbol{b}) = \sigma^2(X'X)^{-1}$ with $(X'X)^{-1}$:

$$\begin{pmatrix} 0.333333 & 0 & 0 & -0.33333 & -0.333335 & 0 \\ 0 & 0.333333 & 0 & 0 & 0 & 0 \\ 0 & 0 & 0.333353 & 0 & 0 & 0 \\ -0.33333 & 0 & 0 & 0.833333 & 0.166676 & 0 \\ -0.333333 & 0 & 0 & 0166676 & 0.833431 & 0 \\ 0 & 0 & 0 & 0 & 0 & 1.333412 \end{pmatrix}$$

In the case of 3 factors and the experimental region of $[-1;+1] \times [-1;+1] \times [-1;+1]$ of $(x_1; x_2, x_3)$, the 13 experimental points are then for a Doehlert design $(x_1; x_2, x_3)$: (1, 0, 0), (0.5, 0.866, 0), (0.5, 0.289, 0.816), (0.5, −0.866, 0), (0.5, −0.289, -0.816), (0, 0, 0), (0, 0.577, −0.816), (0, −0.577, 0.816), (−0.5, 0.866, 0), (−0.5, −0.866, 0), (−0.5, 0.289, 0.816), (−0.5, −0.289, −0.816), (−1, 0, 0).
The 12 outer points lie on a sphere with radius 1 and with centre (0, 0, 0). Factor x_1 has 5 levels, factor x_2 has 7 levels and factor x_3 has 3 levels.

The Doehlert design has the $Var(\boldsymbol{b}) = \sigma^2(X'X)^{-1}$ with $(X'X)^{-1}$:

$$\begin{pmatrix} 1 & 0 & 0 & 0 & -1 & -1 & -1 & 0 & 0 & 0 \\ 0 & 0.25 & 0 & 0 & 0 & 0 & 0 & 0 & 0 & 0 \\ 0 & 0 & 0.25 & 0 & 0 & 0 & 0 & 0 & 0 & 0 \\ 0 & 0 & 0 & 0.25 & 0 & 0 & 0 & 0 & 0 & 0 \\ -1 & 0 & 0 & 0 & 1.5 & 0.83338 & 0.91788 & 0 & 0 & -0.23653 \\ -1 & 0 & 0 & 0 & 0.83338 & 1.50018 & 0.91774 & 0 & 0 & 0.23464 \\ -1 & 0 & 0 & 0 & 0.91788 & 0.91774 & 1.42012 & 0 & 0 & -0.00143 \\ 0 & 0 & 0 & 0 & 0 & 0 & 0 & 1.33341 & -0.4723 & 0 \\ 0 & 0 & 0 & 0 & 0 & 0 & 0 & -0.47225 & 1.66908 & 0 \\ 0 & 0 & 0 & 0 & -0.23653 & 0.23464 & -0.00143 & 0 & 0 & 1.66841 \end{pmatrix}$$

9.3 *D*-optimum and *G*-optimum second order designs

D-optimum designs have already been introduced in Chapter eight.
D-optimal central composite designs need much more observations than the designs discussed so far.

Table 9.4 contains parameters of the smallest *D*- and *G*-optimal designs for $k = 2, \ldots, 9$, but there are smaller designs which are either rotatable, orthogonal or both.

TABLE 9.4
Smallest *D*- and *G*-Optimal Second Order Central Composite Designs.

p	k	n	n_C	n_K	n_S	$\alpha = \sqrt{p}$
2	0	48	8	5	5	1.414
3	0	400	40	27	24	1.732
4	0	180	12	7	7	2.000
5	1	588	28	25	16	2.236
6	1	3584	128	81	72	2.449
7	1	20736	576	245	320	2.646
8	2	900	20	11	11	2.828
9	2	38720	704	243	384	3.000

In general a *D*-optimum design in a hypercube with edges $[a_i; b_i]; i = 1, \ldots p$ can be constructed as follows:
Choose for each i the $p + 1$ support points

$$\begin{aligned} x_1^{(i)} &= a_i \\ x_t^{(i)} &= \frac{b_i - a_i}{2} w_{t-1}^{(p)} + \frac{a_i + b_i}{2} \quad (t = 2, \ldots, p) \\ x_{p+1}^{(i)} &= b_i \end{aligned}$$

$w_j^{(p)}$ is the j-th root of the first derivative of the p-ths Legendre polynomial $P_p(x)$ in the interval $[-1,+1]$, $P_p(x) = \frac{d^p(x^2-1)^p}{2^p p! dx^p}$.
We can obtain these values from Table 8.2 for $a = -1$; $b = 1$. If $N \neq (p+1)n$, the n_i must be chosen as equal as possible.

A list of second order response surface designs for 2–13, 15 and 16 factors over spherical regions, along with their G-efficiencies, can be found at `http://lib.stat.cmu.edu/designs/rbc`

9.4 Comparing the determinant criterion for some examples

The criterion of the D-optimality requires the minimisation of the determinant of the variance-covariance matrix of the estimator of the vector of regression coefficient which equals $\sigma^2(X^TX)^{-1}$. We put w.l.o.g. $\sigma^2 = 1$ and use besides $critI = |(X^TX)^{-1}|$ which has to be minimised, a modified criterion $critII = |(X^TX)|^{\frac{1}{p}}$; which has to be maximised. The resulting design is of course the same in both cases; the second criterion is easier to handle. We also use a criterion proposed by Draper and Lin (1990): $critIII = \frac{|(X^TX)|^{\frac{1}{p}}}{n}$. We now wish to compare these criteria for designs with two and three factors using all designs described above.

9.4.1 Two factors

The $n = 7$ experimental points of (x_1, x_2) are (1, 0), (0.5, 0.866), (0, 0), (−0.5, 0.866), (−1, 0), (−0.5, −0.866) and (0.5, −0.866). The 6 outer points lie on a circle with a radius 1 and centre (0, 0).
The factor x_1 has 5 levels (−1, −0.5, 0, 0.5, 1) and x_2 has 3 levels (−0.866, 0, 0.866).
This Doehlert design has an equally spaced distribution of points over the experimental region, a so-called uniform space filler, where the distances between neighboring experiments are equal. For the Doehlert design $Var(\boldsymbol{b}) = \sigma^2(X'X)^{-1}$ with $(X'X)^{-1}$ equal to

$$\begin{pmatrix} 1 & 0 & 0 & -1 & -1.00006 & 0 \\ 0 & 0.333333 & 0 & 0 & 0 & 0 \\ 0 & 0 & 0.333353 & 0 & 0 & 0 \\ -1 & 0 & 0 & 1.5 & 0.833382 & 0 \\ -1.00006 & 0 & 0 & 0.833382 & 1.500176 & 0 \\ 0 & 0 & 0 & 0 & 0 & 1.333412 \end{pmatrix}$$

For the Doehlert design $critI = |(X^TX)^{-1}| = 1/30.36787 = 0.03293$, and $critIII = \frac{|(X^TX)|^{\frac{1}{p}}}{n} = 0.787244$.
This Doehlert design is rotatable.
If we start with a 2^2 factorial design (1, 1), (1, −1), (−1, 1) and (−1, −1) and we add three points at the x_1 axis (1, 0), (0, 0) and (−1, 0), we have also a design with $n = 7$ points in the region $[-1, 1]$ for both factors. The factor x_1 has 3 levels (−1, 0, 1) and x_2 also has 3 levels (−1, 0, 1).
For this design $(X'X)^{-1}$ is

$$\begin{pmatrix} 1 & 0 & 0 & -1 & 0 & 0 \\ 0 & 0.166667 & 0 & 0 & 0 & 0 \\ 0 & 0 & 0.25 & 0 & 0 & 0 \\ -1 & 0 & 0 & 1.5 & -0.5 & 0 \\ 0 & 0 & 0 & -0.5 & 0.75 & 0 \\ 0 & 0 & 0 & 0 & 0 & 0.25 \end{pmatrix}$$

For this design $critI = |(X^TX)^{-1}| = 1/768 = 0.001302$ and $critIII = \frac{|(X^TX)|^{\frac{1}{p}}}{n} = 3.958973$.
This inscribed design is not rotatable.
The value $critI = |(X^TX)^{-1}|$ *of this design is much smaller than that of the Doehlert design, and therefore the value* $critIII = \frac{|(X^TX)|^{\frac{1}{p}}}{n}$ *of this design is larger than that of the Doehlert design.*
To increase the degrees of freedom for the estimator $\boldsymbol{s}^2$ of σ^2 we can add two centre points (0, 0) to the Doehlert design. The experimental points of (x_1, x_2) are then (1, 0), (0.5, 0.866), (0, 0) three times, (−0.5, 0.866), (−1, 0), (−0.5, −0.866) and (0.5, −0.866). This design remains a regular hexagon with a centre point at (0, 0).
For the Doehlert design $Var(\boldsymbol{b}) = \sigma^2(X'X)^{-1}$ with $(X'X)^{-1}$ equal to

$$\begin{pmatrix} 0.333333 & 0 & 0 & -0.33333 & -0.33335 & 0 \\ 0 & 0.333333 & 0 & 0 & 0 & 0 \\ 0 & 0 & 0.333353 & 0 & 0 & 0 \\ -0.33333 & 0 & 0 & 0.833333 & 0.1666676 & 0 \\ -0.33335 & 0 & 0 & 0.166676 & 0.833431 & 0 \\ 0 & 0 & 0 & 0 & 0 & 1.333412 \end{pmatrix}$$

For this design $critI = |(X^TX)^{-1}| = 1/91.10362 = 0.010977$ and $critIII = \frac{|(X^TX)|^{\frac{1}{p}}}{n} = 1.060536$.
This Doehlert design is rotatable.

The Central Composite Design (CCD) with the star points at −1 and +1 is

called an “Inscribed” Central Composite Design. The circle with the experimental points has the radius of 1; the distance of the star points to the centre (0, 0). This design requires; 5 levels of each factor. Our R-program gives this rotatable and non-orthogonal Central Composite Design for 2 factors over the region $[-1, +1]$ with $n = 9$ runs. For the rotatable and non-orthogonal Central Composite Design with $n = 9$ runs, the experimental points of (x_1, x_2) are the 4 cube points (−0.7071, −0.7071), (−0.7071, 0.7071), (0.7071, −0.7071), (0.7071, 0.7071); the 4 star points (−1, 0), (1, 0), (0, −1), (0, 1) and one centre point (0, 0). This CCD is a regular octagon with the centre point (0, 0). The outer points lie on a circle with radius 0.7071.
For this inscribed CCD the $Var(\boldsymbol{b}) = \sigma^2(X'X)^{-1}$ with $(X'X)^{-1}$ equal to

$$\begin{pmatrix} 1 & 0 & 0 & -1 & -1 & 0 \\ 0 & 0.25 & 0 & 0 & 0 & 0 \\ 0 & 0 & 0.25 & 0 & 0 & 0 \\ -1 & 0 & 0 & 1.375 & 0.875 & 0 \\ -1 & 0 & 0 & 0.875 & 1.375 & 0 \\ 0 & 0 & 0 & 0 & 0 & 1 \end{pmatrix}$$

For this design $critI = |(X^TX)^{-1}| = 1/127.9902 = 0.007813$ and $critIII = \frac{|(X^TX)|^{\frac{1}{P}}}{n} = 1.257031$.
This inscribed CCD is rotatable.

In this design the star points are at the centre of each face of the factorial space. This design requires 3 levels of each factor. The experimental points of (x_1, x_2) are the 4 cube points (1, 1), (1, −1), (−1, 1), (−1, −1); the 4 star points (1, 0), (−1, 0), (0, 1), (0, −1); one centre point (0, 0).
For this face centered CCD $(X'X)^{-1}$ is

$$\begin{pmatrix} 0.555556 & 0 & 0 & -0.33333 & -0.33333 & 0 \\ 0 & 0.166667 & 0 & 0 & 0 & 0 \\ 0 & 0 & 0.166667 & 0 & 0 & 0 \\ -0.33333 & 0 & 0 & 0.5 & 0 & 0 \\ -0.33333 & 0 & 0 & 0 & 0.5 & 0 \\ 0 & 0 & 0 & 0 & 0 & 0.25 \end{pmatrix}$$

For this design $critI = |(X^TX)^{-1}| = 1/5184 = 0.000193$ and $critIII = \frac{|(X^TX)|^{\frac{1}{P}}}{n} = 8$.
This face centered CCD is not rotatable.

According to the determinant criterion and the critIII $= \frac{|(\mathrm{X}^\mathrm{T}\mathrm{X})|^{\frac{1}{\mathrm{P}}}}{\mathrm{n}}$, *the Doehlert design is the worst, then follows the inscribed CCD and the best is the face centered CCD. Also the variance of the estimator of the parameters is for the face centered design the smallest, except for* var(b_0). *But the variance of predicted values at a distance 0.7071 from the centre is smaller for the Doehlert Design than for the face centered Central Composite Design. The Doehlert design has 5 levels for x_1 and 3 levels for x_2. The face centered Central Composite esign has 5 levels for x_1 and x_2. Hence for* $n = 9$ *runs the Doehlert design and the face centered CCD have the preference.*

A rotatable and orthogonal Central Composite Design with $n = 16$ runs is an optimal CCD. The experimental points of (x_1, x_2) are the 4 cube points (−0.7071, −0.7071), (−0.7071, 0.7071), (0.7071, −0.7071), (0.7071, 0.7071); the 4 star points (−1, 0), (1, 0), (0, −1), (0, 1) and one centre point (0, 0) replicated 8 times. This CCD is a regular octagon with the centre point (0, 0). The outer points lie on a circle with radius 0.7071. For this inscribed CCD the $Var(\boldsymbol{b}) = \sigma^2(X'X)^{-1}$ with $(X'X)^{-1}$

$$\begin{pmatrix} 0.125 & 0 & 0 & -0.125 & -0.125 & 0 \\ 0 & 0.25 & 0 & 0 & 0 & 0 \\ 0 & 0 & 0.25 & 0 & 0 & 0 \\ -0.125 & 0 & 0 & 0.5 & 0 & 0 \\ -0.125 & 0 & 0 & 0 & 0.5 & 0 \\ 0 & 0 & 0 & 0 & 0 & 1 \end{pmatrix}$$

For this design $critI = |(X^TX)^{-1}| = 1/1023.921 = 0.000977$ and $critIII = \frac{|(X^TX)|^{\frac{1}{p}}}{n} = 1.999923..$

This inscribed CCD is rotatable.

For this optimal design we have the smallest variance for the predicted values at a distance 0.7071 from the centre (0, 0). But the number of runs is much larger now, $n = 16$.

9.4.2 Three factors

Our R-program gives a non-rotatable and non-orthogonal Central Composite inscribed Design for 3 quantitative factors with only $n = 11$ runs. CADEMO-light[1] gives also a rotatable and non-orthogonal inscribed Central Composite Design with $n = 15$ runs. This design has 8 cube points (±0.5946, ±0.5946, ±0.5946); 6 star points at ±1 on the axis and 1 centre point (0, 0, 0).

A rotatable and orthogonal inscribed Central Composite Design needs $n = 32$ runs. This design has 8 cube points (±0.7071, ±0.7071, ±0.7071); 6 star points twice at ±1 on the axis and 1 centre point (0, 0, 0) 12 times.

The non-rotatable and non-orthogonal inscribed Central Composite Design with $n = 11$ runs has 4 cube points: (−0.8715, −0.8715, 0.8715); (0.8715,

[1]`www.Biomath.de`

−0.8715, −0.8715); (−0.8715, 0.8715, −0.8715); (0.8715, 0.8715, 0.8715); 6 star points: (−1, 0, 0); (1, 0, 0) ; (0, −1, 0) ; (0, 1, 0); (0, 0, −1) ; (0, 0, 1) ; and 1 centre point (0, 0, 0). Each factor has 5 levels.
For this inscribed CCD we have $Var(\boldsymbol{b}) = \sigma^2(X'X)^{-1}$ with $(X'X)^{-1}$

$$\begin{pmatrix} a & 0 & 0 & 0 & b & b & b & 0 & 0 & 0 \\ 0 & 0.5 & 0 & 0 & 0 & 0 & 0 & 0 & 0 & c \\ 0 & 0 & 0.5 & 0 & 0 & 0 & 0 & 0 & c & 0 \\ 0 & 0 & 0 & 0.5 & 0 & 0 & 0 & c & 0 & 0 \\ b & 0 & 0 & 0 & 0.5 & 0 & 0 & 0 & 0 & 0 \\ b & 0 & 0 & 0 & 0 & 0.5 & 0 & 0 & 0 & 0 \\ b & 0 & 0 & 0 & 0 & 0 & 0.5 & 0 & 0 & 0 \\ 0 & 0 & 0 & c & 0 & 0 & 0 & d & 0 & 0 \\ 0 & 0 & c & 0 & 0 & 0 & 0 & 0 & d & 0 \\ 0 & c & 0 & 0 & 0 & 0 & 0 & 0 & 0 & d \end{pmatrix}$$

where $a = 0.4056$, $b = -0.229$, $c = -0.5737$, $d = 1.0917$.
For this design $critI = |(X^TX)^{-1}| = 1/8648.73453 = 0.000115624$ and $critIII = \frac{|(X^TX)|^{\frac{1}{p}}}{n} = 1.866057$.
This CCD is not rotatable.

Draper and Lin (1990) proposed a small composite design for 3 quantitative factors with $n = 11$ runs. The experimental points for (x_1, x_2, x_3) are (1, 0, 0), (−1, 0, 0), (0, 1, 0), (0, −1, 0), (0, 0, 1), (0, 0, −1), (1, −1, 1), (−1, −1, −1), (1, 1, −1), (−1, 1, 1), (0, 0, 0). Each factor has 3 levels.
The small composite design has the $Var(\boldsymbol{b}) = \sigma^2(X'X)^{-1}$ with $(X'X)^{-1}$

$$\begin{pmatrix} a & 0 & 0 & 0 & b & b & b & 0 & 0 & 0 \\ 0 & 0.5 & 0 & 0 & 0 & 0 & 0 & 0 & 0 & 0.5 \\ 0 & 0 & 0.5 & 0 & 0 & 0 & 0 & 0 & 0.5 & 0 \\ 0 & 0 & 0 & 0.5 & 0 & 0 & 0 & 0.5 & 0 & 0 \\ b & 0 & 0 & 0 & c & d & d & 0 & 0 & 0 \\ b & 0 & 0 & 0 & d & c & d & 0 & 0 & 0 \\ b & 0 & 0 & 0 & d & d & c & 0 & 0 & 0 \\ 0 & 0 & 0 & 0.5 & 0 & 0 & 0 & 0.75 & 0 & 0 \\ 0 & 0 & 0.5 & 0 & 0 & 0 & 0 & 0 & 0.75 & 0 \\ 0 & 0.5 & 0 & 0 & 0 & 0 & 0 & 0 & 0 & 0.75 \end{pmatrix}$$

where $a = 0.30435$, $b = -0.13043$, $c = 0.41304$, $d = -0.08696$.
For this design $critI = |(X^TX)|^{-1} = 1/94208 = 1.06148E-05$ and $critIII = \frac{|(X^TX)^{\frac{1}{p}}|}{n} = 4.136534$.

This small CCD is not rotatable.

The small CCD has a smaller determinant criterion and a larger critIII $= \frac{|(X^TX)|^{\frac{1}{p}}}{n}$ *than the inscribed CCD. The variance of the estimator for the parameters is smaller for the simple CCD than for the inscribed CCD, only the variances for* b_1, b_2 *and* b_3 *are the same. Furthermore the simple CCD has only 3 levels for each factor, but the inscribed CCD has 5 levels for each factor. Hence for* $n = 11$ *runs, the small CCD has to be preferred.*

The Doehlert design is characterized by uniformity in space filling; that is, the distances between all neighboring experiments are equal. The 13 experimental points are then for (x_1, x_2, x_3):
(1, 0, 0), (0.5, 0.866, 0), (0.5, 0.289, 0.816), (0.5, −0.866, 0), (0.5, −0.289, −0.816), (0, 0, 0), (0, 0.577, −0.816), (0, −0.577, 0.816), (−0.5, 0.866, 0), (−0.5, −0.866, 0), (−0.5, 0.289, 0.816), (−0.5, −0.289, −0.816), (−1, 0, 0).
The 12 outer points lie on a sphere with radius 1 and with centre (0, 0, 0). Factor x_1 has 5 levels, factor x_2 has 7 levels and factor x_3 has 3 levels.

The Doehlert design has the $Var(\boldsymbol{b}) = \sigma^2(X'X)^{-1}$ with $(X'X)^{-1}$

$$\begin{pmatrix}
1 & 0 & 0 & 0 & -1 & -1 & -1 & 0 & 0 & 0 \\
0 & 0.25 & 0 & 0 & 0 & 0 & 0 & 0 & 0 & 0 \\
0 & 0 & 0.25 & 0 & 0 & 0 & 0 & 0 & 0 & 0 \\
0 & 0 & 0 & 0.25 & 0 & 0 & 0 & 0 & 0 & 0 \\
-1 & 0 & 0 & 0 & 1.5 & 0.83338 & 0.91788 & 0 & 0 & -0.23653 \\
-1 & 0 & 0 & 0 & 0.83338 & 1.50018 & 0.91774 & 0 & 0 & 0.23464 \\
-1 & 0 & 0 & 0 & 0.91788 & 0.91774 & 1.42012 & 0 & 0 & -0.00143 \\
0 & 0 & 0 & 0 & 0 & 0 & 0 & 1.33341 & -0.47225 & 0 \\
0 & 0 & 0 & 0 & 0 & 0 & 0 & -0.47225 & 1.66908 & 0 \\
0 & 0 & 0 & 0 & -0.23653 & 0.23464 & -0.00143 & 0 & 0 & 1.66841
\end{pmatrix}$$

For this design $critI = |(X^TX)^{-1}| = 1/254.3726716 = 0.003931$ and $critIII = \frac{|(X^TX)|^{-\frac{1}{p}}}{n} = 0.487394$.
This Doehlert design is nearly rotatable.

Box and Behnken (1960) proposed the following small design for 3 quantitative factors. Combining 2^2 factorial designs of 2 factors with the zero level of a third factor and a centre point (0, 0, 0) gives the following design. The 13 experimental points for (x_1, x_2, x_3) are (1, 1, 0), (1, −1, 0), (−1, 1, 0), (−1, −1, 0), (1, 0, 1), (1, 0, −1), (−1, 0, 1), (−1, 0, −1), (0, 1, 1), (0, 1, −1), (0, −1, 1), (0, −1, −1), (0, 0, 0).
The Box-Behnken design has the $Var(\boldsymbol{b}) = \sigma^2(X'X)^{-1}$ with $(X'X)^{-1}$

$$\begin{pmatrix} 1 & 0 & 0 & 0 & -0.5 & -0.5 & -0.5 & 0 & 0 & 0 \\ 0 & 0.125 & 0 & 0 & 0 & 0 & 0 & 0 & 0 & 0 \\ 0 & 0 & 0.125 & 0 & 0 & 0 & 0 & 0 & 0 & 0 \\ 0 & 0 & 0 & 0.125 & 0 & 0 & 0 & 0 & 0 & 0 \\ -0.5 & 0 & 0 & 0 & 0.4375 & 0.1875 & 0.1875 & 0 & 0 & 0 \\ -0.5 & 0 & 0 & 0 & 0.1875 & 0.4375 & 0.1875 & 0 & 0 & 0 \\ -0.5 & 0 & 0 & 0 & 0.1875 & 0.1875 & 0.4375 & 0 & 0 & 0 \\ 0 & 0 & 0 & 0 & 0 & 0 & 0 & 0.25 & 0 & 0 \\ 0 & 0 & 0 & 0 & 0 & 0 & 0 & 0 & 0.25 & 0 \\ 0 & 0 & 0 & 0 & 0 & 0 & 0 & 0 & 0 & 0.25 \end{pmatrix}$$

For this design $critI = |(X^TX)^{-1}| = 1/8388608 = 1.19209E - 07$ and $critIII = \frac{|(X^TX)^{\frac{1}{p}}|}{n} = +15.62979$.
The Box-Behnken design is not rotatable.

The Box-Behnken design has a smaller determinant criterion and a larger $\mathrm{critIII} = \frac{|(\mathrm{X^TX})|^{\frac{1}{\mathrm{p}}}}{\mathrm{n}}$ *than the Doehlert design. The variance of the estimator for the parameters is smaller for the Box-Behnken design than for the Doehlert design. Furthermore the Box-Behnken design has only 3 levels for each factor, but the Doehlert design has 5 levels for* x_1*, 7 levels for* x_2 *and 3 levels for* x_3*. Hence for* $n = 13$ *runs, the Box-Behnken design has to be preferred.*

Optimal Central Composite Design with $n = 32$ ***runs.***

The optimal CCD for 3 factors is a rotatable and orthogonal inscribed Central Composite Design with $n = 32$ runs. The experimental points for (x_1, x_2, x_3) are the 8 cube points: (−0.7071, −0.7071, −0.7071), (0.7071, −0.7071, −0.7071), (−0.7071, 0.7071, −0.7071), (0.7071, 0.7071, −0.7071), (−0.7071, −0.7071, 0.7071), (0.7071, −0.7071, 0.7071), (−0.7071, 0.7071, 0.7071), (0.7071, 0.7071, 0.7071; the 6 star points which occur twice: (−1, 0, 0), (1, 0, 0), (0, −1, 0), (0, 1, 0), (0, 0, −1), (0, 0, 1) and the centre point which occur 12 times: (0, 0, 0).
The optimal CCD has the $Var(\boldsymbol{b}) = \sigma^2(X'X)^{-1}$ with $(X'X)^{-1}$:

$$\begin{pmatrix} 0.07813 & 0 & 0 & 0 & -0.0625 & -0.0625 & -0.0625 & 0 & 0 & 0 \\ 0 & 0.125 & 0 & 0 & 0 & 0 & 0 & 0 & 0 & 0 \\ 0 & 0 & 0.125 & 0 & 0 & 0 & 0 & 0 & 0 & 0 \\ 0 & 0 & 0 & 0.125 & 0 & 0 & 0 & 0 & 0 & 0 \\ -0.0625 & 0 & 0 & 0 & 0.25 & 0 & 0 & 0 & 0 & 0 \\ -0.0625 & 0 & 0 & 0 & 0 & 0.25 & 0 & 0 & 0 & 0 \\ -0.0625 & 0 & 0 & 0 & 0 & 0 & 0.25 & 0 & 0 & 0 \\ 0 & 0 & 0 & 0 & 0 & 0 & 0 & 0.5 & 0 & 0 \\ 0 & 0 & 0 & 0 & 0 & 0 & 0 & 0 & 0.5 & 0 \\ 0 & 0 & 0 & 0 & 0 & 0 & 0 & 0 & 0 & 0.5 \end{pmatrix}$$

For this design $critI = |(X^TX)^{-1}| = 1/8387160.073 = 1.1923E - 07$ and

$critIII = \frac{|(X^TX)|^{\frac{1}{p}}}{n} = 6.349239.$
For this optimal design we have the smallest variance for the predicted values at a distance 0.7071 from the centre (0, 0, 0). But the number of runs is much larger now, $n = 32$.

10

Mixture Designs

Mixture designs are a type of factorial design that is used to find the best composition when there is a mixture of ingredients. An important contribution to this kind of designs was made by Scheffé (1958, 1963). A nearly complete overview over the literature of mixture problems is given by Cornell (1973). For further reading, see Becker (1970), Cornell (2002), Draper et al. (2000), Draper and Pukelsheim (1999) and McLean and Anderson (1966), and for a sequential approach, Palasota et al. (1992).

10.1 Introduction

Mixtures are different from other types of factorial experimental design (Chapter seven) because the percentages of the components must add up to 100%. Thus increasing the level of one constituent must reduce the level of at least one other. While in usual factorial designs often ANOVA models are in use in mixture design, regression analysis is mainly used; especially linear, quadratic or cubic response surfaces are assumed in dependence on the mixture components.

In a mixture experiment, the factors are different components of a blend. For example, if one wants to optimise the tensile strength of stainless steel, the factors of interest might be iron, copper, nickel and chromium in the alloy. Other examples would be gasoline, soaps or detergents, beverages, cake mixes, soups a.s.o.

Often there are restrictions on the percentage of the ingredients. For instance, to bake a cake you cannot do it without flour and some liquid (milk or water). So some of the ingredients cannot have a percentage of zero. Then the optimal design must be found under some restrictions.

There are standard mixture designs for fitting standard models, such as *Simplex-Lattice* designs and *Simplex-Centroid* designs. When mixture components are subject to additional constraints, such as a maximum and/or minimum value for each component, designs other than the standard mixture designs, referred to as constrained mixture designs or *Extreme-Vertices* designs, are appropriate.

In mixture experiments, the measured response is assumed to depend only on the relative proportions of the ingredients or components in the mixture and not on the amount of the mixture. The main distinction between mixture experiments and independent variable experiments is that with the former, the input variables or components are non-negative proportionate amounts of the mixture, and if expressed as fractions of the mixture, they must sum to one. In mixture problems, the purpose of the experiment is to model the blending surface with some form of mathematical equation and to find the "best" mixture with respect to a well-defined response variable and an optimality criterion on it.. The usual assumptions made for regression models in general are also made for mixture experiments. In particular, it is assumed that the errors are independent and identically distributed with zero mean and common variance. Another assumption that is made, as with regression models, is that the true underlying response surface is continuous over the region being studied. The underlying model is always a Model I of regression (see Chapter eight) because the proportions are determined by the experimenter.

Planning a mixture experiment typically involves the following steps:

1. Define the objectives of the experiment. Select the mixture components.
2. Identify any constraints on the mixture components to specify the experimental region.
3. Identify the response variable to be measured.
4. Define an optimality criterion for the construction of the design.
5. Propose an appropriate model for modelling the response data as functions of the mixture components.
6. Select an experimental design which suffices points 3 and 4.

Let p be the number of factors occurring in a mixture design after step 2 above.
The unrestricted design region for mixture proportions is a simplex, a regularly sided figure of dimension $p-1$ with p vertices. For example, with two factors, the simplex is the line segment from (0, 1) to (1, 0). With three factors, the simplex would have vertices at (1, 0, 0), (0, 1, 0), and (0, 0, 1). The shapes of the simplex with three variables is a tetrahedron. There is a corresponding simple coordinate system.
We can now consider models for mixture experiments. The usual first order model is

$$E(\boldsymbol{y}) = \beta_0^* + \sum_{i=1}^{p} \beta_i^* x_i, 0 \leq x_i \leq 1; \sum_{i=1}^{p} x_i = 1$$

However, since $\sum_{i=1}^{p} x_i = 1$ for a mixture model, the β_i^*'s will not be uniquely defined. We could eliminate one of the x_i's, but a better approach was suggested by Scheffé (1963). In the equation above multiply β_0^* with $\sum_{i=1}^{p} x_i = 1$ to get $E(\boldsymbol{y}) = \sum_{i=1}^{p}(\beta_0^* + \beta_i^*)x_i, 0 \leq x_i \leq 1; \sum_{i=1}^{p} x_i = 1.$

Relabeling the β_i^*'s, we get the following canonical forms.

Linear:

$$E(\boldsymbol{y}) = \sum_{i=1}^{p} \beta_i \boldsymbol{x}_i, 0 \leq x_i \leq 1; \sum_{i=1}^{p} x_i = 1 \tag{10.1}$$

Analogously we can write a quadratic polynomial in canonical form as:

Quadratic:

$$E(\boldsymbol{y}) = \sum_{i=1}^{p} \beta_i \boldsymbol{x}_i + \sum_{i \leq j}\sum \beta_{ij} \boldsymbol{x}_i \boldsymbol{x}_j, 0 \leq x_i \leq 1; \sum_{i=1}^{p} x_i = 1 \tag{10.2}$$

The linear terms in the canonical mixture polynomials have simple interpretations. Geometrically, the parameter β_i $i = 1, \ldots, p$ in the above equations represents the expected response to the pure mixture $x_i = 1, x_j = 0, i \neq j$ and is the height of the mixture surface at the vertex $x_i = 1$. The portion of each of the above polynomials given by

$$\sum_{i=1}^{p} \beta_i x_i$$

is called the linear blending portion. When blending is strictly additive, then the linear model form above is an appropriate model; otherwise higher polynomials must be used.

Of course there are cubic and other non-linear models possible, but these are seldom used in practice. We restrict ourselves to the linear and the full quadratic models as shown above. It is quite clear that optimal designs depend on the model.

In mixture designs when there are constraints on the component proportions, these are often upper and/or lower bound constraints of the form $L_i \leq x_i \leq U_i, i = 1, 2, \ldots, p$, where L_i is the lower bound for the i-th component and U_i the upper bound for the i-th component. The general form of the constrained mixture problem is

$$x_1 + x_2 + \ldots + x_p = 1$$

$$L_i \leq x_i \leq U_i, \text{ for } i = 1, 2, \ldots, p$$

with $L_i \geq 0$ an $U_i \leq 1$.

Consider the following case in which only the lower bounds in the above equation are imposed, so that the constrained mixture problem becomes

$$x_1 + x_2 + \ldots + x_p = 1$$

$$L_i \leq x_i \leq 1, \text{ for } i = 1, 2, \ldots, p$$

The existence of lower bounds does not affect the shape of the mixture region; it is still a simplex region. In general, this will always be the case if only lower bounds are imposed on any of the component proportions.

Graphically we will demonstrate the case with 3 components for simplicity. The simplex is then a triangle as shown in Figure 10.1.

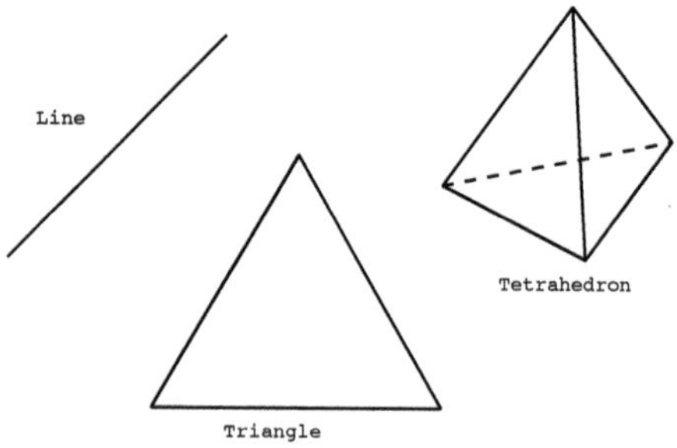

FIGURE 10.1
Simplexes with two, three and four components

These initial trials form the first simplex. The shapes of the simplex in a one, a two or a three variable search space are a line, a triangle or a tetrahedron. A geometric interpretation is difficult with more variables, but the basic mathematical approach outlined below can handle the search for optimum conditions.

10.2 The simplex lattice designs

We first consider unrestricted mixture experiments. If nothing is known about where the optimum of the response surface lies, the simplex defined by the condition $\sum_{i=1}^{p} x_i = 1$ has dimension $p-1$ and p vertices and the support points of the design are allocated equally spaced. If in the case of a polynomial of degree m the support points take for each component x_i one of the $m+1$ equally spaced values $x_i = 0, \frac{1}{m}, \frac{2}{m}, \ldots, \frac{m-1}{m}, 1$, then the set of all possible combinations of the support point is called a (p, m) *simplex lattice*. A pure (p, m) simplex design with p support points at the p vertices (i.e., $m = 1$) is not really a mixture design and will not be discussed further. A $\{p, m\}$ simplex lattice design for p components $x_1, x_2, \ldots, x_p$ consists of $r = \frac{(p+m-1)!}{m/(p-1)!}$ points. If one of those points is in the interior of the simplex (in its centre), the simplex lattice design is called a simplex centroid design.

In Table 10.1 some values of r are given.

TABLE 10.1
Number r of Design Points in $\{p, m\}$ Simplex Lattice Designs.

		m		
		1 (pure simplex)	2	3
	3	3	6	10
	4	4	610	20
p	5	5	15	35
	6	6	21	56
	7	7	28	84

Example 10.1 Let us consider mixtures of 3 components so that the simplex is a triangle. To fit a quadratic response function m should be at least equal to 2. In the case of $m = 3$ we have the following 10 points in the (3,3) simple lattice:

x_1	x_2	x_3
0	0	1
0	0.333	0.667
0	0.667	0.333
0	1	0
0.333	0	0.667
0.333	0.333	0.333
0.333	0.6667	0
0.667	0	0.333
0.667	0.333	0
1	0	0

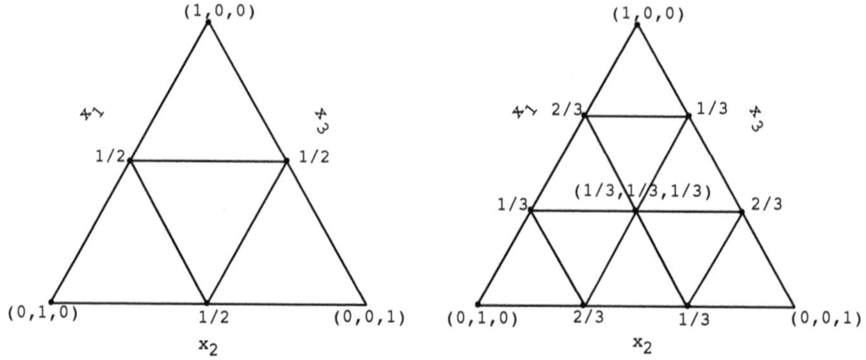

FIGURE 10.2
Simplex lattice designs with $p = 3$ components and $m = 2$ and $m = 3$

The standard simplex lattice design is a boundary-point design; that is, with the exception of the centroid, all the design points are on the boundaries of the simplex.

In Figure 10.2 we show two simplex lattices with $p = 3$ components and a linear, a quadratic and a cubic polynomial as the model equation.

10.3 Simplex centroid designs

A simplex centroid design is a simplex lattice design with an extra point in the centroid of the simplex. An example is given in Figure 10.2

10.4 Extreme vertice designs

An extreme vertice design (McLean and Anderson, 1966) is a mixture design with support points in the vertices of a restricted polygon within a simplex. The restrictions stem from the fact that some portions of all ingredients are necessary for the experimental unit to work.

If we have lower bounds L_i and upper bounds U_i for the first (w.l.o.g.) $q < p$ ingredients of the mixture, we have the conditions

$$x_1 + x_2 + \ldots + x_p = 1$$

$$L_i \leq x_i \leq U_i, \text{ for } i = 1, 2, \ldots, q$$

$$\phi \leq x_i \leq 1, \text{for } i = q + 1, \ldots, p$$

Now the experiment can be designed as follows:

Write all two-level treatment combinations of the lower and upper levels, using the bounds for all but one factor which is left blank ($\leftrightarrow$)

For instance $(L_1, L_2, U_3, \leftrightarrow, U_5, L_6, \ldots, U_p,)$. In this way we generate $p2^{p-1}$ treatment combinations.

10.5 Augmented designs

Augmented mixture designs are basic designs like simplex lattice or simplex centroid designs augmented by some inner points. These designs also use mixtures which include all components. Such designs may appear if optimal mixture designs like D-optimal designs are constructed starting from any basic design.

10.6 Constructing optimal mixture designs with R

The R-program provides the construction of $\{p, m\}$ simplex lattice designs and the corresponding simplex centroid designs. Further for given polygons within a simplex whose coordinates have to be inserted in a table of the program, extreme vertice designs are constructed.

The simplest way to construct a mixture design for a linear model is to choose in the support the extremes of the simplex (simplex lattice design with $m = 1$) or the restricted experimental region and then choose the overall centroid of the region. For a quadratic model we could add the midpoints of boundaries for each component, for instance, a simplex lattice design with $m = 2$ in the case of no restrictions. It is important to check that the restrictions are not contradictory and the experimental region is not empty.

The *sequential simplex method* can be used to handle mixture constraints and possible dependencies between the components. The procedure is as follows. Construct the first simplex with one of the p components left out. Then calculate the component left out by calculating the difference between 1 (or 100%) and the other components. Another way is the use of the *sequential simplex optimisation* (Walters et al. 1991). With two variables the first simplex design is based on three trials; for three variables it is four trials, etc. This number of trials is also the minimum for defining a direction of improvement. Therefore, it is a time-saving and economical way to start an optimisation project. After the initial trials the simplex process is sequential, with the addition and evaluation of one new trial at a time. The simplex searches systematically for the best levels of the control variables. The optimisation process ends when the optimisation objective is reached or when the responses cannot be improved further.

Because the design matrix for any pair $\{p, m\}$ can be chosen in such a way that the mixture design is Φ-optimal, program offers D- and G-optimal mixture designs and restricted mixture designs for any connected design region within the simplex.

10.7 An example

Not to be too complicated let us take an everyday example, namely, baking a cake. A speciality of Renate, the wife of the first author, is "Wolkenkuchen," consisting of $p = 6$ components for the basic batter, which is eventually topped with streusel and chocolate:
flour $= x_1$
sugar $= x_2$
butter $= x_3$
milk $= x_4$
baking powder $= x_5$
eggs $= x_6$
(All variables in g.)
The mixture which we think is an optimal one and is used in the Rasch family is
flour $=$ 500g $= x_1$
sugar $=$ 375g $=x_2$
butter $=$ 300g $=x_3$
milk $=$ 30g $= x_4$
baking powder $=$ 15g $= x_5$
eggs 4 $=$ 280g $= x_6$
Thus the total amount for a usual baking pan in the household is 1500 g.

It is clear that there have to be restrictions: eggs and the milk (to a certain extent, the butter also) are the liquids in the blend, and their contribution cannot be zero. A cake without sugar, flour and baking powder will also not work. We therefore define the following side conditions:

$200 < x_1 < 600; 100 < x_2 < 400; 50 < x_3 < 400; 10 < x_4 < 50; 2 < x_5 < 30; 75 < x_6 < 450$.

When we assume that cakes should always weigh 1.5 kg, then the not-standardised range for each of the ingredients is between 0 and 1500. If we define the restrictions as proportions (between 0 and 1), the standardised restrictions are

$0.1333 < x_1 < 0.4, 0.0667 < x_2 < 0.2667, 0.0333 < x_3 < 0.2667, 0.0067 < x_4 < 0.0333, 0.0013 < x_5 < 0.02, 0.05 < x_6 < 0.3$.

The dependent variables are the lightness of the cake, the sweetness and the texture.

A

Theoretical Background

In this appendix we will give some mathematical tools needed in the text and which are not as commonly known as calculus or elementary matrix algebra. We will also give some definitions from discrete mathematics (combinatorics) which are needed in constructing BIBD and fractional factorial designs, as well as some theorems which may be of interest for a mathematical statistician.

Notation: $I_n = \begin{pmatrix} 1 & 0 & \dots & 0 \\ 0 & 1 & \dots & 0 \\ \dots & \dots & \dots & \dots \\ 0 & 0 & \dots & 1 \end{pmatrix}$ unity matrix of order n;

$$J_n = e_{n,n} = \begin{pmatrix} 1 & 1 & \dots & 1 \\ 1 & 1 & \dots & 1 \\ \dots & \dots & \dots & \dots \\ 1 & 1 & \dots & 1 \end{pmatrix}, \; e_n = \begin{pmatrix} 1 \\ 1 \\ \dots \\ 1 \end{pmatrix}.$$

The indicator function $I_{(a,b)}(x) = \begin{cases} 1, \text{if} x \in (a,b) \\ 0, \text{otherwise} \end{cases}$.

In Chapter eight we need Legendre polynomials in $[-1;1]$. The Legendre polynomial of degree k is defined as $P_k(x) = \frac{1}{2^k k!} \frac{d^k}{dx^k}[(x^2-1)^k], x\varepsilon[-1;1]$.

A.1 Non-central distributions

Non-central distributions play an important role in the determination of sample sizes or more generally in the determination of the size of an experiment. We introduce here the non-cental chi-square, t- and F-distributions.

Definition A.1 Given a family of $N(\mu, I_n)$-distributions of the random variable $\boldsymbol{y}^T = (\boldsymbol{y}_1, \boldsymbol{y}_2, \dots, \boldsymbol{y}_n)$ with $\mu^T = (\mu_1, \mu_2, \dots, \mu_n)$, the family of distributions induced by the measurable function $\boldsymbol{\chi}^2 = \boldsymbol{y}^T\boldsymbol{y}$ is the family of non-central $\boldsymbol{\chi}^2$-distributions with n degrees of freedom and non-centrality parameter $\lambda = \mu^T\mu \neq 0$. For short, we say, $\boldsymbol{\chi}^2 = \boldsymbol{y}^T\boldsymbol{y}$ is CS(n, λ)-distributed. If $\lambda = 0$, the induced family is the family of (central) $\boldsymbol{\chi}^2$-distributions with n degrees of freedom.

Theorem A.1 The density function of the non-central χ^2-distributions with n degrees of freedom and non-centrality parameter $\lambda \neq 0$ is given by

$$n_{n,\lambda}(\chi^2) = \frac{e^{-\frac{1}{2}\chi^2}\cdot(\chi^2)^{\frac{n}{2}-1}}{2^{\frac{n}{2}}\Gamma(\frac{f}{2})}\left[e^{-\frac{\lambda}{2}} + \sum_{j=1}^{\infty}\frac{(\lambda\chi^2)^j\Gamma(\frac{n}{2})e^{-\frac{\lambda}{2}}}{2^{2j}\Gamma(\frac{n}{2}+j)j!}\right], I_{(0,\infty)}(\chi^2)$$

Definition A.2 Let $\boldsymbol{\chi}^2$ be CS$(n,0)$-distributed and $\boldsymbol{x}$ independently of $\boldsymbol{\chi}^2$ be $N(\mu;1)$-distributed. Then the family of distributions induced by the measurable function $t = \frac{\boldsymbol{x}}{\sqrt{\frac{\boldsymbol{\chi}^2}{n}}}$ is the family of non-central t-distributions with n degrees of freedom and non-centrality parameter $\lambda = \mu^2 \neq 0$. For short, we say $\boldsymbol{t}$ is $t(n,\lambda)$-distributed. If $\lambda = 0$, the induced family is the family of (central) t-distributions with n degrees of freedom, for short the family of $t(n)$ distributions.

Theorem A.2 The density function of the non-central t–distributions with n degrees of freedom and non-centrality parameter λ is given by

$$f_{n,\lambda}(t) = \frac{n^{\frac{n}{2}}e^{-\frac{\lambda}{2}}}{\sqrt{\pi}\cdot\Gamma\left(\frac{n}{2}\right)\cdot(n+t^2)^{\frac{n+1}{2}}}\sum_{j=0}^{\infty}\frac{\Gamma\left(\frac{n+j+1}{2}\right)}{j!}\left[\frac{2t}{n+t^2}\right]^{\frac{j}{2}}\cdot\left(\sqrt{\lambda}t\right)^j$$

Definition A.3 Let $\boldsymbol{\chi}_1^2$ be CS(n_1,λ)–distributed and $\boldsymbol{\chi}_2^2$ independently of $\boldsymbol{\chi}_1^2$ be CS$(m_2,0)$-distributed. Then the family of distributions induced by the measurable function $\boldsymbol{F} = \frac{\boldsymbol{\chi}_1^2\cdot m}{\boldsymbol{\chi}_2^2\cdot n}$ is the family of non-central F–distributions with n and m degrees of freedom and non-centrality parameter $\lambda \neq 0$. For short, we say, $\boldsymbol{F}$ is $F(n,m,\lambda)$-distributed. If $\lambda = 0$, the induced family is the family of (central) F–distributions with n and m degrees of freedom, for short, the family of $F(n,m)$–distributions.

Theorem A.3 The density function of the non-central F–distributions with n and m degrees of freedom and non-centrality parameter λ is given by

$$f_{n,m,\lambda}(F) = \sum_{j=0}^{\infty}\frac{e^{-\frac{\lambda}{2}}\lambda^j n^{\frac{n}{2}+j}m^{\frac{m}{2}}\Gamma\left(\frac{n+m}{2}+j\right)}{j!2^j\Gamma\left(\frac{n}{2}+j\right)\Gamma\left(\frac{m}{2}\right)}\frac{F^{n+j-1}}{(m+nF)^{\frac{n}{2}+\frac{m}{2}+j}}I_{(0,\infty)(F)}$$

A.2 Groups, fields and finite geometries

The definitions of this section are needed for the construction of BIBDs in Chapter six.

Definition A.4 A group is a set G of elements with a binary operation $\bullet$ on G that satisfies the following four axioms:

1. *Closure*: For all a, b in G, the result of $a \bullet b$ is also in G.
2. *Associativity*: For all a, b and c in G, $(a \bullet b) \bullet c = a \bullet (b \bullet c)$.
3. *Identity element*: There exists an element e in G such that for all a in $G, e \bullet a = a \bullet e = a$.
4. *Inverse element*: For each a in G, there exists an element b in G such that $a \bullet b = b \bullet a = e$, where e is the identity element.

Definition A.5 A group G is said to be **Abelian**, or **commutative**, if the operation satisfies the commutative law. That is, for all a and b in $G, a \bullet b = b \bullet a$. An **Abelian group** is also called a **module**.
A group table of a finite group is a square matrix with the group elements defining rows and columns and where the entries are the results of the corresponding binary operation between the row element and the column element.

Definition A.6 A **field** is an algebraic structure in which the operations of addition, subtraction, multiplication and division (except division by zero) may be performed, and the same rules hold which are familiar from the arithmetic of ordinary numbers.
A field can be considered *applicable to both the addition and the multiplication of a group.*

Definition A.7 A finite field is called a **Galois field** GF(p), where p is prime number. It is isomorphic to the complete system of residues modulo p.
For finite fields as well for addition and also for multiplication, a group table can be written down.

Definition A.8 A cyclic group is a group that can be generated by a single element g (the group generator). Cyclic groups are Abelian. The generator of a finite cyclic group of order n satisfies $g^n = I$ where I is the identity element. The element g is called the **generator** g of the cyclic group; it is also called a **primitive element**.

In Table A.1, we give the primitive elements for some primes.

The elements of a Galois field GF(p) can be written by help of a primitive element as $0, g^0, g^1, \ldots, g^{p-2}$.
If, for instance, $p = 13$, then $x = 2$ is a primitive element and the elements of GF(13) are $0, 2^0 = 1, 2^1 = 2, 2^2 = 4, 2^3 = 8, 2^4 = 16 \equiv 3 \bmod(13), 2^5 \equiv 6, 2^6 \equiv 12, 2^7 \equiv 11, 2^8 \equiv 9, 2^9 \equiv 5, 2^{10} \equiv 10, 2^{11} \equiv 7$. To each prime number p there is one GF(p^h) containing p^h elements for each positive integer h. The elements of GF(p^h) can be written as polynomials of degree $h-1$ at the most or as powers of a primitive root x of $x^{p^h-1} = 1$ which means it is a root of

TABLE A.1
Some Primitive Elements.

p	Primitive Element	p	Primitive Element
3	2	23	5
5	2	29	2
7	3	31	3
11	2	37	2
13	2	41	6
17	3	43	3
19	2	47	5

this equation, but $x^t \neq 1$ for all divisors t of $p^h - 1$. If we divide $x^{p^h - 1}$ by the least common multiple of all factors $x^t - 1$ in which t is a divisor of $p^h - 1$ and represent all coefficients of the resultant polynomial by their smallest positive residues (mod p), we obtain the **cyclotomic polynomial**. Any irreducible factor of the cyclotomic polynomial is called a **minimum function** $P(x)$. Some minimum functions are given in Table A.2.

TABLE A.2
Mimimum Functions.

p	h	$P(x)$	p	h	$P(x)$	p	h	$P(x)$
2	2	$x^2 + x + 1$	5	2	$x^2 + 2x + 3$	11	2	$x^2 + x + 7$
	3	$x^3 + x^2 + 1$		3	$x^3 + x^2 + 2$		3	$x^3 + x^2 + 3$
	4	$x^4 + x^3 + 1$		4	$x^4 + x^3 + 2x^2 + 2$		4	$x^4 + 4x^3 + 2$
	5	$x^5 + x^3 + 1$		5	$x^5 + x^2 + 2$		5	$x^5 + x^3 + x^2 + 9$
	6	$x^6 + x^5 + 1$		6	$x^6 + x^5 + 2$	13	2	$x^2 + x + 2$
3	2	$x^2 + x + 2$	7	2	$x^2 + x + 3$		3	$x^3 + x^2 + 2$
	3	$x^3 + 2x + 1$		3	$x^3 + x^2 + x + 2$		4	$x^4 + x^3 + 3x^2 + 2$
	4	$x^4 + x + 2$		4	$x^4 + x^3 + x^2 + 3$	17	2	$x^2 + x + 3$
	5	$x^5 + 2x^4 + 1$		5	$x^5 + x^4 + 4$		3	$x^3 + x + 3$
	6	$x^6 + x^5 + 2$		6	$x^6 + x^5 + x^4 + 3$		4	$x^4 + 4x^2 + x + 3$

A minimum function $P(x)$ can be used to generate the elements of GF(p^h). For this we need a function

$$f(x) = a_0 + a_1 x + \ldots + a_{h-1} x^{h-1}$$

with integer coefficients $a_i (i = 0, \ldots, h-1)$ that are elements of GF(p). The function

$$F(x) = f(x) + pq(x) + P(x)Q(x) \tag{A.1}$$

with the minimum function $P(x)$ and any $q(x)$ and $Q(x)$ form a class of functions, the residuals modulo p and $P(x)$. We write

$$F(x) \equiv f(x)(\text{mod}p; P(x)) \tag{A.2}$$

If p and $P(x)$ are fixed and $f(x)$ is variable, $F(x)$ gives rise to p^h classes (functions) which form a Galois field GF(p^h) if and only if p is a prime number and $P(x)$ is a minimum function of GF(p^h).

Definition A.9 Let $a, n, m > 1$ be positive integers and a and m be relatively prime, then a is a quadratic residue modulo m if an integer x exists so that $x^2 \equiv a \bmod(m)$; otherwise a is a quadratic non-residue modulo m.

If $m = p$ is a prime, then the numbers $1, 2, \ldots, p-1$ are relatively prime to p, and we have exactly $\frac{p-1}{2}$ quadratic residues (and $\frac{p-1}{2}$ quadratic non-residues). For instance, if $p = 7$, we have $\frac{p-1}{2} = 3$ and the quadratic residues are $1, 4, 2$ and the quadratic non-residues are $3, 5, 6$. Some quadratic residues are given in Table A.3

Let us introduce the following Legendre symbol:
For an element a of a GF(p^h) with a prime p and some $h \geq 1$, we define
$\varphi = \varphi(a) = \begin{cases} 1, \text{ if } a \text{ is a quadratic residue of } p^h \\ -1 \text{ otherwise} \end{cases}$

Definition A.10 Let p be a prime number. We consider with an integer h the prime power $s = p^h$. Every ordered set $X = (x_0, \ldots, x_n)$ of $n+1$ elements x_i from GF(s) that do not vanish simultaneously is called a point of a **projective geometry** PG(n, s). Two sets $Y = (y_0, \ldots, y_n)$ and $X = (x_0, \ldots, x_n)$ with $y_i = qx_i (i = 0, \ldots, n)$ and any element q from GF(s) that is different from 0 represent the same point. The elements $x_i (i = 0, \ldots, n)$ of X are called **coordinates** of X. All points of a PG(n, s) that satisfy $n - m$ linear independent homogeneous equations $\sum_{i=0}^{n} a_{ji}x_i = 0 \; j = 1, \ldots, n-m; a_{ji} \in$ GF(s) form an ***m*-dimensional subspace** of the PG(n, s). Subspaces with $x_0 = 0$ are called **subspaces in infinity**. In a PG(n, s) we have $Q_n = \frac{s^{n+1}-1}{s-1}$ different points and $Q_m = \frac{s^{m+1}-1}{s-1}$ points in each m-dimensional subspace. The number of m-dimensional subspaces of a PG(n, s) is

$$\varphi(n, m, s) = \frac{(s^{n+1}-1)(s^n-1)\ldots(s^{n-m+1}-1)}{(s^{m+1}-1)(s^m-1)\ldots(s-1)}, (m \geq 0; n \geq m) \tag{A.3}$$

The number of different m-dimensional subspaces of a PG(n, s) that have one fixed point in common is

$$\varphi(n, m, s)\frac{s^{m+1}-1}{s^{n+1}-1}a = \varphi(n-1, m-1, s) \text{ if } m \geq 1.$$

TABLE A.3
Quadratic Residues for Some Small Primes.

p	$(p-1)/2$	$x(\text{mod}p)$	$x^2(\text{mod}p)$	Non-Residue
5	2	± 1	1	2
		± 2	4	3
7	3	± 1	1	3
		± 2	4	5
		± 3	2	6
11	5	± 1	1	2
		± 2	4	6
		± 3	9	7
		± 4	5	8
		± 5	3	10
13	6	± 1	1	2
		± 2	4	5
		± 3	9	6
		± 4	3	8
		± 5	12	10
		± 6	7	11
17	8	± 1	1	3
		± 2	4	5
		± 3	9	6
		± 4	16	7
		± 5	8	10
		± 6	2	11
		± 7	15	12
		± 8	13	14
19	9	± 1	1	2
		± 2	4	3
		± 3	9	8
		± 4	16	10
		± 5	6	12
		± 6	17	13
		± 7	11	14
		± 8	7	15
		± 9	5	18

The number of different m-dimensional subspaces of a PG(n, s) with two different fixed points in common is

$$\varphi(n, m, s)\frac{(s^{m+1}-1)(s^m-1)}{(s^{n+1}-1)(s^n-1)}a = \varphi(n-2, m-2) \text{ if } m \geq 2.$$

Definition A.11 Let p be a prime number. We consider with an integer h

the prime power $s = p^h$. Every ordered set $X^* = (x_1, \ldots, x_n)$ of n elements x_i from GF(s) is called a point of a **Euclidean geometry** EG(n, s). Two sets $Y^* = (y_1, \ldots, y_n)$ and $X^* = (x_1, \ldots, x_n)$ are called identical if and only if $x_i = y_i; i = 1, \ldots, n$. The elements $x_i (i = 1, \ldots, n)$ of X^* are called coordinates of X^*. All points of an EG(n, s) that satisfy $n - m$ non-contradictory and linear independent equations $\sum_{i=0}^{n} a_{ji}x_i = 0\ j = 1, \ldots, n - m; a_{ji} \in \text{GF}(s); x_0 \equiv 1$ form an **m-dimensional subspace** of the EG(n, s). In an EG(n, s) we have s^n points and s^m points in each m-dimensional subspace. The number of m-dimensional subspaces of an EG(n, s) is

$$\varphi(n, m, s) - \varphi(n - 1, m, s)$$

The number of different m-dimensional subspaces of an EG(n, s) that have one fixed point in common is

$$\varphi(n - 1, m - 1, s).$$

The number of different m-dimensional subspaces of an EG(n, s) with two different fixed points in common is

$$\varphi(n - 2, m - 2, s)$$

Theorem A.4 The points of a PG(n, s) for which $x_0 \neq 0$ and the points of an EG(n, s) can be mapped one-to-one onto each other.

A.3 Difference sets

The definitions of this section are needed for the construction of BIBDs in Chapter six—see, for instance, Jungnickel et al. (2006).

Definition A.12 An Abelian group—A—a module—of m elements may be given by $\{1, 2, \ldots, m\}$. We assign symbols $x_1, \ldots, x_s$, where s is a positive integer, to each element x and unite elements with the same subscript j to form the j-th class $K_j (j = 1, \ldots, s)$ of symbols. From the ms symbols we now form t subsets $M_l = \{x_l^1, x_l^2, \ldots, x_l^{p_l}\} (l = 1, \ldots, t)$ containing each $k < v = ms$ different elements so that p_{jl} elements of M_l belong to class K_j, which means that $k = \sum_{j=1}^{s} p_{jl} \forall l$.

A set $M = \{M_1, \ldots, M_t\}$ of t sets M_l is called a (ν, k, ρ, s, t)-**difference set** if

- Exactly r of the kt symbols of M belong (possibly repeatedly) to each of the s classes
- Each element which is different from zero in the module occurs exactly ρ times among the $\sum_l \sum_j p_{jl}(p_{jl}-1)$ differences between the elements of the sets M_l belonging to the same class (**pure differences**)
- Each element which is different from zero of the module occurs exactly ρ times among the $\sum_l \sum_j \sum_{j'} p_{jl}p_{j'l}$ differences between the elements of the sets M_l belonging to different classes (**mixed differences**)

It follows from this definition that $rs = kt$.
In particular, if $s = 1$, the $(\nu, k, \rho, 1, t)$-difference set is called simply a (ν, k, ρ)-difference set.

Example A.1 We take the module $\{0, 1, 2, 3, 4\}$ of residual classes mod 5 and put $s = 2$. We thus have the classes $\{0_1, 1_1, 2_1, 3_1, 4_1\}$ and $\{0_2, 1_2, 2_2, 3_2, 4_2\}$, and $\nu = ms = 5 \cdot 2 = 10$. We form $t = 3$ sets M_1, M_2, M_3 containing $k = 4$ elements each: $\{0_1, 3_1, 0_2, 4_2\}$, $\{0_1, 0_2, 1_2, 3_2\}$, $\{1_1, 2_1, 3_1, 0_2\}$. From the 12 elements in these 3 sets exactly $r = 6$ belong to each of the two classes, $\{0_1, 1_1, 2_1, 3_1\}$ and $\{0_2, 1_2, 3_2, 4_2\}$.
Then the p_{jl} are given as follows.

	$l = 1$	$l = 2$	$l = 3$
$j = 1$	2	1	3
$j = 2$	2	3	1

And there are $2 \times 1 + 2 \times 1 + 3 \times 2 + 3 \times 2 = 16$ pure differences and $2[2 \times 2 + 3 \times 1 + 3 \times 1] = 20$ mixed differences. The value of ρ should therefore be $\rho = 4$. We want to see whether $K = \{M_1, M_2, M_3\}$ is a $(10, 4, 4, 2, 3)$-difference set. The first of the conditions above is satisfied for $r = 6$. We calculate the 16 pure differences mod 5

$$\begin{array}{ll}
0_1 - 3_1 \equiv -3 \equiv 2 & 3_2 - 0_2 \equiv 3 \\
3_1 - 0_1 \equiv 3 & 3_2 - 1_2 \equiv 2 \\
0_2 - 4_2 \equiv 1 & 1_1 - 2_1 \equiv 4 \\
4_2 - 0_2 \equiv 4 & 1_1 - 3_1 \equiv 3 \\
0_2 - 1_2 \equiv 4 & 2_1 - 1_1 \equiv 1 \\
0_2 - 3_2 \equiv 2 & 2_1 - 3_1 \equiv 4 \\
1_2 - 0_2 \equiv 1 & 3_1 - 1_1 \equiv 2 \\
1_2 - 3_2 \equiv 3 & 3_1 - 2_1 \equiv 1
\end{array}$$

The elements of the module which are different from zero occur as differences exactly $\rho = 4$ times. We now form the 20 mixed differences $(0, 0, 1, 4, 3, 2, 4, 1, 0, 0, 4, 1, 2, 3, 1, 4, 2, 3, 3, 2)$. Once again each element of the module appears $\rho = 4$ times. Therefore $K = \{M_1, M_2, M_3\}$ is a $(10, 4, 4, 2, 3)$-difference set.

Example A.2 We take the module $\{0, 1, 2, 3, 4, 5, 6, 7, 8, 9\}$ of residual classes mod 9 and put $s = 1$. We thus have only one class, the module. We form $t = 2$ sets M_1, M_2 containing $k = 4$ elements each: $M_1 = \{0, 1, 2, 4\}; M_2 = \{0, 3, 4, 7\}$. Here we have $r = 7$. The differences in these two sets are

$0 - 1 \equiv 8 \pmod 9$	$1 - 0 \equiv 1 \pmod 9$	$2 - 0 \equiv 2 \pmod 9$	$4 - 0 \equiv 4 \pmod 9$
$0 - 2 \equiv 7 \pmod 9$	$1 - 2 \equiv 8 \pmod 9$	$2 - 1 \equiv 1 \pmod 9$	$4 - 1 \equiv 3 \pmod 9$
$0 - 4 \equiv 5 \pmod 9$	$1 - 4 \equiv 6 \pmod 9$	$2 - 4 \equiv 7 \pmod 9$	$4 - 2 \equiv 2 \pmod 9$
$0 - 3 \equiv 6 \pmod 9$	$3 - 0 \equiv 3 \pmod 9$	$4 - 0 \equiv 4 \pmod 9$	$7 - 0 \equiv 7 \pmod 9$
$0 - 4 \equiv 5 \pmod 9$	$3 - 4 \equiv 8 \pmod 9$	$4 - 3 \equiv 1 \pmod 9$	$7 - 3 \equiv 4 \pmod 9$
$0 - 7 \equiv 2 \pmod 9$	$3 - 7 \equiv 5 \pmod 9$	$4 - 7 \equiv 6 \pmod 9$	$7 - 4 \equiv 3 \pmod 9$

Each of the 8 possible differences occurs exactly $\lambda = \rho = 3$ times. Thus $M_1 = \{0, 1, 2, 4\}; M_2 = \{0, 3, 4, 7\}$ builds up a $(9, 4, 1)$ difference set.

Definition A.13 An Abelian group—A—a module—of m elements may be given by $\{1, 2, \ldots, m\}$. Let $B = \{b_1, \ldots, b_k; 1 \leq k < m\}$ be a subset of A. The A-stabiliser of B is the subgroup $A_B \in A$ consisting of all elements $i \in A; i = 1, \ldots, m$ so that with I is $B + I = B$. B is called short if $A_B \in A$ is not trivial. A collection $\{B_1, \ldots, B_t\}$ of k-subsets of A form a (ν, k, λ)-difference family or a (ν, k, λ)-difference system if every non-identity element of A occurs λ times in $B_1 \cup \ldots \cup B_t$. The sets B_i are called base blocks. A difference family having at least one short block is called partial.

Definition A.14 An Abelian group—A—a module—of $\nu - 1$ elements may be given by $\{0, 1, 2, \ldots, \nu - 2\}$ and let ∞ be a symbol not belonging to A. A collection $B = \{b_1, \ldots, b_k; 1 \leq k < m\}$ of k-subsets of $A \cup \{\infty\}$ is rotational (ν, k, λ)-difference family if every non-identity element of $A \cup \{\infty\}$ occurs λ times in $B_1 \cup \ldots \cup B_t$. The sets B_t are called base blocks.

Definition A.15 An Abelian group—a module—of m elements may be given by $\{1, 2, \ldots, m\}$. We assign symbols $x_1, \ldots, x_s$, where s is a positive integer to each element x and unites elements with the same subscript j to form the j-th class $K_j (j = 1, \ldots, s)$ of symbols. We now form sets $M_l (l = 1, \ldots, t)$ of cardinality $k < ms$ out of different elements taken from the $\nu = ms$ symbols so that $k = \sum_{j=1}^{s} p_{jl}$, where p_{jl} is the number of elements of M_l in the j-th class. Let $M = \{M_1, \ldots, M_t\}$. Analogously we consider t' sets $N_h (h = 1, \ldots, t')$ containing exactly $k - 1$ different elements from among the $\nu = ms$ symbols, and also containing the element ∞. Let $N = \{N_1, \ldots, N_{t'}\}$.
We call $K_\infty = \{M_1, \ldots, M_t, N_1, \ldots, N_{t'}\}$ a (ν, k, ρ, s, t, t')-**difference set (with the element** ∞) if

TABLE A.4
Some (ν, k, λ)-Difference Families.

ν	k	Base Blocks
14	4	(0,1,2,3);(0,2,4,6);(0,4,8,12);(0,8,1,9);(3,6,9,12);(0,1,5,10);(0,2,5,10)
22	6	(0,1,2,5,10,13);(0,1,5,6,8,15);(0,2,3,6,16,18);*(0,2,6,11,13,17)*
22	8	(0,1,5,7,13,17,20,21); (0,2,7,11,13,14,16,17); (0,3,4,12,14,15,17,21)
27	3	(0,1,3); (0,4,11); (0,5,15); (0,6,14); *(0,9,18)*
28	6	(0,4,7,8,16,21);(5,7,8,9,14,20);(7,12,14,16,17,25); (1,4,7,13,14,24);*(2,4,7,16,17,22)*
29	7	(1,7,16,20,23,24,25); (2,3,11,14,17,19,21)
29	8	(0,1,7,16,20,23,24,25); (0,2,3,11,14,17,19,21)
31	5	(0,1,3,7,15); (0,3,9,14,21); (0,4,5,13,15)
33	3	(0,1,3); (0,4,10); (0,5,18); (0,7,19); (0,8,17); *(0,11,22)*
34	4	(0,1,22,24); (0,1,19,25); (0,2,6,29); (0,4,7,20); (0,5,8,20); *(0,8,17,25)*
34	12	(0,5,9,14,15,17,20,25,26,27,28,30); (0,6,7,10,13,17,18,20,22,24,25,26)
39	6	(0,3,4,17,19,32);(0,1,5,12,30,36);(0,3,8,9,25,27); (0,7,10,12,17,21);(0,16,17,19,27,35); (0,2,18,27,28,33);(0,6,13,19,26,32)
41	5	(0,1,4,11,29);(0,2,8,17,22)
41	6	(0,1,10,16,18,37); (0,6,14,17,19,26); (0,2,20,32,33,36); (0,11,12,28,34,38)
43	3	(0,1,3); (0,4,9); (0,6,28); (0,7,23); (0,8,33); (0,11,30); (0,12,26)
43	4	(0,1,6,36); (0,3,18,22); (0,9,11,23); (0,10,12,26); (0,1,6,36); (0,26,27,33); (0,13,35,38); (0,19,28,39)
43	7	(1,4,11,16,21,35,41); (3,5,12,19,20,33,35); (9,13,14,15,17,25,36)
43	8	(0,1,4,11,16,21,35,41); (0,3,5,12,19,20,33,37); (0,9,13,14,15,17,6)
45	3	(0,1,3); (0,4,10); (0,5,28); (0,7,34); (0,8,32); (0,9,29); (0,12,26); *(0,15,30)*
45	15	(1,3,6,13,18,21,22,25,26,27,33,35,36,38,40); (9,10,11,13,16,17,19,23,26,27,28,33,35,38,39)
46	6	(0,1,3,11,31,35); (0,1,4,10,23,29); (0,2,7,15,32,41)
46	10	(3,7,13,16,23,24,25,28,30,42); (2,10,12,18,25,34,40,43,44,45)
49	9	(0,1,3,5,9,14,19,25,37); (0,2,12,13,16,19,34,41,42)

Note: Only those needed in Chapter six are given. Entries of short blocks are in italics.

- Of the kt elements of the set $M = \{M_1, \dots, M_t\}$, exactly $mt' - \lambda$ belong to each of the s classes K_j and of the $(k-1)t'$ finite elements of $N = \{N_1, \dots, N_{t'}\}$ exactly λ occur in each of the s classes K_j, where $\lambda = \frac{\varrho}{s}$ and

- The differences arising from the finite symbols in the $t + t'$ sets $K_\infty = \{M_1, \dots, M_t, N_1, \dots, N_{t'}\}$ are each occurring λ times.

From this definition it follows that $s(mt^{'} - \lambda) = kt$ and $s\lambda = s(k-1)$.

Example A.3 We consider the module of Example A.1 with $m = 1, s = t = t'$. Let $M = \{M_1\} = (0, 1, 4)$ and $N = \{N_1\} = (\infty, 1, 4)$, with $\lambda = 2$ the first condition of definition A.13 is fulfilled.
Further

$0-1 \equiv 4(\text{mod } 5), 0-4 \equiv 1(\text{mod } 5), 1-0 \equiv 1(\text{mod } 5), 1-4 \equiv 2(\text{mod } 5), 4-0 \equiv 4(\text{mod } 5), 4-1 \equiv 3(\text{mod } 5), 1-4 \equiv 2(\text{mod } 5), 4-1 \equiv 3(\text{mod } 5)$. Really, each non-zero element of the module occurs $\lambda = 2$ times. Thus $K_\infty = \{M_1, N_1\} = \{(0,1,4), (\infty, 1, 4)\}$ is a $(3, \rho, 1, 1, 1)$-difference set (with the element ∞).

Definition A.16 Let there be two symbols u and w in $\{0, \ldots, \nu - 1\}$ and in $\{0, \ldots, \nu - 2, \infty\}$, respectively. The **distance** $d(u, w)$ between u and w is defined as

$$d(u, w) = Min\{u - w(\text{mod } \nu*); w - u(\text{mod } \nu*)\} \text{ if } u \neq \infty; w \neq \infty$$

and by $d(u, \infty) = d(\infty, w) = \infty$.
Further, let there be one triple (x, y, z) of distances not all ∞, and let (d_1, d_2, d_3) denote the corresponding ordered triple $(d_1 \leq d_2 \leq d_3)$. Such a triple is called an **admissible triple of type 1** (d_1, d_2, d_3) if all elements are finite and $d_1 + d_2 = d_3$. Such a triple is called an **admissible triple of type 2** $(d_1, d_2, d_3)^*$ if $d_1 + d_2 = -d_3(\text{mod } \nu^*)$ or $(d_1, d_2, d_3) = (d_1, \infty, \infty)$.

Definition A.17 A group divisible design (GDD) with index λ is a triple (V, G, B) where

- $\boldsymbol{V}$ is a finite set of points (treatments) of size ν
- $\boldsymbol{G}$ is a set of subsets of $\boldsymbol{V}$, called groups which partition $\boldsymbol{V}$
- $\boldsymbol{B}$ is a collection of subsets of $\boldsymbol{V}$, called blocks, such that every pair of points from distinct groups occurs in exactly λ blocks
- $|G \cap B| \leq 1, \forall G \in \boldsymbol{G}; B \in \boldsymbol{B}$

Such a GDD is written as $GDD_\lambda(\boldsymbol{V}, \boldsymbol{G}, \boldsymbol{B})$. The type of a $GDD_\lambda(\boldsymbol{V}, \boldsymbol{G}, \boldsymbol{B})$ is the set $H = (|G : G \in \boldsymbol{G}|)$. We use the notation $GD_\lambda(K, H, \nu)$. We are only interested in GDDs with constant block size k and constant group size g; *they are written as* $GDD_\lambda(k, g, \nu)$.

Definition A.18 Let $\boldsymbol{V}$ be the point set of a design and $G \subset \boldsymbol{V}$. A **partial resolution class** is a subset of blocks such that every point of $\boldsymbol{V} \backslash G$ occurs exactly once and the points of G do not occur at all.

Definition A.19 A **frame** is a $GDD_\lambda(\boldsymbol{V}, \boldsymbol{G}, \boldsymbol{B})$ such that

- the block set can be partitioned into a family $\boldsymbol{R}$ of partial resolution classes, and
- each $R \in \boldsymbol{R}$ can be associated with a group $G \in \boldsymbol{G}$ so that R contains every point of $\boldsymbol{V} \backslash G$ exactly once.

In a fashion similar to that of GDDs, we use the notations $F_\lambda(\boldsymbol{V}, \boldsymbol{G}, \boldsymbol{B})$ and $F_\lambda(k, g, \nu)$. For an $F_\lambda(k, g, \nu)$ we say it is a **(k, λ)-frame of type g**. An $F_{k-1}(k, 1, tk+1)$ with a positive integer t is called a **nearly resolvable design** NRD$(k; tk+1)$. For the construction of BIBDs in Chapter six we need uniform frames only.
Necessary conditions for uniform frames are

- $\nu \equiv 0(\text{mod } g)$
- $\lambda(\nu - g) \equiv 0(\text{mod } k-1)$
- $\lambda\nu(\nu - g) \equiv 0(\text{mod } k(k-1))$
- $\nu - g \equiv 0(\text{mod } k)$
- $g\lambda \equiv 0(\text{mod } k-1)$
- $\nu \geq (k+1)g$

If an NRD(k, ν) exists, then $\nu \equiv 0(\text{mod } k)$.

A.4 Hadamard matrices

The definitions in this section are needed for the construction of BIBDs in Chapter six and of fractional factorial designs in Chapter seven.

Definition A.20 A square matix H_n of order n and entries −1 and +1 is said to be an Hadamard matrix if $H_n H_n^T = nI_n$, where I_n is an identity matrix of order n.
We have $H_n H_n^T = H_n^T H_n$ and a necessary condition for the existence of an Hadamard matrix for $n > 2$ is $n \equiv 0(\text{mod } 4)$. The necessary conditions are sufficient for all $n < 201$ (Hedayat and Wallis, 1978). Trivially $H_1 = (1); H_2 = \begin{pmatrix} 1 & 1 \\ 1 & -1 \end{pmatrix}$. An Hadamard matrix remains an Hadamard matrix if we multiply any of its rows or columns by −1. Therefore we can write Hadamard matrices w.l.o.g. in normal form, which means that their first rows and first columns contain only the elements +1. If the first column of an Hadamard matrix contains only the element +1, it is said to be in seminormal form. The Kronecker product of two Hadamard matrices H_{n_1} and H_{n_2}, $H_{n_1} \otimes H_{n_2} = H_{n_1 n_2}$ is an Hadamard matrix of order $n_1 n_2$.

Construction:
If $n \equiv 0(\text{mod } 4)$ and $n - 1 = p$ is a prime, let g a primitive element of GF(p)

TABLE A.5
Some Rotational (ν, k, λ)-Difference Families.

ν	k	Base Blocks
18	6	$(\infty,0,2,8,9,12);(0,1,2,4,7,15);(0,1,3,7,8,12))$
20	8	$(\infty,0,4,5,6,9,16,17);(\infty,0,1,4,6,7,9,11);(0,1,5,6,7,9,15,17);$ $(0,4,5,6,7,10,11,16);(0,1,4,9,11,13,16,17)$
24	4	$(\infty,0,7,10);(1,8,12,22);(2,5,6,11);$ $(3,9,14,18);(4,16,17,19);(13,15,20,21)$
26	6	$\{(0,0);(0,y);(0,2y),(y,y),(2y,4y);(4y,0)\}$ for $y=1,2;$ $\{(\infty,\infty),(0,0),(1,y),(2,2y),(3,3y),(4,4y)\}$ for $y=0,2,4$
27	6	$(\infty,0,7,11,13,21);(0,3,6,14,17,24);(0,9,16,18,20,25);$ $(0,1,2,8,22,23);(\mathit{2},\mathit{5},\mathit{6},\mathit{15},\mathit{18},\mathit{19});$
30	5	$(\infty,4,8,17,18);(0,1,7,24,27);(2,12,15,16,23);$ $(9,11,13,21,26);(3,14,20,22,25);(5,6,10,19,28)$
34	6	$(\infty,0,3,10,12,32);(0,3,16,18,24,32);$ $(0,9,15,23,28,30);(0,3,4,12,31,32);$ $(0,7,13,21,23,30);(\mathit{0},\mathit{4},\mathit{11},\mathit{15},\mathit{22},\mathit{26});(\mathit{0},\mathit{5},\mathit{11},\mathit{16},\mathit{22},\mathit{27});$
34	11	$(\infty,1,6,7,11,14,22,23,25,27,29);(0,1,3,4,6,10,18,23,24,26,32);$ $(0,1,2,6,9,10,13,15,23,27,32);(\mathit{0},\mathit{3},\mathit{6},\mathit{9},\mathit{12},\mathit{15},\mathit{18},\mathit{21},\mathit{24},\mathit{27},\mathit{30})$
36	6	$(\infty,\mathit{0},\mathit{7},\mathit{14},\mathit{21},\mathit{28});(\infty,\mathit{0},\mathit{7},\mathit{14},\mathit{21},\mathit{28});$ $(\infty,\mathit{0},\mathit{7},\mathit{14},\mathit{21},\mathit{28});(4,15,16,20,24,26);(2,3,12,15,18,20)$
36	9	$(\infty,1,10,11,13,15,25,26,29);(0,9,16,17,18,19,22,24,30);$ $(3,4,6,12,21,23,27,31,34);(2,5,7,8,14,20,28,32,33);$
39	13	$(\infty,1,7,8,9,11,12,18,20,23,25,26,30);$ $(0,3,4,6,13,15,21,27,28,29,31,33,37);$ $(1,2,7,10,11,13,14,15,19,22,29,32,34)$
40	6	$(\infty,0,3,19,27,32);(0,6,7,15,18,35);(0,17,19,21,24,33);$ $(0,1,7,8,29,30);(0,5,9,19,23,25);(10,12,15,16,24,27);$ $(\mathit{0},\mathit{5},\mathit{13},\mathit{18},\mathit{26},\mathit{31});(\mathit{0},\mathit{2},\mathit{13},\mathit{15},\mathit{26},\mathit{28})$
40	12	$(\infty,0,2,3,5,9,13,21,24,27,32,33);(1,2,5,6,7,8,9,15,19,24,29,31);$ $(2,3,4,6,14,16,22,27,31,35,36,38);$ $(\mathit{0},\mathit{1},\mathit{3},\mathit{9},\mathit{13},\mathit{14},\mathit{16},\mathit{22},\mathit{26},\mathit{27},\mathit{29},\mathit{35})$
50	8	$\{(0,0),(0,y),(0,2y),(y,0),$ $(2y,2y),(4y,y),(4y,5y),(6y,6y)\}$ for $y=1,2,3;$ $\{(\infty,\infty),(\mathit{0},\mathit{0}),(\mathit{1},y),(\mathit{2},2y),(\mathit{3},3y),(\mathit{4},4y),(\mathit{5},5y),(\mathit{6},6y)\}$ for $y=0,2,3,6$

Note: Only those needed in Chapter six are given. Entries of short blocks are given in italics.

and let $a_0 = 0, a_1 = g^0, a_2 = g^1, \ldots, a_{p-1} = g^{p-2}$ be the elements of the Galois field GF(p). Further for $0 \le j \le p-1$, let $S_j = \{a_j + a_1, a_j + a_2, \ldots, a_j + a_{p-2}\}$ and $e_s^T = (1, 1, \ldots, 1)$ a unit vector with s elements. If $n - 1 = p^t$ is a prime power, we proceed analogously with a minimal polynomial of GF(p^t) in place of g.

Define the elements of $H = (h_{ij})$ as follows:

$h_{ii} = -1, i = 0, \ldots, p-1; h_{ij} = \varphi(a_j - a_i); i \ne j, i, j = 0, \ldots, p-1$ with $\varphi(a)$ the Legendre symbol (see above Definition A.10).

Then H_n can be constructed by $H_n = \begin{pmatrix} 1 & e_p^T \\ e_p & H \end{pmatrix}$.

Example A.4 We construct the Hadamard matrix H_8. We first write the elements of GF(7) using the primitive element (see Table A.1) $g = 3$ as $a_0 = 0; a_1 = 3^0 = 1; a_2 = 3^1 = 3; a_3 = 3^2 = 2; a_4 = 3^3 = 6; a_5 = 3^4 = 4; a_6 = 3^5 = 5$.
We know in general that if for an odd prime $p, r = \frac{p-1}{2}$, the elements $1^2, 2^2, \ldots, r^2$ are quadratic residues (mod p) and all other elements are quadratic non-residues. In our case $p = 7$ just 1, 2, 4 are quadratic residuals (mod 7), so that $\phi(1) = \phi(2) = \phi(4) = 1$ and $\phi(3) = \phi(5) = \phi(6) = -1$.

$$\text{We obtain } H = \begin{pmatrix} -1 & 1 & 1 & -1 & 1 & -1 & -1 \\ -1 & -1 & 1 & 1 & -1 & 1 & -1 \\ -1 & -1 & -1 & 1 & 1 & -1 & 1 \\ 1 & -1 & -1 & -1 & 1 & 1 & -1 \\ -1 & 1 & -1 & -1 & -1 & 1 & 1 \\ 1 & -1 & 1 & -1 & -1 & -1 & 1 \\ 1 & 1 & -1 & 1 & -1 & -1 & -1 \end{pmatrix}$$

Thus the Hadamard matrix we are looking for is

$$H_8 = \begin{pmatrix} 1 & 1 & 1 & 1 & 1 & 1 & 1 & 1 \\ 1 & -1 & 1 & 1 & -1 & 1 & -1 & -1 \\ 1 & -1 & -1 & 1 & 1 & -1 & 1 & -1 \\ 1 & -1 & -1 & -1 & 1 & 1 & -1 & 1 \\ 1 & 1 & -1 & -1 & -1 & 1 & 1 & -1 \\ 1 & -1 & 1 & -1 & -1 & -1 & 1 & 1 \\ 1 & 1 & -1 & 1 & -1 & -1 & -1 & 1 \\ 1 & 1 & 1 & -1 & 1 & -1 & -1 & -1 \end{pmatrix}$$

R-program for constructing Hadamard matrices.

The program provides the construction of Hadamard matrices. Up to $n = 100$ all Hadamard matrices can be calculated; for larger n there are some matrices missing. The program uses the following algorithms (up to $n = 100$):

$H_2 = \begin{pmatrix} 1 & 1 \\ 1 & -1 \end{pmatrix}$; $H_4 = H_2 \otimes H_2, H_8 = H_4 \otimes H_2, H_{16} = H_8 \otimes H_2, H_{24} = H_{12} \otimes H_2, H_{32} = H_{16} \otimes H_2, H_{40} = H_{20} \otimes H_2, H_{48} = H_{24} \otimes H_2, H_{56} = H_{28} \otimes H_2, H_{64} = H_{32} \otimes H_2, H_{72} = H_{36} \otimes H_2, H_{80} = H_{40} \otimes H_2, H_{88} = H_{44} \otimes H_2, H_{96} = H_{48} \otimes H_2$.

Further

$H_{12} = \begin{pmatrix} 1 & e_{11}^T \\ e_{11} & H \end{pmatrix}$ from GF(11) with $g = 2$, $H_{20} = \begin{pmatrix} 1 & e_{19}^T \\ e_9 & H \end{pmatrix}$ from GF(19) with $g = 2$, $H_{28} = \begin{pmatrix} 1 & e_{27}^T \\ e_{27} & H \end{pmatrix}$ from GF(27) with $g^3 = g + 2$, $H_{44} = \begin{pmatrix} 1 & e_{43}^T \\ e_{43} & H \end{pmatrix}$ from GF(43) with $g = 3$, $H_{60} = \begin{pmatrix} 1 & e_{59}^T \\ e_{59} & H \end{pmatrix}$ from GF(59) with $g = 2$, $H_{84} = \begin{pmatrix} 1 & e_{83}^T \\ e_{83} & H \end{pmatrix}$ from GF(83) with $g = 2$.

For $n = 36, 52, 68, 76, 92$ and 100, special methods are used. For instance, for $n = 36$ the following algorithm is programmed:

$$H_{36} = \begin{pmatrix} 1 & e_{35}^T \\ e_{35} & H \end{pmatrix} \text{ with } H = \begin{pmatrix} h_1 & h_2 & \dots & h_{34} & h_{35} \\ h_{35} & h_1 & \dots & h_{33} & h_{34} \\ \dots & \dots & \dots & \dots & \dots \\ h_3 & h_4 & \dots & h_1 & h_2 \\ h_2 & h_3 & \dots & h_{35} & h_1 \end{pmatrix} \text{ where}$$

$h_1 = -1, h_2 = +1, h_3 = -1, h_4 = +1, h_5 = +1, h_6 = +1, h_7 = -1, h_8 = -1, h_9 = -1, h_{10} = +1, h_{11} = +1, h_{12} = +1, h_{13} = +1, h_{14} = +1, h_{15} = -1, h_{16} = +1, h_{17} = +1, h_{18} = +1, h_{19} = -1, h_{20} = -1, h_{21} = +1, h_{22} = -1, h_{23} = -1, h_{24} = -1, h_{25} = -1, h_{26} = +1, h_{27} = -1, h_{28} = +1, h_{29} = -1, h_{30} = +1, h_{31} = +1, h_{32} = -1, h_{33} = -1, h_{34} = +1, h_{35} = -1.$

R-program: We are using function `hadamard(n)` from package `survey` which returns the smallest Hadamard matrix with more than n rows. Therefore to get an Hadamard matrix with 12 rows let us try

```
> hadamard(11)
```

A.5 Existence and non-existence of non-trivial BIBD

We give some theorems about the existence of non-trivial BIBD; most of the proofs can be found in Rasch and Herrendörfer (1985) and Dey (1986). Other sources are shown in parentheses.

Theorem A.5 (Shrikande, 1950)
For the existence of a symmetric BIBD with ν even and $k > 1$, $r - \lambda$ must be a perfect square.
Hence there are no symmetrical BIBD for the following triples (ν, k, λ) satisfying the necessary conditions for $\nu < 50$: $(22, 7, 2); (34, 12, 4); (46, 10, 2)$. Further the non-existence for the parameters (29, 8, 2) and (43, 15, 5) was shown by Bose (1942) with ν is odd.

Theorem A.6 (Calinski and Kageyama, 2003)
In a symmetric BIBD with parameters $\nu = b, r = k, \lambda$, any two blocks have exactly λ treatments in common.

Theorem A.7 A necessary condition for the existence of a symmetric BIBD with parameters $\nu = b, r = k, \lambda$ is that the diophantic equation

$$x^2 = (r - \lambda)y^2 + (-1)^{\frac{\nu-1}{2}} \lambda z^2 \tag{A.4}$$

has a solution in integers not all zero.

Theorem A.8 If u is the square-free part of $(r-\lambda)$ and t the square-free part of λ, then a necessary condition for the existence of a symmetric BIBD with parameters $\nu = b$ odd, $r = k$ and λ is that (using the Legendre symbol) $\varphi = \left([-1]^{\frac{\nu-1}{2}} \cdot t|p\right) = 1$ for all odd primes p dividing u but not t, for all odd primes p dividing both u and t, $\varphi = -\left([-1]^{\frac{\nu-1}{2}} \cdot u_0 \cdot t_0|p\right) = 1; u_0p = u; t_0p = t$. We are looking for the smallest designs for given ν and k. Therefore we give only the results for the minimal λ from some theorems of Hanani (1961, 1975).

Theorem A.9 Let

$$\lambda\nu(\nu_1) \equiv 0[\text{mod } k(k-1)] \tag{A.5}$$

Then for:
$k = 3$ a BIBD exists for all λ and ν [which means that $\lambda(\nu-1) \equiv 0(\text{mod } 2)$ and $\lambda\nu(v-1) \equiv 0(\text{mod } 6)$
$k = 4$ a BIBD exists for all λ and ν (which means that $\lambda(\nu-1) \equiv 0(\text{mod } 3)$ and $\lambda\nu(v-1) \equiv 0(\text{mod } 12)$)
$k = 5$ a BIBD exists for all λ and ν, satisfying (A.5) with the exception of $\nu = 15, \lambda = 2$
$k = 6$ a BIBD exists for all λ and ν, satisfying (A.5) with the exception of $\nu = 21, \lambda = 2$
$k = 7$ a BIBD exists for all ν and $\lambda \equiv 0, 6, 7, 12, 18, 24, 30, 35, 36(\text{mod } 42)$ or, for $\lambda > 30$ but not divisible by 2 or 3.

Theorem A.10 For the following parameters no BIBD exists:
$\nu = 15, b = 21, r = 7, k = 5, \lambda = 2$
$\nu = 21, b = 28, r = 8, k = 6, \lambda = 2$
$\nu = 36, b = 45, r = 10, k = 8, \lambda = 2$
$\nu = b = 22; r = k = 7; \lambda = 2$
$\nu = b = 29; r = k = 8; \lambda = 2$
$\nu = b = 34; r = k = 12; \lambda = 4$
$\nu = b = 43; r = k = 7; \lambda = 1$
$\nu = b = 43; r = k = 15; \lambda = 5$
$\nu = b = 46; r = k = 10; \lambda = 2$
$\nu = b = 52; r = k = 18; \lambda = 6$
$\nu = b = 53; r = k = 13; \lambda = 3$
$\nu = b = 58; r = k = 19; \lambda = 6$
$\nu = b = 61; r = k = 21; \lambda = 7$
$\nu = b = 67; r = k = 12; \lambda = 2$
$\nu = b = 77; r = k = 20; \lambda = 5$
$\nu = b = 92; r = k = 14; \lambda = 2$

$\nu = b = 93; r = k = 24; \lambda = 6$
$\nu = b = 103; r = k = 18; \lambda = 3$
$\nu = b = 106; r = k = 15; \lambda = 2$
$\nu = b = 106; r = k = 21; \lambda = 4$
$\nu = b = 137; r = k = 17; \lambda = 2$
$\nu = b = 141; r = k = 21; \lambda = 3$

Note that these parameters fulfilled the three necessary conditions (6.4), (6.5) and (6.6) for BIBD, but the BIBD does not exist.

There is the following asymptotic result due to Wilson (1972 a, b; 1975):
For a given k and λ, there exists a finite V such that a BIBD with parameters (ν, k, λ) exists for all $\nu > V$ whenever (6.4) and (6.5) are satisfied. For a given k and $\lambda = 1$, there exists a finite W such that an RBIBD with parameters $(w, k, \lambda = 1)$ exists for all $w > W$ whenever (6.4) and (6.5) are satisfied.

Theorem A.11 (Calinski and Kageyama 2003)
For $\nu > 6$, the parameters of a BIBD satisfy the inequality

$$\frac{2\lambda[\nu r - b - (\nu - 1)](b - 3)}{r(r-1)(\nu r - b - \nu + 3)} \geq 1.$$

A.6 Conference matrices

Definition A.21 A conference matrix C of order n is a square matrix with zero diagonal and the off-diagonal elements equal to either 1 or -1, satisfying $CC^T = (n-1)I_n$. A conference matrix with zero main diagonal and $+1$ in all elements of the first row and column outside the main diagonal is called normalised.
The name *conference matrix* stems from the fact that Belevitch (1950) used such matrices in conference telephony.

Theorem A.12 If $n = p^h + 1$ with an odd prime p and h a positive integer such that $p^h \equiv 1 \mathrm{mod}(4)$, we denote the elements of GF(p^h) by $\alpha_0, \alpha_1, \ldots, \alpha_{n-2}$. Then the matrix $\begin{pmatrix} 0 & e_{n-1}^T \\ e_{n-1} & G \end{pmatrix}$ with $G = (g_{ij}); i, j = 0, \ldots, n-2$ and $g_{ii} = 0; i = 0, \ldots, n-2; g_{ij} = \varphi(\alpha_j - \alpha_i); i \neq j = 0, \ldots, n-2$ (with $\varphi(\alpha)$) the Legendre symbol (see above Definition A.10) is a conference matrix of order n.

Theorem A.13 A necessary condition for the existence of a conference matrix C of order n is that $n = 1 + \prod_{i=1}^{m} p_i^{h_i}$ with an odd prime p_i and h_i a positive integer such that $p^{h_i} \equiv 1 \mathrm{mod}(4); i = 1, 2, \ldots, m$.

Theorem A.14 If q is an odd prime power with $q \equiv 1 \bmod (4)$ then a conference matrix of order $q+1$ exists (Paley, 1933).

For the construction of conference matrices, the following theorem can be used.

Theorem A.15 If A is a square matrix of order n with elements 0, 1 and –1 and B is a square matrix of order n with elements 1 and –1 such that $AB = BA$ and $AA^T + BB^T = (2n-1)I_n$, then the matrix $C = \begin{pmatrix} A & B \\ B^T & -A^T \end{pmatrix}$ is a conference matrix of order $2n$.

Conference matrices of order $2n$ exist (E) or do not exist (NE) for the following values of n: (Ionin and Kharagani, 2006):
$E : n = 8, 9, 10, 12, 13, 14, 15, 16, 18, 19, 20, 21, 22, 23, 24, 25, 26, 30, 31, 32$
$NE : n = 11, 17, 29, 39.$

Example A.5 The matrix $\begin{pmatrix} 0 & 1 \\ 1 & 0 \end{pmatrix}$ is a symmetric conference matrix of order 2, the matrix

$$\begin{pmatrix} 0 & 1 & 1 & 1 \\ -1 & 0 & 1 & -1 \\ -1 & -1 & 0 & 1 \\ -1 & 1 & -1 & 0 \end{pmatrix}$$ is a conference matrix of order 4, and

$$\begin{pmatrix} 0 & 1 & 1 & 1 & 1 & 1 \\ 1 & 0 & 1 & -1 & -1 & 1 \\ 1 & 1 & 0 & 1 & -1 & -1 \\ 1 & -1 & 1 & 0 & 1 & -1 \\ 1 & -1 & -1 & 1 & 0 & 1 \\ 1 & 1 & -1 & -1 & 1 & 0 \end{pmatrix}$$ is a symmetric normalized conference matrix of order 6.

Example A.6 Let x be the least common multiple of $s_1, s_2, \ldots, s_p$ and $M_c = \{c_1, c_2, \ldots, c_p\}$ be an (additive) module. Consider the bilinear form $[c, t] = \sum_{i=1}^{p} c_i t_i \frac{x}{s_i} \bmod x$. In dependence on $c \in M_c; t \in B$ we call $B^0 = \{c | c \in M_c; [c, t]\} = 0 \forall t \in B$ the **annihilator** of B or the **defining contrasts** subgroup.

References

Abel, R. J. R. (1994). Forty-three balanced incomplete block designs. *J. Comb. Theory A 65*, 252–267.

Anderson, D. A. and A. M. Thomas (1979). Resolution iv fractional factorial designs for the general asymmetric factorial. *Commun. Stat. A Theory 8*, 931–941.

Atkinson, A., A. Donev, and T. Randall (2007). *Optimum Experimental Designs, with SAS.* Oxford, New York: Oxford University Press.

Bailey, R. A. (2008). *Design of Comparative Experiments.* Cambridge: Cambridge University Press.

Bartlett, M. S. (1933). On the theory of statistical regression. *Proc. Roy. Soc. Edinburgh 53*, 260–283.

Bechhofer, R. E. (1954). A single sample multiple-decision procedure for ranking means of normal populations with known variances. *Ann. Math. Stat. 25*, 16–39.

Bechhofer, R. E., J. Kiefer., and M. Sobel (1968). *Sequential Identification and Ranking Procedures.* Chicago: The University of Chicago Press.

Becker, N. G. (1970). Mixture designs for a model linear in the proportions. *Biometrika 57*, 329–338.

Belevitch, V. (1950). Theory of 2n-terminal networks with applications to conference telephony. *Electr. Commun. 27*, 231–244.

Beth, T., D. Jungnickel, and H. Lenz (1999). *Design Theory* (2nd ed.), Volume I. Cambridge: Cambridge University Press.

Bhattacharya, K.N. (1945). On a new symmetrical balanced incomplete block design. *Bull. Calcutta Math. Soc. 36*, 91–96.

Bisgaard, S. (1994). Blocking generators for small 2^{k-p} designs. *J. Qual. Technol. 26*, 288–296.

Bock, J. (1998). *Bestimmung des Stichprobenumfangs.* München, Wien: Oldenbourg.

Bose, R. C. (1939). On the construction of balanced incomplete block designs. *Ann. Eugenics 9*, 353–399.

Bose, R. C. (1941). Discussion on the mathematical theory of design of experiments. *Sankhya 5*, 170–174.

Bose, R. C. (1942). On some new series of balanced incomplete block designs. *Bull. Calcutta Math. Soc 34*, 17–31.

Bowman, K. O. (1972). Tables of the sample size requirement. *Biometrika 59*, 234.

Bowman, K. O. and M. A. Kastenbaum (1975). Sample size requirement: Single and double classification experiments. In *Selected Tables in Mathematical Statistics* Volume III, pp. 111–120. Providence, RI: American Mathematical Society.

Box, G. E. P. and D. W. Behnken (1960). Some new three level design for the study of quantitative variables. *Technometrics 2*, 455–475.

Bratcher, T. L., M. A. Moran, and W. J. Zimmer (1970). Tables of sample size in the analysis of variance. *J. Qual. Technol. 2*, 156–164.

Brown, W. J. and D. J. Murdoch (2008). *A First Course in Statistical Computing with R.* Cambridge, UK: Cambridge University Press, Section 1.1, 2.

Calinski, T. and S. Kageyama (2000). *Block Designs: A Randomization Approach*, Volume I of *Analysis. Lecture Notes in Statistics 150.* New York: Springer.

Calinski, T. and S. Kageyama (2003). *Block Designs: A Randomization Approach*, Volume II of *Design. Lecture Notes in Statistics 170.* New York: Springer.

Casagrande, J. T., M. C. Pike, and P. G. Smith (1978). An improved approximate formula for calculating sample sizes for comparing two binomial distributions. *Biometrics 34*, 483–486.

Chen, J., D. X. Sun, and C. F. J. Wu (1993). A catalogue of two-level and three-level fractional factorial designs with small runs. *Int. Stat. Rev. 61*, 131–145.

Cochran, W. G. and G. M. Cox (1957). *Experimental Designs* (2nd ed.). New York: Wiley.

Colbourn, C. J. and J. H. Dinitz (Eds.) (2006). *Handbook of Combinatorial Designs* (2nd ed.). Boca Raton, FL: Chapman & Hall.

Collens, R. J. (1976). Constructing BIBDs with a computer. *Ars Combinatoria 2*, 187–231.

Cornell, J. A. (1973). Experiments with mixtures: A review. *Technometrics 15*, 437–456.

Cornell, J. A. (2002). *Experiments with Mixtures: Designs, Models and the Analysis of Mixture Data* (3rd ed.). New York: Wiley.

Cox, D. R. (1958). *Planning of Experiments.* New York: Wiley.

Cox, D. R. (2009). Randomization in experimental design. *Int. Stat. Rev. 77*, 415–429.

Dalgaard, P. (2008). *Introductory Statistics with R (Statistics and Computing).* New York: Springer.

Das Gupta, P. (1968). Tables of the non-centrality parameter of the F-test as a function of power. *Sankhya B 30*, 73–82.

Dey, A. (1986). *Theory of Block Designs.* New York: Wiley.

Dey, A. and R. Mukerjee (1999). *Fractional Factorial Plans.* New York: Wiley.

Doehlert, D. H. (1970). Uniform shell designs. *J. Roy. Stat. Soc., Ser. C (Applied Statistics) 19*, 231–239.

Domröse, H. and D. Rasch. (1987). Robustness of selection procedures. *Biometrical J. 28*, 541–553.

Draper, N. R., B. Heiligers, and F. Pukelsheim (2000). Kiefer-ordering of simplex designs for second-degree mixture models with four or more ingredients. *Ann. Stat. 28*, 578–590.

Draper, N. R. and D. K. T. Lin (1990). Small response-surface designs. *Technometrics 32*, 187–194.

Draper, N. R. and F. Pukelsheim (1999). Kiefer ordering of simplex designs for first- and second-degree mixture models. *J. Stat. Plan. Infer. 79*, 325–348.

Dynkin, E. B. (1963a). The optimal choice of the instant for stopping a Markov process. *Dokl. Akad. Nauk. USSR 150*, 238–240.

Dynkin, E. B. (1963b). The optimal choice of the instant for stopping a Markov process. *Soviet Math. Dokl. (English translation) 4*, 627–629.

Eisenberg, B. and B. K. Ghosh (1991). The sequential probability ratio test. In: *Handbook of sequential analysis* pp. 47–65. eds: B.K. Ghosh and P.K. Sen, New York: Marcel Dekker.

Eisenhart, C. (1947). The assumptions underlying the analysis of variance. *Biometrics, 3*, 1–21.

Ermakov, S. M. and A. A. Zhiglyavskii (1987). *Mathematical Theory of Experiments.* Moskov: Nauka (in Russian).

Fisher, R. A. (1926). The arrangement of field experiments. *J. Min. Agric. G. Br. 33*, 503–513.

Fisher, R. A. (1940). An examination of the different possible solutions of a problem in incomplete block designs. *Ann. Eugenics 10*, 52–75.

Fisher, R. A. (1942). The theory of confounding in factorial experiments in relation to the theory of groups. *Ann. Eugenics 11*, 341–353.

Fleiss, J. L. (1981). *Statistical Methods for Rates and Proportions* (2nd ed.). New York: Wiley.

Flood, M. R. (1958). Letter. Copy of this letter in Martin Gardner papers at Stanford University. Stanford, CA: Stanford University libraries: *Archives, series, box 5, folder 19*.

Fox, J. (2008). *Social Organization of the R Project. Use R!* Dortmund, Germany: http://www.statistik.uni-dortmund.de/useR-2008/slides/Fox.pdf.

Fox, M. (1956). Charts of the power of the F-test. *Ann. Math. Stat. 27*, 484–497.

Franklin, M. F. (1985). Constructing tables of minimum aberration p^{n-m} designs. *Technometrics 7*, 165–142.

Gardner, M. (1960). Mathematical games. *Scientific American 202*, 135–178.

Gardner, M. (1966). Chapter 3, Problem 3. *New Mathematical Diversions from Scientific American*, New York: Simon & Schuster.

Ghosh, B. K. and P. K. Sen (1991). *Handbook of Sequential Analysis.* New York: M. Dekker.

Graybill, A. F. (1961). *An Introduction to Linear Statistical Models.* New York: McGraw Hill.

Grömping, U. (2009). *CRAN Task View: Design of Experiments DoE & Analysis of Experimental Data.* http://cran.r-project.org/web/views/ExperimentalDesign.html.

Guiard, V. (1996). Different definitions of Δ-correct selection for the indifference zone formulation. *J. Stat. Plan. Infer. 54*, 175–199.

Gupta, S. S. (1956). *On a decision rule for a problem in ranking means.* Ph. D. thesis, Inst. of Statist. Mimeo Series No 150, University of North Carolina, Chapel Hill, NC.

Gupta, S. S. and S. Panchapakesan (1972). On multiple procedures,. *J. Math and Phys. Sciences 6*, 1–72.

Gupta, S. S. and S. Panchapakesan (1979). *Multiple Decision Procedures: Theory and Methodology of Selecting and Ranking Populations.* New York: Wiley.

Hall, M. (1986). *Combinatorial Theory* (2nd ed.). New York: Wiley.

Hammond, B., J. Lemen, and R. Dudek et al. (2006). Results of a 90-day safety assurance study with rats fed grain from corn rootworm-protected corn. *Food Chem. Tox. 44*, 147–160.

Han, C. and K. Chaloner (2003). D- and C-optimal designs for exponential regression models used in viral dynamics and other applications. *J. Stat. Plan. Infer. 115*, 585–601.

Hanani, H. (1961). The existence and construction of balanced incomplete block designs. *Ann. Math. Stat. 32*, 361–386.

Hanani, H. (1975). Balanced incomplete block designs and related designs. *Discrete Math. 11*, 275–289.

Hartley, H. O. (1959). Smallest composite designs for quadratic response surfaces. *Biometrics 15*, 611–624.

Hedayat, A. S. and P. W. M. John (1974). Resistant and susceptible BIB designs. *Ann. Statist. 2*, 148–158.

Hedayat, A. S. and W. D. Wallis (1978). Hadamard matrices and their applications. *Ann. Statist. 6*, 1184–1238.

Herrendörfer, G., D. Rasch, and K. Schmidt et al. (1997). *Determination of the Size of an Experiment for the F-test in the Analysis of Variance—Mixed Models.* Pasadena, CA: Proceedings of the Second World Congress of IASC.

Herrendörfer, G. and J. Schmidt (1978). Estimation and test for the mean in a model II of analysis of variance. *Biom. J. 20*, 355–361.

Hinkelmann, K. and O. Kempthorne (1994). *Design and Analysis of Experiments.* Volume 1, *Introduction to Experimental Design*, New York: Wiley.

Hinkelmann, K. and O. Kempthorne (2005). *Design and Analysis of Experiments.* Volume 2, *Advanced Experimental Design*, New York: Wiley.

Hochberg, Y. and A. C. Tamhane (1987). *Multiple Comparison Procedures.* New York: Wiley.

Hurvich, C. M. and C. Tsai (1989). Regression and time series model selection in small samples. *Biometrika 76*, 297–307.

Hwang, F. K. and S. Lin (1974). A direct method to construct triple systems. *J. Comb. Theory 17*, 84–94.

Ionin, Y. J. and H. Kharagani (2006). Balanced general weighing matrices and conference matrices. In C. Colbourn and J. H. Dinitz (Eds.), *Handbook of Combinatorial Designs*, pp. 306–312. Boca Raton, FL: Chapman & Hall.

Jennrich, R. L. (1969). Asymptotic properties of non-linear least squares estimators. *Ann. Math. Statist. 40*, 633–643.

John, J. A. (1987). *Cyclic Designs.* London: Chapman & Hall.

John, J. A. and E. R. Williams (1995). *Cyclic and Computer Generated Designs.* London: Chapman & Hall.

Jungnickel, D., A. Pott, and K. W. Smith (2006). *Difference Sets.* In *Handbook of Combinatorial Design* (2nd ed.), pp. 419–435. eds.: C. Colbourn and J. H. Dinitz, Boca Raton, FL: Chapman & Hall.

Kageyama, S. (1983). A resolvable solution of BIBD 18.51.17.6.5. *Ars Combin. 15*, 315–316.

Kastenbaum, M. A., D. G. Hoel, and K. O. Bowman (1970a). Sample size requirements for one-way analysis of variance. *Biometrika 57*, 421–430.

Kastenbaum, M. A., D. G. Hoel, and K. O. Bowman (1970b). Sample size requirements for randomized block designs. *Biometrika 57*, 573–577.

Kempthorne, O. (1952). *The Design of Experiments.* New York: John Wiley & Sons.

Kenneth, L. J. (1977). Appropriate sample sizes for selecting a population with the largest correlation coefficient from among k bivariate normal populations. *Edu. Psych. Meas. 37*, 61–66.

Khuri, A. I. and J. A. Cornell (1987). *Response Surfaces: Design and Analysis.* New York: Marcel Dekker.

Khursheed, A., R. M. Haseeb, and H. Solomon (1976). Selection of largest multiple correlation coefficients: Exact sample size case. *Ann. Statist. 4*, 614–620.

Kirkman, T. P. (1947). On a problem in combinatorics. *Cambridge Dublin Math. J. 2*, 191–204.

Lehmann, E. L. (1959). *Testing Statistical Hypotheses.* New York: Wiley.

Lehmann, E. L. and J. P. Romang (2008). *Testing Statistical Hypotheses.* New York: Wiley.

Lehmer, E. (1944). Inverse tables of probabilities of errors of the second kind. *Ann. Math. Stat. 15*, 388–398.

Lenth, R. V. (1986). Computing non-central Beta probabilities. *Appl. Statistics 36*, 241–243.

Linek, V. and E. B. Wantland (1998). Coloring BIBDs with block size 4. *J. Comb. Designs. 6*, 403–410.

Listing, J. and D. Rasch (1996). Robustness of subset selection procedures. *J. Stat. Plan. Infer. 54*, 291–305.

Mathon, R. and A. Rosa (1990). Tables of parameters of BIBDs with $r \leq 41$ including existence, enumeration and resolvability: An update. *Ars Combin. 30*, 65–96.

McFarland, R. L. (1973). A family of difference sets in non-cyclic groups. *J. Comb. Theory A 15*, 1–10.

McLean, R. A. and V. L. Anderson (1966). Extreme vertices designs of mixture experiments. *Technometrics 8*, 447–454.

McLean, R. A. and V. L. Anderson (1984). *Applied Fractional Factorial Designs.* New York: M. Dekker.

Melas, V. B. (2006). *Functional Approach to Optimal Experimental Design. Lecture Notes in Statistics*, Volume 184. New York: Springer.

Miescke, K. J. and D. Rasch (1996). Special issue on 40 years of statistical selection theory. Parts i & ii. *J. Stat. Plan. Infer. 5*(3).

Montgomery, D. C. (2005). *Design and Analysis of Experiments* (6th ed.). New York: Wiley.

Myers, R. H. (1976). *Response Surface Methodology.* Ann Arbor, MI: Edwards Brothers.

Palasota, J. A., I. Leonidou, and J. M. Palasota et al. (1992). Sequential simplex optimization in a constrained simplex mixture space in liquid chromatography. *Analytica Chimica Acta* (270), 101–106.

Paley, R. E. A. C. (1933). On orthogonal matrices. *J. Math. Phys. 12*, 811–820.

Pukelsheim, F. (1993). *Optimal Design of Experiments.* New York: Wiley.

Raghavarao, D. (1971). *Constructions and Combinatorial Problems in Design of Experiments.* New York: Wiley.

Raghavarao, D., W. T. Federer, and S. J. Schwager (1985). Characteristics for distinguishing balanced incomplete block designs with repeated blocks. *J Stat. Plan. Infer. 13*, 151–163.

Rasch, D. (1990). Optimum experimental design in nonlinear regression. *Commun. Statist.—Theory Meth. 19*, 4789–4806.

Rasch, D. (1995). *Mathematische Statistik.* Berlin, Heidelberg: Joh. Ambrosius Barth—Wiley.

Rasch, D. (1996). Software for selection procedures. *J. Stat. Plan. Infer. 54*, 345–358.

Rasch, D. (1998). Determination of the size of an experiment. In *MODA5. Advances in Model-Oriented Data Analysis and Experimental Design*, pp. 205–212. eds.: A. C. Atkinson, L. Pronzato, and H. P. Wynn, Heidelberg, Germany: Physika Verlag.

Rasch, D. and V. Guiard (2004). The robustness of parametric statistical methods. *Psy. Sci. 46*, 175–208.

Rasch, D. and G. Herrendörfer (1982). *Statistische Versuchsplanung.* Berlin: Deutscher Verlag der Wissenschaften.

Rasch, D. and G. Herrendörfer (1985). *Experimental Designs — Sample Size Determination and Block Designs.* Dordrecht, Boston, Lancaster, Tokyo: Reidel.

Rasch, D., E. M. T. Hendrix, and E. P. J. Boer (1997a). Replication–free optimal designs in regression analysis. *Comp. Stat. 12*, 19–52.

Rasch, D., E. M. T. Hendrix, and P. Kraan (1996). Recent results in computing replication–free designs in logistic regression. In *SoftStat '95 – Advances in Statistical Software*, pp. 497–507. eds.: F. Faulbaum and W. Bandilla, Heidelberg, Germany: Physika Verlag.

Rasch, D., G. Herrendörfer, and J. Bock et al. (2008). *Verfahrensbibliothek Versuchsplanung und –auswertung* (2nd rev. ed. with CD-ROM). München, Wien: Oldenbourg Verlag.

Rasch, D., K. D. Kubinger, and K. Moder (2009). The two-sample t-test: Pre-testing its assumptions does not pay off. *Statistical Papers.* DOI: 10.1007/s00362-009-0224-x.

Rasch, D., B. Spangl, and M. Wang (2011), Minimal experimental size in the three-way, ANOVA cross-classification model with approximate F-tests. *J. Stat. Plan. Inter.*, in press.

Rasch, D., F. Teuscher, and V. Guiard (2007a). How robust are tests for two indepenedent samples? *J. Statist. Plan. Infer. 137*, 2706–2720.

Rasch, D., L. R. Verdooren, and J. I. Gowers (2007b). *Fundamentals in the Design and Analysis of Experiments and Survey.* München: Oldenbourg.

Rasch, D., M. Wang, and G. Herrendörfer (1997b). Determination of the size of an experiment for the F-test in the analysis of variance. Model i. In *Advances in Statistical Software 6.* The 9th Conference on the Scientific Use of Statistical Software, Heidelberg.

Rizvi, H. and H. Solomon (1973). Selection of largest multiple correlation coefficient; asymptotic case. *J. Amer. Statist. Assoc. 69*, 184–188.

Sahai, H. and A. Khurshid (1996). Formulae and tables for the determination of sample sizes and power in clinical trials for testing differences in proportions for the two-sample design: A review. *Stat. in Med. 15*, 1–21.

Scheffé, H. (1958). Experiments with mixtures. *J. Roy. Statist. Soc. 20*, 344–360.

Scheffé, H. (1959). *The Analysis of Variance.* New York: Wiley.

Scheffé, H. (1963). The simplex centroid design for experiments with mixtures. *J. Roy. Statist. Soc. 25*, 235–263.

Schneider, B. (1994). Sequentialverfahren. Unpublished manuscript.

Shah, K. R. and B. K. Sinha (1989). *Theory of Optimal Designs. Lecture Notes in Statistics.* New York: Springer.

Shrikande, S. S. (1950). The impossibility of certain symmetrical balanced incomplete block designs. *Ann. Math. Stat. 21*, 106–111.

Stein, C. (1945). A two-sample test of a linear hypothesis whose power is independent of the variance. *Ann. Math. Statist 16*, 243–258.

Street, A. P. and D. J. Street (1987). *Combinatorics of Experimental Design.* Oxford: Clarendon Press.

Takeuchi, K. (1962). A table of difference sets generating balanced incomplete block designs. *Rev. Inst. Int. Statist. 30*, 361–366.

Tang, P. C. (1938). The power function of the analysis of variance tests with table and illustrations of their use. *Stat. Res. Mem. 2*, 126–149.

Thomas, E. (2006). *Feldversuchswesen.* Stuttgart: Ulmer.

Tiku, M. L. (1967). Tables of the power of F-test. *J. Amer. Stat. Assoc. 62*, 525–539.

Tiku, M. L. (1972). More tables of the power of the F-test. *J. Amer. Stat. Assoc. 67*, 709–710.

Turiel, T. P. (1988). A computer program to determine defining contrasts and factor combinations for two-level fractional factorial designs of resolution III, IV and V. *J. Qual. Techn. 20*, 267–272.

Vanstone, S. A. (1975). A note on construction of BIBDs. *Utilitas Math 7*, 321–322.

Venables, W. N. and B. D. Ripley (2003). *Modern Applied Statistics with S* (4th ed.). New York: Springer.

Wald, A. (1943). Sequential analysis of statistical data. Technical report, Columbia University. Theory Statist. Res. Group Rep. 75, New York.

Wald, A. (1944). On cumulative sums of random variables. *Ann. Math. Statist. 15*, 283–296.

Wald, A. (1947). *Sequential Analysis.* New York: Wiley.

Wallis, W. D., A. P. Street, and J. D. Wallis (2008). *Combinatorics: Room Squares, Sum-free Sets, Hadamard Matrices.* New York: Springer.

Walters, F. H., L. R. Parker, and S. L. Morgan et al. (1991). *Sequential Simplex Optimization.* Boca Raton, FL: CRC Press.

Wang, M. (2002). *Sample Size and Efficiency for Hypotheses Testing in ANOVA Model.* Berlin: Logos.

Wang, M., D. Rasch, and R. Verdooren (2005). Determination of the size of a balanced experiment in mixed ANOVA models using the modified approximate F-test. *J. Stat. Plan. Infer. 132*, 183–201.

Welch, B.L. (1947). The generalisation of Student's problem when several, different population variances are involved. *Biometrika 34*, 28–35.

Whitehead, J. (1997). *The Design and Analysis of Sequential Clinical Trials* (2nd rev. ed.). New York: Wiley.

Wijsman, R. A. (1991). *Stopping Times: Termination, Moments, Distribution.* In *Handbook of Sequential Analysis.* pp. 69–120. eds.: B. K. Ghosh and P. K. Sen, New York: M. Dekker.

Williams, E. R. (2002). *Experimental Design and Analysis for Tree Improvement.* Harwood: Matheson.

Wilson, R. W. (1972a). An existence theory for pairwise balanced designs. I. Composition theorems and morphisms. *J. Comb. Theory A 13*, 220–245.

Wilson, R. W. (1972b). An existence theory for pairwise balanced designs. II. The structure of PBD-closed sets and the existence conjecture. *J. Comb. Theory A 13*, 246–273.

Wilson, R. W. (1975). An existence theory for pairwise balanced designs. III. Proof of the existence conjecture. *J. Comb. Theory A 18*, 71–79.

Wu, C. F. J. and M. Hamada (2000). *Experiments: Planning, Analysis and Parameter Design Optimization.* New York: Wiley.

Yates, F. (1935). Complex experiments. Supplement to *J. Roy. Statist. Soc. B2 2*(2) 181–247.

Yates, F. (1936). Incomplete randomised blocks. *Ann. Eugenics 7*, 121–140.

Yates, F. (1937). The design and analysis of factorial experiments. Technical report, Technical Communication No. 35 of the Commonwealth Bureau of Soils (alternatively attributed to the Imperial Bureau of Soil Science), Harpenden, UK.

Young, L. J. (1994). Computation of some exact properties of Wald's SPRT when sampling from a class of discrete distributions. *Biom. J. 36*, 627–637.

Index

Milton Keynes UK
Ingram Content Group UK Ltd.
UKHW021629071024
449327UK00020BA/1242